Crystal Structures of Metal Complexes

Crystal Structures of Metal Complexes

Editor

Hiroshi Sakiyama

MDPI • Basel • Beijing • Wuhan • Barcelona • Belgrade • Manchester • Tokyo • Cluj • Tianjin

Editor
Hiroshi Sakiyama
Department of Science
Faculty of Science
Yamagata University
Yamagata
Japan

Editorial Office
MDPI
St. Alban-Anlage 66
4052 Basel, Switzerland

This is a reprint of articles from the Special Issue published online in the open access journal *Molecules* (ISSN 1420-3049) (available at: www.mdpi.com/journal/molecules/special_issues/crystal_structure).

For citation purposes, cite each article independently as indicated on the article page online and as indicated below:

LastName, A.A.; LastName, B.B.; LastName, C.C. Article Title. *Journal Name* **Year**, *Volume Number*, Page Range.

ISBN 978-3-0365-6453-1 (Hbk)
ISBN 978-3-0365-6452-4 (PDF)

© 2023 by the authors. Articles in this book are Open Access and distributed under the Creative Commons Attribution (CC BY) license, which allows users to download, copy and build upon published articles, as long as the author and publisher are properly credited, which ensures maximum dissemination and a wider impact of our publications.

The book as a whole is distributed by MDPI under the terms and conditions of the Creative Commons license CC BY-NC-ND.

Contents

Preface to "Crystal Structures of Metal Complexes" vii

Ryusei Hoshikawa, Ryoji Mitsuhashi, Eiji Asato, Jianqiang Liu and Hiroshi Sakiyama
Structures of Dimer-of-Dimers Type Defect Cubane Tetranuclear Copper(II) Complexes with Novel Dinucleating Ligands
Reprinted from: *Molecules* **2022**, *27*, 576, doi:10.3390/molecules27020576 1

Franz A. Mautner, Florian Bierbaumer, Ramon Vicente, Saskia Speed, Ánnia Tubau and Mercè Font-Bardía et al.
Magnetic and Luminescence Properties of 8-Coordinate Holmium(III) Complexes Containing 4,4,4-Trifluoro-1-Phenyl- and 1-(Naphthalen-2-yl)-1,3-Butanedionates
Reprinted from: *Molecules* **2022**, *27*, 1129, doi:10.3390/molecules27031129 15

Mezna Saleh Altowyan, Saied M. Soliman, Jamal Lasri, Naser E. Eltayeb, Matti Haukka and Assem Barakat et al.
A New Pt(II) Complex with Anionic s-Triazine Based *NNO*-Donor Ligand: Synthesis, X-ray Structure, Hirshfeld Analysis and DFT Studies
Reprinted from: *Molecules* **2022**, *27*, 1628, doi:10.3390/molecules27051628 31

Ryoji Mitsuhashi, Yuya Imai, Takayoshi Suzuki and Yoshihito Hayashi
Selective Formation of Intramolecular Hydrogen-Bonding Palladium(II) Complexes with Nucleosides Using Unsymmetrical Tridentate Ligands
Reprinted from: *Molecules* **2022**, *27*, 2098, doi:10.3390/molecules27072098 43

Sabrina Grenda, Maxime Beau and Dominique Luneau
Synthesis, Crystal Structure and Magnetic Properties of a Trinuclear Copper(II) Complex Based on P-Cresol-Substituted Bis(α-Nitronyl Nitroxide) Biradical
Reprinted from: *Molecules* **2022**, *27*, 3218, doi:10.3390/molecules27103218 57

Richard F. D'Vries, Germán E. Gomez and Javier Ellena
Highlighting Recent Crystalline Engineering Aspects of Luminescent Coordination Polymers Based on F-Elements and Ditopic Aliphatic Ligands
Reprinted from: *Molecules* **2022**, *27*, 3830, doi:10.3390/molecules27123830 73

Sifani Zavahir, Hamdi Ben Yahia, Julian Schneider, DongSuk Han, Igor Krupa and Tausif Altamash et al.
Fluorescent Zn(II)-Based Metal-Organic Framework: Interaction with Organic Solvents and CO_2 and Methane Capture
Reprinted from: *Molecules* **2022**, *27*, 3845, doi:10.3390/molecules27123845 87

Masahiro Mikuriya, Yuko Naka, Ayumi Inaoka, Mika Okayama, Daisuke Yoshioka and Hiroshi Sakiyama et al.
Mixed-Valent Trinuclear Co^{III}-Co^{II}-Co^{III} Complex with 1,3-Bis(5-chlorosalicylideneamino)-2-propanol
Reprinted from: *Molecules* **2022**, *27*, 4211, doi:10.3390/molecules27134211 101

Kennedy Mawunya Hayibor, Yukinari Sunatsuki and Takayoshi Suzuki
Selective Formation of Unsymmetric Multidentate Azine-Based Ligands in Nickel(II) Complexes
Reprinted from: *Molecules* **2022**, *27*, 6788, doi:10.3390/molecules27206788 115

Izabela Pospieszna-Markiewicz, Marta A. Fik-Jaskółka, Zbigniew Hnatejko, Violetta Patroniak and Maciej Kubicki
Synthesis and Characterization of Lanthanide Metal Ion Complexes of New Polydentate Hydrazone Schiff Base Ligand
Reprinted from: *Molecules* **2022**, *27*, 8390, doi:10.3390/molecules27238390 . . . . **125**

Alexander G. Medvedev, Oleg Yu. Savelyev, Dmitry P. Krut'ko, Alexey A. Mikhaylov, Ovadia Lev and Petr V. Prikhodchenko
Speciation of Tellurium(VI) in Aqueous Solutions: Identification of Trinuclear Tellurates by ^{17}O, ^{123}Te, and ^{125}Te NMR Spectroscopy
Reprinted from: *Molecules* **2022**, *27*, 8654, doi:10.3390/molecules27248654 . . . . **139**

Preface to "Crystal Structures of Metal Complexes"

Since I was a student, I have been working on crystal structure analysis of metal complexes. I cut single crystals with a razor blade while looking at them under a microscope, encapsulated single crystals in thin glass tubes, and centered crystals in the beam path of the apparatus. I was very happy every time when a single crystal structure was solved, and I was always impressed that the new metal complexes I had synthesized were contained in the crystal in regular rows.

The crystal structure reveals many things, such as the coordination geometry around metals and intermolecular interactions. Furthermore, quantum calculations based on the crystal structure can provide information on electronic states, which can be related to various physicochemical properties.

This book contains 11 papers published in a Special Issue of *Molecules* entitled "Crystal Structures of Metal Complexes". I will be very happy if readers will be interested in the crystal structures of metal complexes.

Hiroshi Sakiyama
Editor

Article

Structures of Dimer-of-Dimers Type Defect Cubane Tetranuclear Copper(II) Complexes with Novel Dinucleating Ligands

Ryusei Hoshikawa [1], Ryoji Mitsuhashi [2], Eiji Asato [3], Jianqiang Liu [4] and Hiroshi Sakiyama [1,*]

[1] Department of Science, Faculty of Science, Yamagata University, 1-4-12 Kojirakawa, Yamagata 990-8560, Yamagata, Japan; hosiryu@yahoo.com
[2] Institute of Liberal Arts and Science, Kanazawa University, Kakuma, Kanazawa 920-1192, Ishikawa, Japan; mitsuhashi@staff.kanazawa-u.ac.jp
[3] Department of Chemistry, Biology and Marine Science, Faculty of Science, University of the Ryukyus, Nishihara 903-0213, Nakagami-gun, Okinawa, Japan; asato@sci.u-ryukyu.ac.jp
[4] Guangdong Provincial Key Laboratory of Research and Development of Natural Drugs, School of Pharmacy, Guangdong Medical University, Guangdong Medical University Key Laboratory of Research and Development of New Medical Materials, Dongguan 523808, China; jianqiangliu2010@126.com
* Correspondence: saki@sci.kj.yamagata-u.ac.jp

Citation: Hoshikawa, R.; Mitsuhashi, R.; Asato, E.; Liu, J.; Sakiyama, H. Structures of Dimer-of-Dimers Type Defect Cubane Tetranuclear Copper(II) Complexes with Novel Dinucleating Ligands. *Molecules* **2022**, *27*, 576. https://doi.org/10.3390/molecules27020576

Academic Editor: Catherine Housecroft

Received: 24 December 2021
Accepted: 14 January 2022
Published: 17 January 2022

Publisher's Note: MDPI stays neutral with regard to jurisdictional claims in published maps and institutional affiliations.

Copyright: © 2022 by the authors. Licensee MDPI, Basel, Switzerland. This article is an open access article distributed under the terms and conditions of the Creative Commons Attribution (CC BY) license (https://creativecommons.org/licenses/by/4.0/).

Abstract: Only a limited number of multinucleating ligands can stably maintain multinuclear metal structures in aqueous solutions. In this study, a water-soluble dinucleating ligand, 2,6-bis{[N-(carboxylatomethyl)-N-methyl-amino]methyl}-4-methylphenolate ((sym-cmp)$^{3-}$), was prepared and its copper(II) complexes were structurally characterized. Using the single-crystal X-ray diffraction method, their dimer-of-dimers type defect cubane tetranuclear copper(II) structures were characterized for [Cu$_4$(sym-cmp)$_2$Cl$_2$(H$_2$O)$_2$] and [Cu$_4$(sym-cmp)$_2$(CH$_3$O)$_2$(CH$_3$OH)$_2$]. In the complexes, each copper(II) ion has a five-coordinate square-pyramidal coordination geometry. The coordination bond character was confirmed by the density functional theory (DFT) calculation on the basis of the crystal structure, whereby we found the bonding and anti-bonding molecular orbitals. From the cryomagnetic measurement and the magnetic analysis, overall antiferromagnetic interaction was observed, and this magnetic behavior is also explained by the DFT result. Judging from the molar conductance and the electronic spectra, the bridging chlorido ligand dissociates in water, but the dinuclear copper(II) structure was found to be maintained in an aqueous solution. In conclusion, the tetranuclear copper(II) structures were crystallographically characterized, and the dinuclear copper(II) structures were found to be stabilized even in an aqueous solution.

Keywords: tetranuclear copper(II) complex; dinucleating ligand; dimer-of-dimers type; crystal structure; magnetic properties; density functional theory (DFT)

1. Introduction

Copper is an essential trace element [1,2], and we humans cannot live without it. In fact, a 70 kg adult human body contains ~0.11 g of copper [1]. Humans need oxygen for cellular respiration to extract energy from food, and for cellular respiration, cytochrome *c* oxidase requires iron and copper to bind and activate oxygen [1–3]. In addition, toxic superoxide is produced daily together with cellular respiration, and superoxide dismutase (SOD) requires copper and zinc to decompose superoxide [1–3]. These are just a few examples of copper enzymes, and various copper proteins and copper enzymes play important roles in life. Some of the copper proteins have two or more copper ions at the active site and have functions that cannot be achieved by one copper ion. For the purpose of artificially realizing the function of such multi-copper proteins, many multinucleating ligands have been developed to stabilize the multinuclear metal complex structures.

2,6-Bis[bis(2-pyridylmethyl)aminomethyl]-4-methylphenol (H(bpmp)) is one of the most well-known acyclic dinucleating ligands, providing two N_3O coordination sites [4]. N,N'-(2-Hydroxy-5-methyl-1,3-xylylene)bis[N-(carboxymethyl)glycine] (H_5(5-Me-hxta)) is another well-known acyclic dinucleating ligand, possessing two NO_3 coordination sites [5,6]. Both ligands, (bpmp)$^-$ and (5-Me-hxta)$^{5-}$, are end-off type acyclic dinucleating ligands with a phenolato moiety as a bridging group and are suitable for incorporating various dinuclear metal cores. In addition, metal complexes with these ligands and their derivatives [7] are stable in aqueous solutions, while Schiff base ligands are often hydrolyzed in aqueous solutions. When (bpmp)$^-$ or (5-Me-hxta)$^{5-}$ incorporates two octahedral metal ions, two coordination sites will be available for substrate incorporation in catalytic reactions. In this study, for the purpose of increasing the number of available coordination sites, a novel dinucleating ligand, 2,6-bis{[N-(carboxylatomethyl)-N-methylamino]methyl}-4-methylphenolate ((sym-cmp)$^{3-}$) was synthesized (Figure 1). The ligand (sym-cmp)$^{3-}$ has a bridging phenolato moiety and two NO_2 coordination sites and is expected to incorporate two metal ions. This paper reports the crystal structures of dimer-of-dimers type tetranuclear copper(II) complexes with (sym-cmp)$^{3-}$.

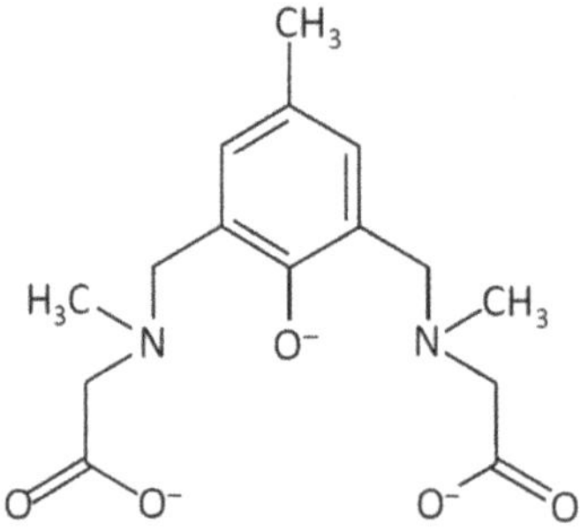

Figure 1. Chemical structure of (sym-cmp)$^{3-}$.

2. Results and Discussion

2.1. Preparation

2.1.1. Preparation of a Dinucleating Ligand

A dinucleating ligand, 2,6-bis{[N-(carboxylatomethyl)-N-methyl-amino]methyl}-4-methylphenolate ((sym-cmp)$^{3-}$) was synthesized via a Mannich reaction from p-cresol and sarcosine. The ligand was obtained as a sodium salt and recrystallized from ethanol. The ligand was characterized by IR, elemental analysis, ^{1}H and ^{13}C NMR, and electrospray ionization (ESI) mass spectrometry. In ^{1}H NMR, five singlet signals characteristic for (sym-cmp)$^{3-}$ were obtained (Figure S1), and in ^{13}C NMR, nine characteristic signals were obtained (Figure S2). In ESI-mass spectra, the main peak at $m/z = 309$ was assigned to [H_2(sym-cmp)]$^-$ (Figure S3), and its elemental composition ($C_{15}H_{21}N_2O_5$) was confirmed by the isotope pattern (Figure S4). Judging from the small molar conductance value (19 S·cm^2·mol^{-1}) in water, the sodium ions are considered to be tightly incorporated in the ligand.

2.1.2. Preparation of Copper(II) Complexes

With the dinucleating ligand (sym-cmp)$^{3-}$, tetranuclear copper(II) complexes were prepared as dimers of dinuclear copper(II) units. Using the copper(II) chloride, a chlorido complex, [Cu_4(sym-cmp)$_2Cl_2(H_2O)_2$]·$2H_2O$ (**1**), was obtained, while a methoxido derivative, [Cu_4(sym-cmp)$_2$(CH_3O)$_2$(CH_3OH)$_2$]·$2C_3H_7OH$·$2CH_3OH$ (**2**), was obtained by using the copper(II) nitrate. Purification of **2** was very difficult, and the crude product often contains nitrate ions. So, complex **2** was characterized only by the single-crystal X-ray diffraction method. Structural details will be described in the following crystallographic section (Section 2.2) and the theoretical calculation section (Section 2.4).

2.2. Crystal Structures of Copper(II) Complexes

2.2.1. Crystal Structures of [Cu$_4$(sym-cmp)$_2$Cl$_2$(H$_2$O)$_2$]·2.4CH$_3$OH·1.8H$_2$O (**1'**)

Single crystals of **1'** were obtained by recrystallization of **1** from methanol. (Note that **1** is a dried sample, while **1'** is a sample in a crystalline state where drying was prevented.) Although **1** was considered to contain two water molecules as solvent of crystallization per tetranuclear copper(II) unit from the elemental analysis, solvents of crystallization of **1'** were empirically determined as 2.4 methanol and 1.8 water molecules. The crystal structure of the tetranuclear copper(II) complex [Cu$_4$(sym-cmp)$_2$Cl$_2$(H$_2$O)$_2$] and its tetranuclear bridging structure are shown in Figure 2, and selected atomic distances and angles are summarized in Tables 1 and 2.

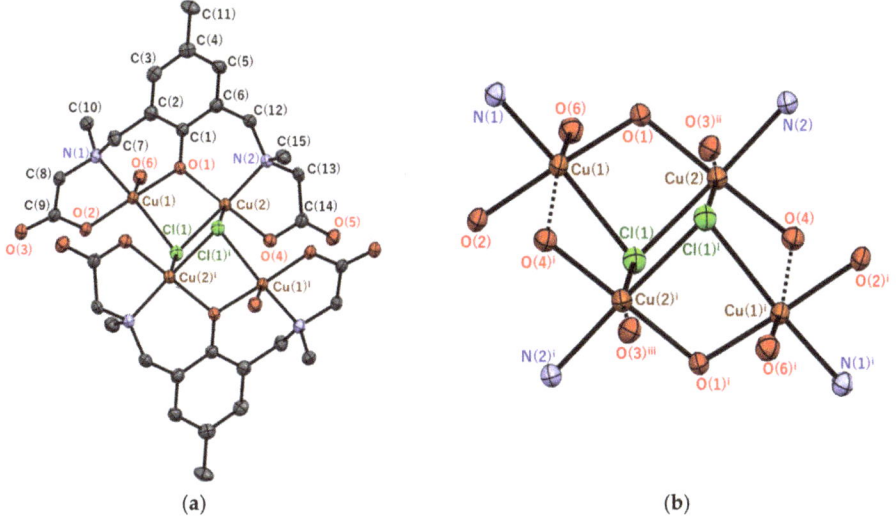

Figure 2. Molecular structures of (**a**) [Cu$_4$(sym-cmp)$_2$Cl$_2$(H$_2$O)$_2$] and (**b**) Cu$_4$Cl$_2$N$_4$O$_{10}$ core in **1'** with atom labeling. Hydrogen atoms are omitted for clarity. Thermal ellipsoids are drawn at the 50% probability level. Symmetry code: [i] ($-x + 1/2, -y + 1/2, -z + 1$), [ii] ($-x + 1/2, y + 1/2, -z + 3/2$), [iii] ($x, -y, z - 1/2$).

Table 1. Selected distances for **1'**.

Atom–Atom [1]	Distance/Å	Atom–Atom [1]	Distance/Å
Cu(1)–Cl(1)	2.3638(14)	Cu(1)–O(1)	1.937(3)
Cu(1)–O(2)	1.926(3)	Cu(1)–O(6)	2.214(4)
Cu(1)–N(1)	1.998(4)	Cu(1)–O(4)[i]	3.208(4)
Cu(2)–Cl(1)	2.3113(14)	Cu(2)–O(1)	1.942(3)
Cu(2)–O(4)	1.936(3)	Cu(2)–N(2)	2.003(4)
Cu(2)–Cl(1)[i]	2.8012(14)	Cu(2)–O(3)[ii]	2.804(4)
Cu(1)···Cu(2)	3.1274(8)	Cu(1)···Cu(1)[i]	5.8824(12)
Cu(1)···Cu(2)[i]	3.8024(9)	Cu(2)···Cu(2)[i]	3.7248(13)

[1] Symmetry code: [i] ($-x + 1/2, -y + 1/2, -z + 1$), [ii] ($-x + 1/2, y + 1/2, -z + 3/2$).

Table 2. Selected angles for **1'**.

Atom–Atom–Atom [1]	Angle/°	Atom–Atom–Atom [1]	Angle/°
Cl(1)–Cu(1)–O(1)	83.38(11)	Cl(1)–Cu(1)–O(2)	94.87(10)
Cl(1)–Cu(1)–O(6)	94.72(10)	Cl(1)–Cu(1)–N(1)	161.10(12)
Cl(1)–Cu(1)–O(4)^i	77.29(7)	O(1)–Cu(1)–O(2)	173.59(15)
O(1)–Cu(1)–O(6)	90.83(14)	O(1)–Cu(1)–N(1)	94.06(15)
O(1)–Cu(1)–O(4)^i	92.89(12)	O(2)–Cu(1)–O(6)	95.46(14)
O(2)–Cu(1)–N(1)	85.61(15)	O(2)–Cu(1)–O(4)^i	80.71(12)
O(6)–Cu(1)–N(1)	104.06(15)	O(6)–Cu(1)–O(4)^i	170.73(12)
N(1)–Cu(1)–O(4)^i	84.16(13)	Cl(1)–Cu(2)–O(1)	84.68(11)
Cl(1)–Cu(2)–O(4)	95.38(11)	Cl(1)–Cu(2)–N(2)	172.22(13)
Cl(1)–Cu(2)–Cl(1)^i	86.96(5)	Cl(1)–Cu(2)–O(3)^{ii}	84.35(8)
O(1)–Cu(2)–O(4)	175.93(15)	O(1)–Cu(2)–N(2)	94.72(16)
O(1)–Cu(2)–Cl(1)^i	89.05(11)	O(1)^i–Cu(2)–O(3)^{ii}	130.04(9)
O(4)–Cu(2)–N(2)	84.66(15)	O(4)–Cu(2)–Cl(1)^i	95.02(11)
O(4)–Cu(2)–O(3)^{ii}	88.47(12)	N(2)–Cu(2)–Cl(1)^i	100.79(12)
N(2)–Cu(2)–O(3)^{ii}	87.87(14)	Cl(1)^i–Cu(2)–O(3)^{ii}	170.91(8)
Cu(1)–Cl(1)–Cu(2)	83.96(5)	Cu(1)–O(1)–Cu(2)	107.46(17)
Cu(1)–Cl(1)–Cu(2)^i	94.44(4)	Cu(2)–Cl(1)–Cu(2)^i	93.04(5)

[1] Symmetry code: ^{i} $(-x + 1/2, -y + 1/2, -z + 1)$, ^{ii} $(-x + 1/2, y + 1/2, -z + 3/2)$.

The tetranuclear copper(II) complex [Cu$_4$(*sym*-cmp)$_2$Cl$_2$(H$_2$O)$_2$] is centrosymmetric (Figure 2a) and considered as a dimer-of-dimers type tetranuclear copper(II) complex, possessing the defect cubane tetranuclear copper(II) core (Figure 2b). Each dinucleating ligand, (*sym*-cmp)$^{3-}$, incorporates two copper(II) ions bridged by one phenolic oxygen of the dinucleating ligand and by one chlorido ligand. If we consider only the typical coordination bonds, each copper(II) ion has five-coordinate square-pyramidal coordination geometry, and Cu(1) and Cu(2) ions are surrounded by NO$_3$Cl and NO$_2$Cl$_2$ donor sets, respectively. The distortion parameter τ defined as τ = (θ − φ)/60 × 100% is calculated as 20.9% for Cu(1) and 6.1% for Cu(2), where θ° and φ° are the largest and the second largest bond angles around each copper atom, respectively. The parameter τ is 100% if the coordination geometry is purely trigonal–bipyramidal, while τ is 0% if the geometry is purely square–pyramidal. Therefore, both coordination geometries are considered to be square–pyramidal. Each of the apical bond distances (Cu(1)–O(6) = 2.222(3) Å and Cu(2)–Cl(1)^i = 2.8011(13) Å) is longer than the other basal Cu–O (1.925(3)–1.943(3) Å) and Cu–Cl (2.3124(13)–2.3630(12) Å) bond lengths, respectively. This can be explained by the Jahn–Teller effect [8], typical for the copper(II) complexes with d^9 electronic configuration.

The apical Cu–Cl distance (Cu(2)–Cl(1)^i = 2.8011(13) Å) may seem to be slightly too long for the coordination bond; however, the covalent bond character was confirmed by the density functional theory (DFT) calculation (Section 2.4). Therefore, the chlorido ligand is definitely bridging three copper(II) ions, forming the defect cubane tetranuclear copper(II) core structure. On the other hand, two more weak coordination bonds were found by DFT calculations (dashed bonds in Figure 2b). One is the Cu(1)–O(4)^i bond (3.209(3) Å), and the other is the Cu(2)–O(3)^{ii} bond (2.804(4) Å) between Cu(2) and an oxygen atom in a neighboring tetranuclear copper(II) complex. Here, the weak coordination bond refers to a bond with less covalency than the typical coordination bond, where the overlap of atomic orbitals involved in the bond is smaller. When the weak coordination bonds are also taken into account, the coordination geometries around the two copper(II) ions are both octahedral.

Tetranuclear copper(II) complexes with similar tetranuclear copper(II) cores are reported [9–11], and their magnetic properties were analyzed. In two of them, all three adjacent copper(II) pairs are doubly bridged [9,10], while in the rest of them, two of the adjacent copper(II) pairs are doubly-bridged ones, but one pair is singly-bridged [11]. In a precise sense, this type of core structure is often called the stepped cubane, but in the case of **1'**, the core structure can be included in the defect cubane.

2.2.2. Crystal Structures of [Cu$_4$(sym-cmp)$_2$(CH$_3$O)$_2$(CH$_3$OH)$_2$]·2C$_3$H$_7$OH·2CH$_3$OH (2)

Single crystals of **2** were obtained by slow diffusion of 2-propanol to a methanolic solution of the crude product. The crystal structure of the tetranuclear copper(II) complex [Cu$_4$(sym-cmp)$_2$(CH$_3$O)$_2$(CH$_3$OH)$_2$] and its tetranuclear bridging structure are shown in Figure 3, and selected atomic distances and angles are summarized in Tables 3 and 4.

The basic skeletal structure of the tetranuclear copper(II) complex [Cu$_4$(sym-cmp)$_2$(CH$_3$O)$_2$(CH$_3$OH)$_2$] in **2** is very similar to the complex structure in **1′**. That is, the bridging chlorido and water ligands in **1′** are replaced with methoxido and methanol ligands, respectively. The [Cu$_4$(sym-cmp)$_2$(CH$_3$O)$_2$(CH$_3$OH)$_2$] complex is centrosymmetric (Figure 3a) and considered a dimer-of-dimers type tetranuclear copper(II) complex, possessing the defect cubane tetranuclear copper(II) core (Figure 3b). A pair of copper(II) ions incorporated into one dinucleating ligand are bridged by one phenolic oxygen of the dinucleating ligand and by one methoxido ligand. Both types of copper(II) ions have five-coordinate square-pyramidal coordination geometries with NO$_4$ donor atoms, if we consider only the typical coordination bonds. The distortion parameter τ was 16.5% for Cu(1) and 11.0% for Cu(2). The apical bond distances (Cu(1)–O(7) = 2.331(3) Å and Cu(2)–O(3)^i = 2.305(3) Å) are longer than the other basal Cu–O distances (1.930(3)–1.999(3) Å). The longer apical distances are consistent with the d^9 electronic configuration of the copper(II) centers discussed in Section 2.2.1. When weak coordination bonds were also taken into account, another bond, Cu(1)–O(5)^i (2.856(3) Å), was found in the DFT calculation, and the coordination geometry around Cu(1) became octahedral. In contrast, no covalent nature was observed between Cu(2) and adjacent O(6)^{ii} in another complex.

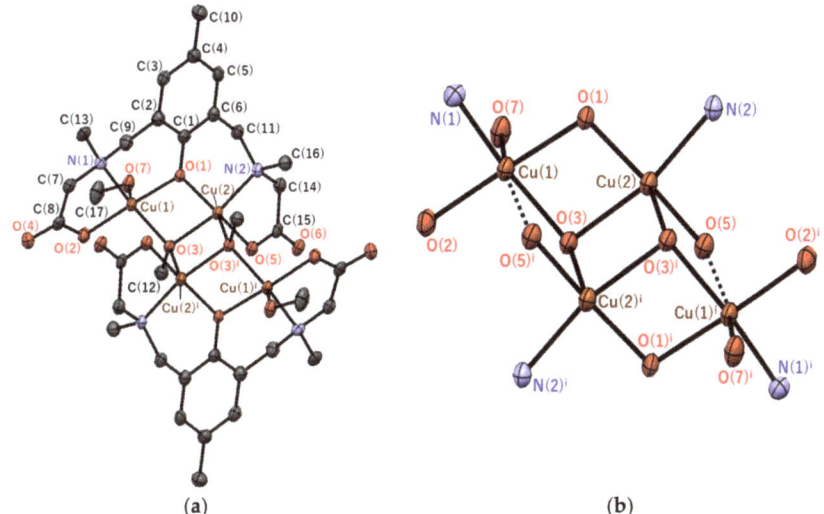

Figure 3. Molecular structures of (**a**) [Cu$_4$(sym-cmp)$_2$(CH$_3$O)$_2$(CH$_3$OH)$_2$] and (**b**) Cu$_4$N$_4$O$_{10}$ core in **2** with atom labeling. Hydrogen atoms are omitted for clarity. Thermal ellipsoids are drawn at the 50% probability level. Symmetry code: ^{i} ($-x + 1, -y + 1, -z + 1$).

Table 3. Selected distances for **2**.

Atom–Atom [1]	Distance/Å	Atom–Atom [1]	Distance/Å
Cu(1)–O(1)	1.930(3)	Cu(1)–O(2)	1.933(3)
Cu(1)–O(3)	1.981(3)	Cu(1)–O(7)	2.331(3)
Cu(1)–N(1)	2.008(4)	Cu(1)–O(5)^i	2.856(3)
Cu(2)–O(1)	1.930(3)	Cu(2)–O(3)	1.999(3)
Cu(2)–O(5)	1.936(3)	Cu(2)–N(2)	2.019(4)
Cu(2)–O(3)^i	2.305(3)	Cu(2)···O(6)^{ii}	3.3478(3)
Cu(1)···Cu(2)	3.0138(8)	Cu(1)···Cu(1)^i	5.5123(12)
Cu(1)···Cu(2)^i	3.3743(9)	Cu(2)···Cu(2)^i	3.2483(11)

[1] Symmetry code: ^{i} $(-x+1, -y+1, -z+1)$, ^{ii} $(-x+2, -y+1, -z+1)$.

Table 4. Selected angles for **2**.

Atom–Atom–Atom [1]	Angle/°	Atom–Atom–Atom [1]	Angle/°
O(1)–Cu(1)–O(2)	174.14(13)	O(1)–Cu(1)–O(3)	79.22(13)
O(1)–Cu(1)–O(7)	88.52(13)	O(1)–Cu(1)–N(1)	93.40(14)
O(1)–Cu(1)–O(5)^i	89.28(12)	O(2)–Cu(1)–O(3)	100.38(14)
O(2)–Cu(1)–O(7)	97.33(13)	O(2)–Cu(1)–N(1)	85.53(14)
O(2)–Cu(1)–O(5)^i	84.96(11)	O(3)–Cu(1)–O(7)	90.78(13)
O(3)–Cu(1)–N(1)	164.26(14)	O(3)–Cu(1)–O(5)^i	76.54(11)
O(7)–Cu(1)–N(1)	103.00(14)	O(7)–Cu(1)–O(5)^i	167.32(11)
N(1)–Cu(1)–O(5)^i	89.59(13)	O(1)–Cu(2)–O(3)	78.76(13)
O(1)–Cu(2)–O(5)	170.63(13)	O(1)–Cu(2)–N(2)	93.49(14)
O(1)–Cu(2)–O(3)^i	96.27(13)	O(3)–Cu(2)–O(5)	99.19(13)
O(3)–Cu(2)–N(2)	164.05(15)	O(3)–Cu(2)–O(3)^i	82.25(13)
O(5)–Cu(2)–N(2)	86.18(14)	O(5)–Cu(2)–O(3)^i	92.48(12)
N(2)–Cu(2)–O(3)^i	112.69(13)	Cu(1)–O(1)–Cu(2)	102.66(15)
Cu(1)–O(3)–Cu(2)	98.45(14)	Cu(1)–O(3)–Cu(2)^i	103.63(13)
Cu(2)–O(3)–Cu(2)^i	97.75(12)		

[1] Symmetry code: ^{i} $(-x+1, -y+1, -z+1)$.

2.3. Magnetic Properties

The cryomagnetic behavior for complex **1** was measured for the purpose of confirming the electronic configuration of the ground state and revealing the exchange interactions between the copper(II) ions. The $\chi_M T$ versus T plot is shown in Figure 4a. The observed $\chi_M T$ product at 300 K was 1.63 cm^3·K·mol^{-1}, which was close to the spin-only value for the four independent $S = 1/2$ magnetic centers (1.50 cm^3·K·mol^{-1}). The $\chi_M T$ value decreased on cooling to 1.9 K (0.026 cm^3·K·mol^{-1}), suggesting a strong antiferromagnetic interaction between copper(II) ions. For the magnetic analysis, the method of Hatfield and Inman [12] was used to obtain the magnetic susceptibility equation. In this study, the Hamiltonian $H = -J_1 (S_{A1} \cdot S_{B1} + S_{A2} \cdot S_{B2}) - J_2 S_{B1} \cdot S_{B2} - J_3 (S_{A1} \cdot S_{B2} + S_{A2} \cdot S_{B1}) - J_4 S_{A1} \cdot S_{A2}$ was used (see Figure 4b). The magnetic susceptibility equations (Equations (1)–(11)) used in this study are as follows:

$$\chi_M = \frac{Ng^2\beta^2}{kT} \frac{10\exp(-\frac{A}{kT}) + 2\exp(-\frac{B}{kT}) + 2\exp(-\frac{C}{kT}) + 2\exp(-\frac{D}{kT})}{5\exp(-\frac{A}{kT}) + 3\exp(-\frac{B}{kT}) + 3\exp(-\frac{C}{kT}) + 3\exp(-\frac{D}{kT}) + \exp(-\frac{E}{kT}) + \exp(-\frac{F}{kT})} (1-\rho) + \frac{Ng^2\beta^2}{kT}\rho + 4\,\text{TIP}, \quad (1)$$

$$A = -\frac{K}{2} - Q, \quad (2)$$

$$B = -\frac{K}{2} + Q, \quad (3)$$

$$C = \frac{K}{2} - \sqrt{L^2 + P^2}, \quad (4)$$

$$D = \frac{K}{2} + \sqrt{L^2 + P^2}, \quad (5)$$

$$E = \frac{K}{2} + Q - \sqrt{K^2 + 3L^2 - 2KQ + Q^2}, \quad (6)$$

$$F = \frac{K}{2} + Q + \sqrt{K^2 + 3L^2 - 2KQ + Q^2}, \quad (7)$$

$$K = \frac{J_2 + J_4}{2}, \quad (8)$$

$$L = \frac{J_1 - J_3}{2}, \quad (9)$$

$$P = \frac{J_2 - J_4}{2}, \quad (10)$$

$$Q = \frac{J_1 + J_3}{2}, \quad (11)$$

where TIP and ρ are the temperature-independent paramagnetism per copper and the paramagnetic impurity with $S = 1/2$, respectively.

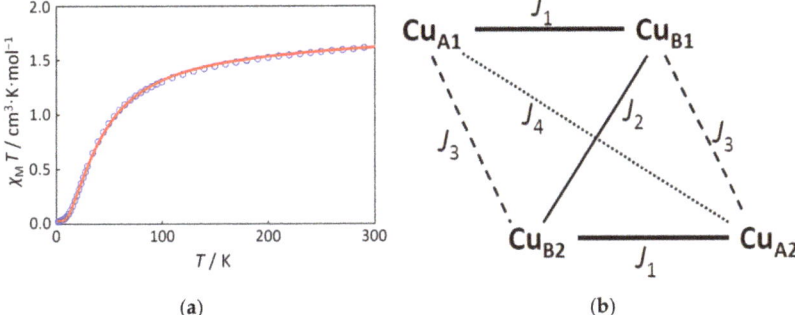

Figure 4. (a) The $\chi_M T$ versus T plot for **1**. The observed data (○) and the theoretical curve (—) with the best-fitting parameter set (J_1, J_2, J_3, J_4, g, TIP, ρ) = (−47.9 cm^{-1}, −38.5 cm^{-1}, 15.3 cm^{-1}, 0 cm^{-1} (fixed), 2.10, 60 × 10^{-6} cm^3·mol^{-1} (fixed), 0.0196); (**b**) Interactions in the centrosymmetric tetranuclear copper(II) core.

As the result, the best fitting parameter set was found to be (J_1, J_2, J_3, J_4, g, TIP, ρ) = (−47.9 cm^{-1}, −38.5 cm^{-1}, 15.3 cm^{-1}, 0 cm^{-1} (fixed), 2.10, 60 × 10^{-6} cm^3·mol^{-1} (fixed), 0.0196) with a good discrepancy factor ($R\chi = 7.2 \times 10^{-5}$). Overall, the magnetic interaction is antiferromagnetic, but the strongest antiferromagnetic interaction is considered to occur between the copper(II) ions in the same dinucleating ligand, expressed as J_1 (=−47.9 cm^{-1}), bridged by one phenolato oxygen and one chlorido chlorine atoms. The second strongest antiferromagnetic interaction occurs between the copper(II) ions bridged by two chlorido ligands, expressed as J_2 (=−38.5 cm^{-1}). The third interaction, considered to be ferromagnetic, was between the copper(II) ions bridged by one chlorido ligand, expressed as J_3 (=15.3 cm^{-1}). This order, $|J_1| > |J_2| > |J_3|$, is consistent with the order of Cu···Cu distances. That is, the shorter the distance, the stronger the interaction, although this is a rough estimation. From the viewpoint of the molecular orbital theory, the magnetic orbitals should be on the basal planes of square–pyramidal coordination geometries around copper(II) ions. Using the local coordinates, each magnetic orbital is expressed as $d_{x^2-y^2}$, assuming the local z-axis to the apical direction and the local x- and y-axes to the donor atom directions in the basal plane. Since the two local magnetic orbitals of the pair of copper(II) ions in the same dinucleating ligand are almost in the same plane, the strongest antiferromagnetic interaction is expected between Cu$_{A1}$ and Cu$_{B1}$ in Figure 4b, expressed as J_1. In this way, the obtained interaction parameters, J_1, J_2, and J_3, can be reasonably understood. Other obtained parameters are also reasonable for copper(II) complexes.

2.4. Density Functional Theory (DFT) Calculation

In order to confirm the bonding nature around copper(II) ions in the complexes, density functional theory (DFT) calculations were conducted. In particular, the Cu(2)–Cl(1)^i distance (2.8011(13) Å) in **1′** seems to be long, and whether this bond is a coordination bond or an ionic bond should be clarified based on the molecular orbital theory. As a result of the DFT calculation, bonding and anti-bonding orbitals were observed for the Cu(2)–Cl(1)^i bond (Figure 5), indicating the covalent nature of the bond. Note that the coordination bond is the same as the covalent bond from the viewpoint of molecular orbital theory [13], although they were long considered to be different. In **1′**, the Cu(2)–Cl(1)^i bond is formed using the d_{z^2} atomic orbital of Cu(2) atom and p_z atomic orbital of Cl(1)^i, where the z direction is defined as the local apical direction around the Cu(2) atom. In this way, the five typical coordination bonds around each copper(II) ion were confirmed by the DFT calculations for both complexes **1′** and **2**. In addition, molecular orbitals with bonding and anti-bonding characters were found for each of the three weak coordination bonds (Cu(1)–O(4)^i and Cu(2)–O(3)^{ii} in **1′** and Cu(1)–O(5)^i in **2**) discussed in Section 2.2, although the overlap of atomic orbitals involved was smaller than that of typical coordination bonds.

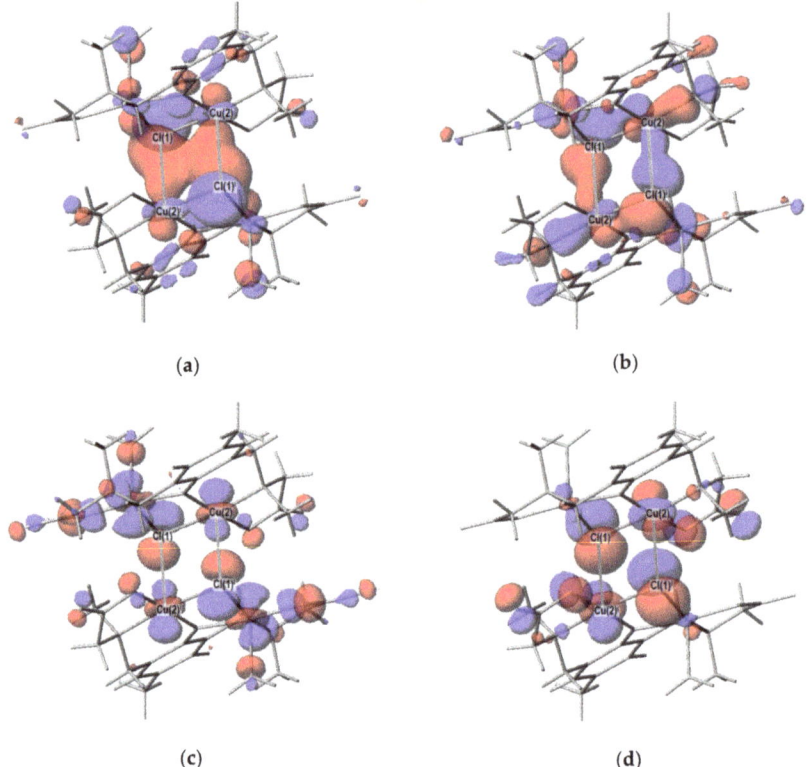

Figure 5. Molecular orbitals in **1′** with respect to the Cu(2)–Cl(1)^i coordination bond: (**a**) *gerade* bonding orbital; (**b**) *ungerade* bonding orbital; (**c**) *gerade* anti-bonding orbital; (**d**) *ungerade* anti-bonding orbital.

Another purpose of the DFT calculation is to confirm the magnetic orbitals possessing the unpaired electrons. The doubly degenerate highest occupied molecular orbitals and doubly degenerate lowest unoccupied molecular orbitals were all found to be based on the local $d_{x^2-y^2}$ atomic orbitals of copper(II) ions orientated along the four donor atoms in the basal plane of each copper(II) coordination polyhedron. In addition, the strong antiferromagnetic interaction (J_1) between Cu(1) and Cu(2) was understood by the phenolato and chlorido bridges, considering the overlaps of the local magnetic orbitals and bridging atomic orbitals.

2.5. Electronic Spectra and Structure in Aqueous Solution

The electronic spectra of complex **1** were measured in water (Figure 6). Judging from the molar conductance in water (see Section 3.2), the chlorido ligands were thought to be dissociated in water to break the tetranuclear copper(II) structure. However, judging from the green color of the solution, copper(II) ions were thought to be remain in the dinucleating ligand (*sym*-cmp)$^{3-}$. The spectra are shown in Figure 6. The first band at around 15,000 cm^{-1} can be assigned to the *d–d* band. As a result of the Gaussian curve fitting, the first band was found to consist of two absorption components at around 14,200 cm^{-1} and 16,400 cm^{-1}. The intensity of the first component is slightly larger than that of the second one, and this pattern is typical of trigonal–bipyramidal copper(II) complexes. Under the D_{3h} symmetry, the first and the second components are assigned to $^2A_{1'} \to {}^2E'$ and $^2A_{1'} \to {}^2E''$, respectively. The electronic spectra of the complexes containing trigonal–bipyramidal [CuCl$_5$]$^{3-}$ anions were investigated earlier [14], and the positions of the components of **1** are found to be reasonable, considering the ligand-field strengths. This spectral feature of **1** suggests that all the copper(II) ions have almost the same structures, incorporated in the dinucleating ligand to form [Cu$_2$(*sym*-cmp)(H$_2$O)$_4$]$^+$ species in aqueous solutions (Figure 7a). Generally, additional chlorido bridge or hydroxido bridge is expected to be formed between copper(II) ions; however, in this case, such an anionic bridging ligand is not so favorable because of the negative charge on the carboxylate side chains of the (*sym*-cmp)$^{3-}$ ligand. This idea is consistent with the facile dissociation of the chlorido ligands in aqueous solutions confirmed by the conductivity measurement described previously.

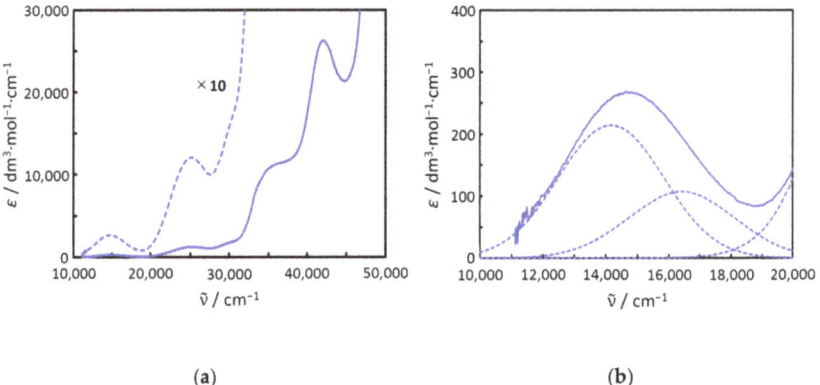

Figure 6. Electronic spectra of **1′** in water: (**a**) spectra in the range of 10,000–50,000 cm^{-1}; (**b**) the first band with the Gaussian spectral components.

The proposed species in an aqueous solution, [Cu$_2$(*sym*-cmp)(H$_2$O)$_4$]$^+$, was confirmed on the basis of the DFT calculation assuming the water environment. Judging from the electronic spectra, the coordination geometry was trigonal–bipyramidal. When the [Cu$_2$(*sym*-cmp)(H$_2$O)$_4$]$^+$ was structurally optimized with two additional water molecules, the proposed trigonal–bipyramidal structure was successfully reproduced as shown in Figure 7b. The distortion parameter τ around the copper(II) ions fell in the range of 85.4–86.0%, which is considered to be trigonal–bipyramidal. This DFT result is strong evidence of the proposed structure. The proposed structure, including two copper(II) ions with four water molecules, is in concordance with the previously proposed structures for dinuclear cobalt(II) and nickel(II) complexes with a related dinucleating ligand in aqueous solutions [7].

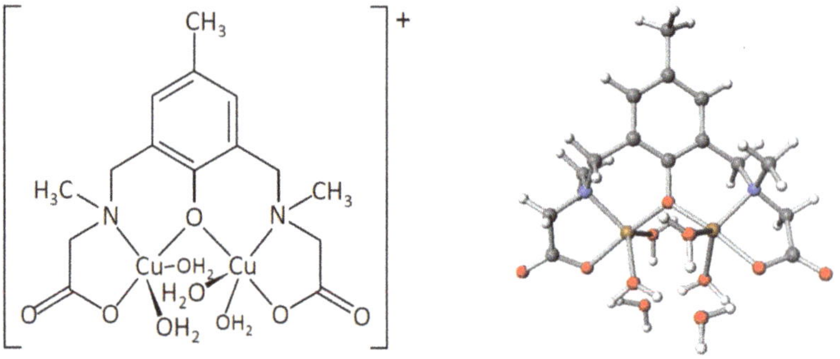

Figure 7. Proposed dinuclear copper(II) structure in aqueous solution: (**a**) chemical structure of $[Cu_2(sym\text{-}cmp)(H_2O)_4]^+$; (**b**) DFT-based optimized structure of $\{[Cu_2(sym\text{-}cmp)(H_2O)_4]\cdot 2H_2O\}^+$.

2.6. Structures of Metal Complexes and Ligand Design

Similar to the popular $(5\text{-}Me\text{-}hxta)^{5-}$ ligand, possessing four carboxylate chelating side chains, the $(sym\text{-}cmp)^{3-}$ ligand in this study possesses two carboxylate chelating side chains. Although the number of side chains is reduced, the ligand is still capable of holding two copper(II) ions in an aqueous solution. With the reduction of the ligand-occupying sites, a larger substrate is expected to be incorporated at the coordination sites. On the other hand, changes in the ligand charge can lead to a variety of metal complex structures. For example, 2,6-bis[(2-hydroxyethyl)methylaminomethyl]-4-methylphenolate $((sym\text{-}hmp)^-)$ [15] and 4-chloro-2,6-bis[(2-hydroxyethyl)methylaminomethyl] phenolate $((sym\text{-}hcp)^-)$ [16] each have a skeletal structure very similar to $(sym\text{-}cmp)^{3-}$ ligand; however, the obtained complex structures are different. The $(sym\text{-}hmp)^-$ and $(sym\text{-}hcp)^-$ ligands form 2:2 (ligand:metal) metal complexes, $[M_2(sym\text{-}hmp)_2]$ (M = Mg(II) [17], Mn(II) [16], Co(II) [15,18,19], Ni(II) [20,21], Zn(II) [21]), while the $(sym\text{-}cmp)^{3-}$ ligand forms 2:4 (ligand:metal) metal complexes, $[M_4(sym\text{-}hmp)_2X_2Y_2]$ (M = Cu(II); X = Cl$^-$, CH$_3$O$^-$; Y = H$_2$O, CH$_3$OH), as presented in this paper. This difference is thought to be caused by the difference in ligand charge. Concludingly, the skeletal structure, charge, bulkiness, etc., of the ligand can give rise to various metal complex structures. The knowledge of creating various controlled structures will also enable the development of metal-organic frameworks (MOFs) and is expected to be useful in various applications [22–29], beyond the use of molecular complexes as homogeneous catalysts.

3. Materials and Methods

3.1. Measurements

Elemental analyses (C, H, and N) were performed at the Elemental Analysis Service Centre of Kyushu University. Copper(II) ions were quantified by titration with ethylenediaminetetraacetic acid in the presence of hydrochloric acid, using murexide as an indicator. IR spectra were recorded on a Jasco FT/IR-4100 FT-IR spectrometer. ^{1}H and ^{13}C NMR spectra (400 MHz) on a Bruker-Biospin AV 400 NMR spectrometer in D$_2$O, electrospray ionization (ESI) mass spectra on a Waters Quattro micro API mass spectrometer in methanol, and electronic spectra on Jasco V-560 (200–800 nm) and Hitachi 330 (800–2000 nm) spectrophotometers. Molar conductance was measured in H$_2$O on a DKK AOL-10 conductivity meter at room temperature. Magnetic susceptibility measurements were performed with a Quantum Design MPMS-7 SQUID magnetometer in the temperature range from 1.9 to 300 K with a static field of 5 kOe. The polycrystalline samples were ground into fine powders in an agate mortar. The sample was wrapped with aluminum foil. Data were corrected for paramagnetism of the aluminum foil. The susceptibilities were corrected for the diamagnetism of the samples by means of Pascal's constants.

3.2. Materials

All the chemicals were commercial products and were used as supplied. Methanol, ethanol, copper(II) nitrate–water (1/3), copper(II) chloride–water (1/2), paraformaldehyde, *p*-cresol, sodium hydroxide, lithium hydroxide–water (1/1), phosphorus pentoxide, 2-propanol, ethylenediaminetetraacetic acid, and hydrochloric acid were supplied by Nacalai Tesque Inc. (Kyoto, Japan). Sarcosine and murexide were supplied by Tokyo Chemical Industry Co., Ltd. (Tokyo, Japan).

3.3. Preparations

Disodium 2,6-bis{[*N*-(carboxylatomethyl)-*N*-methyl-amino]methyl}-4-methylphenol—water (1/3) (Na$_2$H(*sym*-cmp)·3H$_2$O). To an aqueous solution (20 mL) containing *p*-cresol (5.41 g, 50 mmol), NaOH (6.10 g, 153 mmol), sarcosine (8.95 g, 100 mmol), and paraformaldehyde (3.00 g, 100 mmol) were added ethanol (20 mL) and the resulting solution was refluxed for 1 week. Ethanol and water were removed by evaporation to give Na$_2$H(*sym*-cmp) as a colorless powder. Yield 14.25 g (70%). (Found: C, 44.00; H, 6.45; N, 7.00; Calc. for C$_{15}$H$_{20}$N$_2$Na$_2$O$_5$·3H$_2$O: C, 44.10; H, 6.40; N, 6.85). Selected IR data [$\tilde{\nu}$/cm^{-1}] using KBr disk (Figure S5): 3325, 2985, 2955, 2830, 1585, 1420, 1405, 1365, 1330, 1245, 855, 770, 715, 665. ^{1}H NMR in D$_2$O: δ 2.13 (s, 3 H), 2.27 (s, 6 H), 3.13 (s, 4 H), 3.67 (s, 4 H), 6.94 (s, 2 H). ^{13}C NMR in D$_2$O: δ 19.29, 40.97, 56.74, 59.86, 122.77, 127.10, 130.81, 155.36, 177.46. ESI mass spectrum in MeOH: *m*/*z* 309, [H$_2$(*sym*-cmp)]$^-$; 331, [NaH(*sym*-cmp)]$^-$. Molar conductance in H$_2$O [Λ/S·cm^2·mol^{-1}] 19.

[Cu$_4$(*sym*-cmp)$_2$Cl$_2$(H$_2$O)$_2$]·2H$_2$O **1**. To a methanolic solution (5 mL) of copper(II) chloride—water (1/2) (0.34 g, 2.0 mmol) was added a methanolic solution (5 mL) of Na$_2$H(*sym*-cmp)·3H$_2$O (0.38 g, 0.93 mmol), and the resulting solution was stirred for 30 min to give the precipitation of green powder. Recrystallized from methanol, washed with methanol, and dried in vacuo over P$_2$O$_5$. Yield 0.30 g (64%) (Found: C, 35.20; H, 4.60; N, 5.50; Cu, 24.70; Calc. for C$_{30}$H$_{42}$Cl$_2$Cu$_4$N$_4$O$_{12}$·2H$_2$O: C, 35.60; H, 4.60; N, 5.55; Cu, 25.10). Selected IR data [$\tilde{\nu}$/cm^{-1}] using KBr disk (Figure S6): 3050-3700, 3010, 2970, 2920, 2865, 2815, 1630, 1475, 1385, 1195, 870, 545, 465. Molar conductance in H$_2$O [Λ/S·cm^2·mol^{-1}] 250 (1.1 × 10^{-3} mol·dm^{-3}), 250 (5.7 × 10^{-4} mol·dm^{-3}), 290 (1.1 × 10^{-4} mol·dm^{-3}).

[Cu$_4$(*sym*-cmp)$_2$(CH$_3$O)$_2$(CH$_3$OH)$_2$]·2C$_3$H$_7$OH·2CH$_3$OH **2**. To a methanolic solution (5 mL) of copper(II) nitrade—water (1/3) (0.24 g, 0.99 mmol) was added a methanolic solution (5 mL) of Na$_2$H(*sym*-cmp)·3H$_2$O (0.19 g, 0.47 mmol), and the resulting solution was refluxed for 2 h to give the precipitation of white powder. After filtration, the addition of 2-propanol (5 mL) resulted in the precipitation of dark-green powder. Recrystallized from methanol/2-propanol to give dark-green crystals. Yield 0.09 g (31%).

3.4. Crystallography

Crystallographic data are summarized in Table 5. Single crystals of **1'** suitable for X-ray analysis were obtained from a methanolic solution of **1**. Single crystals of **2** were obtained by slow diffusion of 2-propanol to a methanolic solution of the crude product. Single-crystal X-ray diffraction data were obtained with a Rigaku XtaLAB AFC11 diffractometer with graphite-monochromated Mo Kα radiation (λ = 0.71073 Å). A single crystal was mounted with a glass capillary and flash-cooled with a cold N$_2$ gas stream. Data were processed using the CrysAlisPro software packages. The structure was solved by intrinsic phasing methods using the SHELXT [30] software packages and refined on F2 (with all independent reflections) using the SHELXL [31] software packages. The non-hydrogen atoms were refined anisotropically, and hydrogen atoms were refined using the riding model. Complex **2** was refined as a two-component twin with only the non-overlapping reflections of component 1 and was refined using the hklf 5 routine with all reflections of component 1 (including the overlapping ones). The Cambridge Crystallographic Data Centre (CCDC) deposition numbers are included in Table 5.

Table 5. Crystallographic data and refinement parameters of **1'** and **2**.

Complex	1'	2
Empirical formula [1]	$C_{16.2}H_{27.6}ClCu_2N_2O_{8.1}$	$C_{21}H_{38}Cu_2N_2O_9$
Formula weight [1]	542.53	589.61
Crystal system	Monoclinic	triclinic
Space group	$C2/c$	$P\bar{1}$
$a/Å$	27.0693(14)	8.5827(6)
$b/Å$	13.2690(5)	13.0679(8)
$c/Å$	13.1356(7)	13.2862(7)
$\alpha/°$	90	115.021(6)
$\beta/°$	100.422(5)	102.208(5)
$\gamma/°$	90	115.021(6)
$V/Å^3$	4640.2(4)	1263.54(15)
Z [1]	8	2
Crystal dimensions/mm	$0.070 \times 0.050 \times 0.030$	$0.130 \times 0.057 \times 0.038$
T/K	100	100
$\lambda/Å$	0.71073	0.71073
$\rho_{calcd}/g \cdot cm^{-3}$	1.553	1.645
μ/mm^{-1}	1.990	1.734
$F(000)$	2229	616
$2\theta_{max}/°$	55	55
No. of reflections measured	9520	16014
No. of independent reflections	9520 (R_{int} = 0.0623)	5771 (R_{int} = 0.0770)
Data/restraints/parameters	9520/3/296	5771/45/341
$R1$ [2] $[I > 2.00 \sigma(I)]$	0.0575	0.0674
$wR2$ [3] (all reflections)	0.1664	0.1635
Goodness of fit indicator	1.020	0.992
Highest peak, deepest hole/e $Å^{-3}$	1.787, −0.647	1.669, −1.068
CCDC deposition number	2130618	2130619

[1] Based on dinuclear unit, [2] $R1 = \Sigma||Fo| - |Fc||/\Sigma|Fo|$, [3] $wR2 = [\Sigma(w(Fo^2 - Fc^2)^2)/\Sigma w(Fo^2)^2]^{1/2}$.

3.5. Computation

Magnetic analyses and magnetic simulation were conducted using the MagSaki(TetraW 9.2.0Cu) programs of the MagSaki series. DFT computations were performed using the GAMESS program [32,33] on Fujitsu PRIMERGY CX2550/CX2560 M4 (ITO super computer system) at Kyushu University. Calculations were performed with LC-BOP/6-31G [34]. When considering the solvent effect, the polarizable continuum model (PCM) method was used.

4. Conclusions

A water-soluble dinucleating ligand, (*sym*-cmp)$^{3-}$, was prepared, and two dimer-of-dimers type tetranuclear copper(II) complexes with (*sym*-cmp)$^{3-}$ were prepared. The structures of the complexes were crystallographically characterized, and [Cu$_4$(*sym*-cmp)$_2$Cl$_2$(H$_2$O)$_2$] and [Cu$_4$(*sym*-cmp)$_2$(CH$_3$O)$_2$(CH$_3$OH)$_2$] complexes were found to have the defect cubane tetranuclear copper(II) core structures. In the complexes, each copper(II) ion has a five-coordinate square-pyramidal coordination geometry, and the coordination bonds were confirmed by the DFT calculation, whereby we found the bonding and anti-bonding molecular orbitals. The cryomagnetic measurement was conducted to find the overall antiferromagnetic interaction in the tetranuclear copper(II) structure. The observed magnetic data were successfully simulated with the tetranuclear model to find reasonable magnetic parameters. Judging from the molar conductance and the electronic spectra, the tetranuclear structure was found to be broken in an aqueous solution, but the dinuclear copper(II) structure, [Cu$_2$(*sym*-

cmp)(H$_2$O)$_4$]$^+$, was considered to be maintained in an aqueous solution. This proposed structure was supported by DFT calculation.

Supplementary Materials: The following supporting information can be downloaded. Figure S1: ^{1}H NMR of Na$_2$H(sym-cmp)·3H$_2$O., Figure S2: ^{13}C NMR of Na$_2$H(sym-cmp)·3H$_2$O, Figure S3: ESI-mass spectra of Na$_2$H(sym-cmp)·3H$_2$O, Figure S4: Isotope pattern for Na$_2$H(sym-cmp)·3H$_2$O: (a) observed for m/z 309; (b) theoretical for m/z 309; (c) observed for m/z 331; (d) theoretical for m/z 331., Figure S5: IR spectra of of Na$_2$H(sym-cmp)·3H$_2$O, Figure S6: IR spectra of 1.

Author Contributions: Conceptualization, R.H. and H.S.; methodology, R.H., R.M., E.A. and H.S.; software, H.S.; validation, R.H. and H.S.; formal analysis, R.H., R.M., E.A. and H.S.; investigation, R.H., R.M., E.A. and H.S.; resources, R.H., R.M., J.L. and H.S.; data curation, R.H. and H.S.; writing—original draft preparation, R.H. and H.S.; writing—review and editing, R.H., R.M., E.A., J.L. and H.S.; visualization, R.H. and H.S.; supervision, H.S.; project administration, H.S. All authors have read and agreed to the published version of the manuscript.

Funding: This research received no external funding.

Institutional Review Board Statement: Not applicable.

Informed Consent Statement: Not applicable.

Data Availability Statement: The crystallographic data are available from the Cambridge Crystallographic Data Centre (CCDC). Other data not presented in Supplementary Materials are available on request from the corresponding author.

Acknowledgments: The magnetic measurements and the single-crystal X-ray measurement were conducted at the Institute of Molecular Science, supported by the Nanotechnology Platform Program (Molecule and Material Synthesis).

Conflicts of Interest: The authors declare no conflict of interest.

Sample Availability: Not applicable.

References

1. Kaim, W.; Schwederski, B. *Bioinorganic Chemistry: Inorganic Elements in the Chemistry of Life*; Wiley: Chichester, UK, 1991.
2. Cowan, J.A. *Inorganic Biochemistry*; VCH: New York, NY, USA, 1993.
3. Lippard, S.J.; Berg, J.M. *Principles of Bioinorganic Chemistry*; University Science Books: Mill Valley, CA, USA, 1994.
4. Suzuki, M.; Kanatomi, H.; Murase, I. Synthesis and properties of binuclear cobalt(II) oxygen adduct with 2,6-bis[bis(2-pyridylmethyl)aminomethyl]-4-methylphenol. *Chem. Lett.* **1981**, *10*, 1745–1748. [CrossRef]
5. Murch, B.P.; Boyle, P.D.; Que, L., Jr. Structures of binuclear and tetranuclear iron(III) complexes as models for ferritin core formation. *J. Am. Chem. Soc.* **1985**, *107*, 6728–6729. [CrossRef]
6. Murch, B.P.; Bradley, F.C.; Boyle, P.D.; Papaefthymiou, V.; Que, L., Jr. Iron-oxo aggregates. Crystal structures and solution characterization of 2-hydroxy-1,3-xylylenediaminetetraacetic acid complexes. *J. Am. Chem. Soc.* **1987**, *109*, 7993–8003. [CrossRef]
7. Kazama, A.; Wada, A.; Sakiyama, H.; Hossain, M.J.; Nishida, Y. Synthesis of water-soluble dinuclear metal complexes [metal = cobalt(II) and nickel(II)] and their behavior in solution. *Inorg. Chim. Acta* **2008**, *361*, 2918–2922. [CrossRef]
8. Jahn, H.A.; Teller, E. Stability of polyatomic molecules in degenerate electronic states I—Orbital degeneracy. *Proc. R. Soc. Lond. Ser. A-Math. Phys. Sci.* **1937**, *161*, 220–235.
9. Tandon, S.S.; Thompson, L.K.; Bridson, J.N.; Bubenik, M. Synthesis and structural and magnetic properties of mononuclear, dinuclear, and tetranuclear copper(II) complexes of a 17-membered macrocyclic ligand (HM3), capable of forming endogenous phenoxide and pyridazino bridges. X-ray crystal structures of [Cu2(M3)(µ2-OMe)(NO3)2], [Cu4(M3)2(µ3-OMe)2(µ2-Cl)2Cl2], [Cu4(M3)2(µ3-OEt)2(µ2-N3)2(N3)2](MeOH), [Cu4(M3)2(µ3-OMe)2(NCS)4](DMF), and [Cu(M3)(NCS)2]. *Inorg. Chem.* **1993**, *32*, 4621–4631.
10. Koikawa, M.; Yamashita, H.; Tokii, T. Crystal structures and magnetic properties of tetranuclear copper(II) complexes of N-(2-hydroxymethylphenyl)salicylideneimine with a defective double-cubane core. *Inorg. Chim. Acta* **2004**, *357*, 2635–2642. [CrossRef]
11. Li, X.; Cheng, D.; Lin, J.; Li, Z.; Zheng, Y. Di-, tetra-, and hexanuclear hydroxy-bridged copper(II) cluster compounds: Syntheses, structures, and properties. *Cryst. Growth Des.* **2008**, *8*, 2853–2861. [CrossRef]
12. Hatfield, W.E.; Inman, G.W. Spin-spin coupling in magnetically condensed complexes. IX. Exchange coupling constants for tetranuclear Schiff's base complexes of copper(II). *Inorg. Chem.* **1969**, *8*, 1376–1378. [CrossRef]
13. The IUPAC Compendium of Chemical Terminology (Gold Book Version 2.3.3). Available online: https://goldbook.iupac.org/ (accessed on 24 December 2021).

14. Allen, G.C.; Hush, N.S. Reflectance spectrum and electronic states of the $CuCl_5^{3-}$ ion in a number of crystal lattices. *Inorg. Chem.* **1967**, *6*, 4–8. [CrossRef]
15. Tone, K.; Sakiyama, H.; Mikuriya, M.; Yamasaki, M.; Nishida, Y. Magnetic behavior of dinuclear cobalt(II) complexes assumed to be caused by a paramagnetic impurity can be explained by tilts of local distortion axes. *Inorg. Chem. Commun.* **2007**, *10*, 944–947. [CrossRef]
16. Sakiyama, H.; Kato, M.; Sasaki, S.; Tasaki, M.; Asato, E.; Koikawa, M. Synthesis and magnetic properties of a dinuclear manganese(II) complex with two manganese(II) ions of C_2-twisted octahedral geometry. *Polyhedron* **2016**, *111*, 32–37. [CrossRef]
17. Sakiyama, H.; Takahata, S.; Kashimoto, N.; Mitsuhashi, R.; Mikuriya, M. Crystal structure of a dinuclear magnesium(II) complex with 4-chloro-2,6-bis[(2-hydroxyethyl)methylaminomethyl]phenolate. *X-ray. Struct. Anal. Online* **2017**, *33*, 75–76. [CrossRef]
18. Deutsch, M.; Claiser, N.; Gillet, J.-M.; Lecomte, C.; Sakiyama, H.; Tone, K.; Souhassou, M. d-Orbital orientation in a dimer cobalt complex: Link to magnetic properties? *Acta Cryst.* **2011**, *B67*, 324–332. [CrossRef]
19. Ridier, K.; Gillon, B.; Gukasov, A.; Chaboussant, G.; Cousson, A.; Luneau, D.; Borta, A.; Jacquot, J.-F.; Checa, R.; Chiba, Y.; et al. Polarized neutron diffraction as a tool for mapping molecular magnetic anisotropy: Local susceptibility tensors in Co^{II} complexes. *Chem. Eur. J.* **2016**, *22*, 724–735. [CrossRef]
20. Sakiyama, H.; Tone, K.; Yamasaki, M.; Mikuriya, M. Electronic spectrum and magnetic properties of a dinuclear nickel(II) complex with two nickel(II) ions of C_2-twisted octahedral geometry. *Inorg. Chim. Acta* **2011**, *365*, 183–189. [CrossRef]
21. Sakiyama, H.; Chiba, Y.; Tone, K.; Yamasaki, M.; Mikuriya, M.; Krzystek, J.; Ozarowski, A. Magnetic properties of a dinuclear nickel(II) complex with 2,6-bis[(2-hydroxyethyl)methylaminomethyl]-4-methylphenolate. *Inorg. Chem.* **2017**, *56*, 138–146. [CrossRef]
22. Pan, Y.; Rao, C.Y.; Tan, X.L.; Ling, Y.; Singh, A.; Kumar, A.; Li, B.H.; Liu, J.Q. Cobalt-seamed C-methylpyrogallol[4]arene nanocapsules-derived magnetic carbon cubes as advanced adsorbent toward drug contaminant removal. *Chem. Eng. J.* **2021**, 133857. [CrossRef]
23. Zhong, Y.Y.; Chen, C.; Liu, S.; Lu, C.Y.; Liu, D.; Pan, Y.; Sakiyama, H.; Muddassir, M.; Liu, J.Q. A new magnetic adsorbent of eggshell-zeolitic imidazolate framework for highly efficient removal of norfloxacin. *Dalton Trans.* **2021**, *50*, 18016–18026. [CrossRef]
24. Sun, Y.M.; Jiang, X.D.; Liu, Y.W.; Liu, D.; Chen, C.; Lu, C.Y.; Zhuang, S.Z.; Kumar, A.; Liu, J.Q. Recent advances in Cu(II)/Cu(I)-MOFs based nano-platforms for developing new nano-medicines. *J. Inorg. Biochem.* **2021**, *225*, 111599. [CrossRef]
25. Liu., Y.W.; Zhou, L.Y.; Dong, Y.; Wang, R.; Pan, Y.; Zhuang, S.Z.; Liu, D.; Liu, J.Q. Recent developments on MOF-based platforms for antibacterial therapy. *RSC Med. Chem.* **2021**, *12*, 915–928. [CrossRef]
26. Ding, Q.J.; Liu, Y.W.; Shi, C.C.; Xiao, J.F.; Dai, W.; Liu, D.; Chen, H.Y.; Li, B.H.; Liu, J.Q. Applications of ROS-InducedZr-MOFs platform in multimodal synergistic therapy. *Mini-Rev. Med. Chem.* **2021**, *21*, 1718–1733. [CrossRef]
27. Qiu, Y.Z.; Tan, G.J.; Fang, Y.Q.; Liu, S.; Zhou, Y.B.; Kumar, A.; Trivedi, M.; Liu, D.; Liu, J.Q. Biomedical applications of metal–organic framework (MOF)-based nano-enzymes. *New J. Chem.* **2021**, *45*, 20987–21000. [CrossRef]
28. Wang, J.; Rao, C.Y.; Lu, L.; Zhang, S.L.; Muddassir, M.; Liu, J.Q. Efficient photocatalytic degradation of methyl violet using two new 3D MOFs directed by different carboxylate spacers. *Cryst. Eng. Comm.* **2021**, *23*, 741–747. [CrossRef]
29. Liu, J.Q.; Luo, Z.D.; Pan, Y.; Singh, A.K.; Trivedi, M.; Kumar, A. Recent developments in luminescent coordination polymers: Designing strategies, sensing application and theoretical evidences. *Coord. Chem. Rev.* **2020**, *406*, 213145. [CrossRef]
30. Sheldrick, G.M. A short history of SHELX. *Acta Cryst. Sect. A* **2008**, *64*, 112–122. [CrossRef]
31. Sheldrick, G.M. Crystal structure refinement with SHELXL. *Acta Cryst. Sect. C* **2015**, *71*, 3–8. [CrossRef]
32. Schmidt, M.W.; Baldridge, K.K.; Boatz, J.A.; Elbert, S.T.; Gordon, M.S.; Jensen, J.H.; Koseki, S.; Matsunaga, N.; Nguyen, K.A.; Su, S.; et al. General atomic and molecular electronic structure system. *J. Comput. Chem.* **1993**, *14*, 1347–1363. [CrossRef]
33. Gordon, M.S.; Schmidt, M.W. *Advances in Electronic Structure Theory*; Elsevier: Amsterdam, The Netherlands, 2005.
34. Tawada, Y.; Tsuneda, T.; Yanagisawa, S.; Yanai, T.; Hirao, K. A long-range-corrected time-dependent density functional theory. *J. Chem. Phys.* **2004**, *120*, 8425–8433. [CrossRef]

Article

Magnetic and Luminescence Properties of 8-Coordinate Holmium(III) Complexes Containing 4,4,4-Trifluoro-1-Phenyl- and 1-(Naphthalen-2-yl)-1,3-Butanedionates

Franz A. Mautner [1,*], Florian Bierbaumer [1], Ramon Vicente [2], Saskia Speed [2], Ánnia Tubau [2], Mercè Font-Bardía [3], Roland C. Fischer [4] and Salah S. Massoud [5,6,*]

[1] Institute of Physical and Theoretical Chemistry, Graz University of Technology, Stremayrgasse 9, A-8010 Graz, Austria; bierbaumerflorian97@gmail.com
[2] Departament de Química Inorgànica i Orgànica, Universitat de Barcelona, Martí i Franquès 1-11, E-08028 Barcelona, Spain; rvicente@ub.edu (R.V.); saskia.speed@qi.ub.es (S.S.); anniatubau@ub.edu (Á.T.)
[3] Departament de Mineralogia, Cristallografia i Dipòsits Minerals and Unitat de Difracció de R-X, Centre Científic i Tecnològic de la Universitat de Barcelona (CCiTUB), Universitat de Barcelona, Solé i Sabarís 1–3, 08028 Barcelona, Spain; mercefont@ub.edu
[4] Institute of Inorganic Chemistry, Graz University of Technology, Stremayrgasse 9, A-8010 Graz, Austria; roland.fischer@tugraz.at
[5] Department of Chemistry, University of Louisiana at Lafayette, P.O. Box 43700, Lafayette, LA 70504, USA
[6] Department of Chemistry, Faculty of Sciences, Alexandria University, Moharam Bey, Alexandria 21511, Egypt
* Correspondence: mautner@tugraz.at (F.A.M.); ssmassoud@louisiana.edu (S.S.M.); Tel.: +43-316-873-32270 (F.A.M.); +1-337-482-5672 (S.S.M.); Fax: +43-316-873-8225 (F.A.M.); +1-337-482-5676 (S.S.M.)

Citation: Mautner, F.A.; Bierbaumer, F.; Vicente, R.; Speed, S.; Tubau, Á.; Font-Bardía, M.; Fischer, R.C.; Massoud, S.S. Magnetic and Luminescence Properties of 8-Coordinate Holmium(III) Complexes Containing 4,4,4-Trifluoro-1-Phenyl- and 1-(Naphthalen-2-yl)-1,3-Butanedionates. *Molecules* 2022, 27, 1129. https://doi.org/10.3390/molecules27031129

Academic Editor: Hiroshi Sakiyama

Received: 29 December 2021
Accepted: 4 February 2022
Published: 8 February 2022

Publisher's Note: MDPI stays neutral with regard to jurisdictional claims in published maps and institutional affiliations.

Copyright: © 2022 by the authors. Licensee MDPI, Basel, Switzerland. This article is an open access article distributed under the terms and conditions of the Creative Commons Attribution (CC BY) license (https://creativecommons.org/licenses/by/4.0/).

Abstract: A new series of mononuclear Ho^{3+} complexes derived from the β-diketonate anions: 4,4,4-trifluoro-1-phenyl-1,3-butanedioneate (btfa$^-$) and 4,4,4-trifluoro-1-(naphthalen-2-yl)-1,3-butanedionate (ntfa$^-$) have been synthesized, [Ho(btfa)$_3$(H$_2$O)$_2$] (**1a**), [Ho(ntfa)$_3$(MeOH)$_2$] (**1b**), (**1**), [Ho(btfa)$_3$(phen)] (**2**), [Ho(btfa)$_3$(bipy)] (**3**), [Ho(btfa)$_3$(di-^{t}bubipy)] (**4**), [Ho(ntfa)$_3$(Me$_2$bipy)] (**5**), and [Ho(ntfa)$_3$(bipy)] (**6**), where phen is 1,10-phenantroline, bipy is 2,2′-bipyridyl, di-^{t}bubipy is 4,4′-di-*tert*-butyl-2,2′-bipyridyl, and Me$_2$bipy is 4,4′-dimethyl-2,2′-bipyridyl. These compounds have been characterized by elemental microanalysis and infrared spectroscopy as well as single-crystal X-ray diffraction for **2–6**. The central Ho^{3+} ions in these compounds display coordination number 8. The luminescence-emission properties of the pyridyl adducts **2–6** display a strong characteristic band in the visible region at 661 nm and a series of bands in the NIR region (excitation wavelengths (λ_{ex}) of 367 nm for **2–4** and 380 nm for **5** and **6**). The magnetic properties of the complexes revealed magnetically uncoupled Ho^{3+} compounds with no field-induced, single-molecule magnet (SMMs).

Keywords: lanthanides; holmium; X-ray; diketones; magnetic properties; luminescence

1. Introduction

The luminescent emissions of lanthanides in general, and specifically holmium complexes, have been known for decades, as they play crucial roles in research and have a wide range of useful applications [1–28]. Compared to other lanthanides, holmium was proved to serve as a good candidate to make quantum computers, where one bit of data can be stored on a single holmium atom set on a bed of magnesium oxide [23,24]. In addition, Ho is used to generate the strongest artificial magnetic fields when placed within high-strength magnets [25]; Ho-dopped yttrium iron garnet is used in optical insulators, microwave equipment, and in solid-state lasers [26], and is one of the colorant's sources for yellow and red colors in glass and cubic zirconia [27].

Lanthanide ions, Ln^{3+} and their complexes, are known to exhibit narrow and characteristic *f–f* transitions of luminescent emissions that span from ultraviolet (UV) to visible and near-infrared (NIR) regions [1–3,29–32]. The *f–f* transitions in Ln^{3+} complexes are weak, but this process is enhanced via effective energy transfer from ligands or linker electrons to the central metal ions "antenna effect", from which the emission occurs [1–3,20,21,32–35]. Most of the investigated complexes, such as Eu^{3+} and Tb^{3+}, emit red or green luminescent light, respectively [36–39], but other Ln^{3+} complexes, such as those containing Yb^{3+}, Nd^{3+}, and Pr^{3+} metal ions, exhibit luminescence in the near-IR region [40–43].

The lanthanide cations (Ln^{3+}) as hard Lewis acids exhibit a strong binding affinity for O-donor ligands such as β-diketone compounds (HL) [43–49]. Typically electrically neutral tris complexes, $Ln(L)_3$ are most likely to be formed [40–55], but in some cases, the anionic tetrakis complexes, $(Cat^+)[Ln(L)_4]^-$ are also formed [49,56–59]. The two categories of these compounds exhibit good luminescent properties [40–59]. The luminescence efficiency of the β-diketonato complexes can be enhanced by the appropriate choice of the substituents on the β-diketone ligand because, in this way, the ligands' triplet levels can be tuned to provide efficient energy transfer between the diketonato ligand and the lanthanide ion [60–63]. This has been observed when aromatic and fluorinated alkyl groups are incorporated into the β-diketone skeletons. This helps in reducing the nonradiative quenching of lanthanide luminescence [40–42,50–63]. In the anionic $(Cat^+)[Ln(L)_4]$ complexes, additional tuning of the photophysical properties is possible by changing the counterion, Cat^+, which in turn changes the structure of the complex and, in particular, the local coordination geometry of the metal ion [56–59].

The rare-earth complexes with fluorinated-β-diketones (HL), such as L = 4,4,4-trifluoro-1-phenyl-1,3-butanedionate (btfa) and 4,4,4-trifuoro-1-(naphthalen-2-yl)-butane-1,3-dionate (L = ntfa) anions, have been extensively investigated. The structure formulas of H(btfa) and H(ntfa) are shown in Scheme 1. Among the Ln(III)-btfa complexes, half of them are for Eu(III) compounds [52–57,59,62–64], whereas the rest are for Dy(III) [50,51,60], Er(III) [55,61], Tb(III) [62], and Gd(III) [56,63]. In addition, small numbers were reported for Sm(III) [58], Pr(III) [42], and Ho(III) [65]. No structural results were found for La(III), Ce(III), Nd(III), Yb(III), nor Lu(III). In case of Ln(III)-ntfa, less structures were reported compared to the corresponding Ln(III)-btfa compounds, where most were obtained with Eu(III) [38–41,55,56,60–66], Gd(III) [43,53,55–57,66], and Pr(III) [42,45,48,65], some with Dy(III) [50,60,66] and Er(III) [55,65], as well as Tb(III) [62,66]. To the best of our knowledge, few structures were characterized with La(III) [49], Nd(III) [43], Ho(III) [65], and Sm(III) [65], but no structures for Ce(III), Yb(III), nor Lu(III) were found.

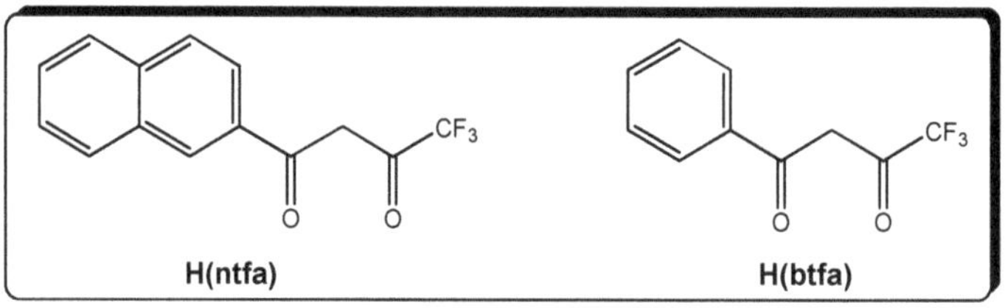

Scheme 1. Structures of the β-diketones used in this study.

As part of a long project to explore the coordination properties and the physicochemical properties of the less-studied Ln^{3+} ions with the β-diketones, Ho(btf) and Ho(ntfa), the following studies were undertaken and devoted for the interaction of these two compounds with Ho^{3+} ions in the presence of different polypyridyl ligands.

2. Materials and Methods

2.1. Materials and Physical Measurements

4,4,4-Trifluoro-1-(phenyl)butane-1,3-dione, 4,4,4-trifluoro-1-(naphthalen-2-yl)-butane-1,3-dione, 4,4'-di-*tert*-butyl-2,2'-bipyridine, 5,5'-dimethyl-2,2'-bipyridine, and 2,2'-bipyridine were purchased from TCI, and the other chemicals were of analytical grade quality. Infrared spectra of solid complexes were either recorded on a Bruker Alpha P (platinum-ATR-cap) spectrometer (Bruker AXS, Madison, WI, USA) or a Thermo Scientific Nicolet IS5 spectrophotometer. Elemental microanalyses were carried out with an Elementar Vario EN3 analyzer (Langenselbold, Germany) at the Serveis Científics i Tecnològics of the Universitat de Barcelona. PXRD patterns were recorded with a Bruker D8 Advance powder diffractometer (Cu-Kα radiation) (Bruker AXS, Madison, WI, USA).

Solid-state fluorescence spectra of compounds **2–6** were recorded on a Horiba Jobin Yvon SPEX Nanolog fluorescence spectrophotometer (Fluorolog-3 v3.2, HORIBA Jovin Yvon, Cedex, France) equipped with a three-slit, double-grating excitation and emission monochromator with dispersions of 2.1 nm/mm (1200 grooves/mm) at room temperature. The steady-state luminescence was excited by unpolarized light from a 450 W xenon CW lamp and detected at an angle of 22.5° for solid-state measurement by a red-sensitive Hamamatsu R928 photomultiplier tube. Near Infra-red (NIR) spectra were recorded at an angle of 22.5° using a liquid-nitrogen-cooled, solid indium/gallium/arsenic detector (900–1600 nm). The instrument was adjusted to obtain the highest background-to-noise ratio with a band pass of 2 for the visible and 10 for the NIR measurements. The sample was mounted between two quartz plates. Spectra were corrected for both the excitation source light intensity variation (lamp and grating) and the emission spectral response (detector and grating).

The magnetic susceptibility and magnetization measurements were performed with a Quantum Design MPMS-XL SQUID magnetometer at the Magnetic Measurements Unit of the University of Barcelona. Pascal's constants were used to estimate the diamagnetic corrections, which were subtracted from the experimental susceptibilities to give the corrected molar magnetic susceptibilities.

2.2. Synthesis of the Complexes

2.2.1. [Ho(btfa)$_3$(H$_2$O)$_2$] (**1a**)

To a methanol solution (10 mL) containing NaOH (6 mmol, 0.240 g), Hbtfa was added in an amount of 6 mmol, 0.130 g, and HoCl$_3$·6H$_2$O was added in an amount of 2 mmol, 0.759 g. The solution was stirred for 1 h at room temperature, then 80 mL of deionized water was added to the reaction mixture and stirred overnight. The light pink precipitate, which was obtained, was filtrated and dried in a desiccator overnight (yield: 1.194 g, 71%). Anal. Calcd. for C$_{30}$H$_{22}$F$_9$HoO$_8$ (846.4 g/mol): C, 42.6; H, 2.6%. Found: C, 42.5; H, 2.7%. Selected IR bands (cm^{-1}): 3658 (m), 3462 (br), 1609 (s), 1575 (s), 1527 (m), 1488 (m), 1464 (m), 1329 (s), 1283 (s), 1245 (m), 1182 (s), 1144 (s), 1071(m), 945 (m), 777 (m), 694 (m), 631(m), 580 (m).

2.2.2. [Ho(ntfa)$_3$(MeOH)$_2$] (**1b**)

A methanolic solution (10 mL) of Ho(NO$_3$)$_3$ 5H$_2$O (281 mg, 0.64 mmol) and a methanolic solution (20 mL) of 4,4,4-trifluoro-1-(2-naphthyl)-1,3-butanedione (515 mg, 1.93 mmol) with 1M NaOH (2.0 mL) were dissolved. After 20 min of stirring, the 4,4,4-trifluoro-1-(2-naphthyl)-1,3-butanedione solution was added to the Ho(NO$_3$)$_3$ 5H$_2$O solution. After 3 h of stirring, 30 mL of deionized water was added to complete the reaction. The mixture was stirred for 12 h at ambient temperature and then filtered. The obtained white powder was re-crystallized from MeOH and dried at 60 °C for 30 min (yield: 509 mg, 81%). Characterization: Anal. Calcd. for: C$_{44}$H$_{32}$F$_9$HoO$_8$ (1018.62 g/mol): C, 51.9; H, 3.2%. Found: C, 51.8; H, 3.1%. Selected IR bands (ATR-IR, cm^{-1}): 3448 (m, br), 1602 (s), 1594 (m), 1568 (m), 1529 (m), 1458 (w), 1356 (w), 1285 (s), 1251 (m), 1184 (s), 1124 (s), 1073 (w), 958 (w), 865 (w), 824 (w), 794 (s), 762 (w), 684 (w).

2.2.3. [Ho(btfa)$_3$(L)] (**2**: L = phen; **3**: L = bipy; **4**: L = di-^{t}Bubipy)

A general method was used to prepare the complexes **2–4**. An ethanol solution (15 mL) containing bipyridyl derivatives (1 mmol, **2**: 0.180 g 1,10-phenanthroline; **3**: 0.156 g 2,2′-bipyridine; **4**: 0.846 g 4,4′-di-*tert*-butyl-2,2′-bipyridine) was added to another ethanol solution (15 mL) containing [Ho(btfa)$_3$(H$_2$O)$_2$] (1 mmol, 0.846 g). The solution was stirred for 30 min and then left to stand at room temperature. Single light pink crystals suitable for

X-ray diffraction were obtained within a week. These were collected by filtration and dried with air.

[Ho(btfa)$_3$(phen)] (**2**) (yield: 38%). Characterization: Anal. Calcd. for C$_{42}$H$_{26}$F$_9$HoN$_2$O$_6$ (990.58 g/mol): C, 50.9; H, 2.6; N, 2.8%. Found: C, 50.7; H, 2.5; N, 2.8%. Selected IR bands (cm^{-1}): 1610 (s), 1574 (s), 1522 (s), 1483 (m), 1476 (m), 1319 (m), 1291 (s), 1246 (m), 1178 (s), 1134 (s) 1077 (m), 846 (m), 763 (s), 770 (s), 631 (m), 580 (m).

[Ho(btfa)$_3$(bipy)] (**3**) (yield: 80%). Characterization: Anal. Calcd. for C$_{40}$H$_{26}$F$_9$HoN$_2$O$_6$ (966.56 g/mol): C, 49.7; H, 2.7; N, 2.9%. Found: C, 49.7; H, 2.5; N, 2.8%. Selected IR bands (cm^{-1}): 1606 (s), 1569 (s), 1533 (m), 1472 (m), 1320 (m), 1279 (s), 1242 (m), 1177 (s), 1122 (s), 1067 (m), 1016 (m), 947 (m), 758 (s), 688 (s), 624 (s).

[Ho(btfa)$_3$(di-^{t}bubipy)] (**4**) (yield: 23%). Characterization: Anal. Calcd. for C$_{48}$H$_{42}$F$_9$HoN$_2$O$_6$ (1078.77 g/mol): C, 53.4; H, 3.9; N, 2.6%. Found: C, 53.3; H, 3.7; N, 2.7%. Selected IR bands (cm^{-1}): 2971 (w), 1612 (s), 1576 (m), 1539 (m), 1479 (m), 1403 (w), 1321 (m), 129 (s), 1248 (m), 1181 (s), 1127 (s), 1075 (m), 1026 (w), 948 (w), 848 (w), 766 (s), 699 (s), 635 (s), 580 (s).

2.2.4. [Ho(ntfa)$_3$(5,5'-Me$_2$bipy)] (**5**)

[Ho(ntfa)$_3$(MeOH)$_2$] (127 mg, 0.125 mmol) and 5,5'-Dimethyl-2,2'-dipyridyl (28 mg, 0.15 mmol) were dissolved in 30 mL ethanol/acetone (3:1). The solution was stirred for approximately for 2 h. The mixture was filtered, and the mother liquor was left in an open atmosphere. After two weeks, pink crystals of **5** were obtained from the mother liquor (yield: 43 mg, 30%). Characterization: Anal. Calcd. for: C$_{54}$H$_{36}$F$_9$HoN$_2$O$_6$ (1144.78 g/mol): C, 56.7; H, 3.2; N, 2.4%. Found: C, 56.6; H, 3.1; N, 2.5%. Selected IR bands (ATR-IR, cm^{-1}): 1738 (w), 1608 (s), 1590 (m), 1566 (m), 1526 (m), 1506 (m), 1476 (w), 1384 (w), 1353 (w), 1284 (s), 1217 (w), 1183 (m), 1131 (s), 1073 (w), 956 (m), 935 (w), 862 (w), 790 (s), 748 (m), 681 (m), 569 (m), 517 (w), 467 (m), 416 (w).

2.2.5. [Ho(ntfa)$_3$(bipy)] (**6**)

[Ho(ntfa)$_3$(MeOH)$_2$] (124 mg, 0.122 mmol) was dissolved in 15 mL ethanol/acetone (4:1). 2,2´-bipyridyl (28 mg, 0.18 mmol) was dissolved in 15 mL ethanol/acetone (4:1). The solutions were combined and stirred approximately for 2 h. The mixture was filtered, and the mother liquor was left in an open atmosphere. After ten days, light pink crystals of **6** were obtained from the mother liquor (yield: 37 mg, 29%). Characterization of solvent-free compound: Anal. Calcd. for: C$_{52}$H$_{32}$F$_9$HoN$_2$O$_6$ (1116.73 g/mol): C, 55.9; H, 2.9; N, 2.5%. Found: C, 55.8; H, 2.8; N, 2.6%. Selected IR bands (ATR-IR, cm^{-1}): 1610 (s), 1591 (m), 1568 (m), 1528 (m), 1507 (m), 1460 (m), 1437 (w), 1387 (w), 1354 (w), 1286 (s), 1188 (m), 1121 (s), 1075 (w), 958 (m), 865 (w), 790 (s), 760 (m), 682 (w), 568 (m), 518 (w), 470 (m), 414 (w).

2.3. X-Ray Crystal Structure Analysis

Single crystals of **2–4** were set up in air on a Bruker-AXS D8 VENTURE diffractometer with a CMOS detector of **5** and **6** on a Bruker-AXS APEX II diffractometer (Bruker-AXS; Madison, WI, USA). The crystallographic data and details of the refinement are listed in Table 1. All the structures were refined by the least-squares method. Intensities were collected with multilayer monochromated Mo-Kα radiation. Lorentz polarization and absorption corrections were made in all the samples [67,68]. The structures were solved by direct methods using the SHELXS-97 computer program and refined by full-matrix least-squares method using the SHELXL-2014 computer program [69,70]. The non-hydrogen atoms were located in successive difference Fourier syntheses and refined with anisotropic thermal parameters on F^2. Isotropic temperature factors were assigned as 1.2 or 1.5 times the respective parent for hydrogen atoms. For **6**, a SQUEEZE treatment was used to eliminate disordered solvent molecules. Further programs used: Mercury [71] and PLATON [72]. CCDC 2120112-2120116 contains the supplementary crystallographic data for **2–6**, respectively.

Table 1. Crystal data and details of the structure determination of compounds 2–6.

Compound	2	3	4	5	6
Empirical formula	$C_{42}H_{26}F_9HoN_2O_6$	$C_{40}H_{26}F_9HoN_2O_6$	$C_{48}H_{42}F_9HoN_2O_6$	$C_{54}H_{36}F_9HoN_2O_6$	$C_{52}H_{32}F_9HoN_2O_6$
Formula mass	990.58	966.56	1078.76	1144.78	1116.73
System	Monoclinic	Monoclinic	Triclinic	Orthorhombic	Orthorhombic
Space group	P21/c	P21/n	P-1	Pca21	Pna21
a (Å)	9.6058(7)	11.0408(10)	12.3569(16)	20.2138(9)	20.7013(6)
b (Å)	36.627(2)	22.6440(18)	13.6076(18)	11.7503(5)	10.9059(3)
c (Å)	10.7464(7)	15.2463(13)	14.3853(18)	19.5852(7)	42.3027(10)
α (°)	90	90	92.478(5)	90	90
β (°)	92.932(3)	101.972(3)	99.883(5)	90	90
γ (°)	90	90	105.233(5)	90	90
V (Å3)	3776.0(4)	3728.8(6)	2289.3(5)	4651.8(3)	9550.5(4)
Z	4	4	2	4	8
μ (mm^{-1})	2.192	2.218	1.815	1.792	1.744
D_{calc} (Mg/m^3)	1.742	1.722	1.565	1.635	1.553
θ max (°)	26.420	34.495	27.171	28.998	28.000
Data collected	91277	103169	50752	93718	264468
Unique refl./R_{int}	7737/0.0836	15617/0.0776	10082/0.0390	12315/0.0812	23060/0.0515
Parameters/Restraints	542/0	523/0	601/0	651/1	1262/19
Goodness-of-fit on F^2	1.120	1.050	1.131	1.012	1.165
R1/wR2 (all data)	0.0615/0.1342	0.0466/0.0809	0.0445/0.1042	0.0374/0.0632	0.0466/0.1060

3. Results and Discussion

3.1. Synthesis and IR Spectra of the Complexes

The precursor complexes [Ho(btfa)$_3$(H$_2$O)$_2$] (**1a**) and [Ho(ntfa)$_3$(MeOH)$_2$] (**1b**) were prepared by the reaction of methanolic solutions containing Ho(III) salts, beta-dicetonate molecules (Hbtfa) or (Hntfa), respectively, and NaOH in the stoichiometric ratio 1:3:3, followed by stirring the resulting solution in H$_2$O. The PXRD pattern confirmed that **1a** is isostructural with [La(btfa)$_3$(H$_2$O)$_2$] [47] and **1b** is isostructural with [Pr(nfa)$_3$(MeOH)$_2$] [48]. The interaction of [Ho(btfa)$_3$(H$_2$O)$_2$] with poly-pyridyl compounds phen, bipy, and di-^{t}bubipy in EtOH afforded the light-pink crystalline adducts [Ho(btfa)$_3$(phen)] (**2**), [Ho(btfa)$_3$-(bipy)] (**3**), and [Ho(btfa)$_3$(di-^{t}bubipy)] (**4**), respectively, whereas the interaction of [Ho(ntfa)$_3$-(MeOH)$_2$] with poly-pyridyl compounds 5,5'-Me$_2$bipy and bipy in ethanol/acetone mixtures afforded the crystalline adducts [Ho(ntfa)$_3$(5,5'-Me$_2$bipy)] (**5**) and [Ho(ntfa)$_3$(bipy)] (**6**) with reasonable yields (38–80%). The approach used here for the synthesis of complexes **2–6** is similar to that successfully employed in similar Ln(III) (Ln = La, Pr, Nd) mono bipyridyl adducts [47–49]. The isolated complexes were structurally characterized by elemental microanalyses and by IR spectroscopy, as well as by single-crystal X-ray crystallography for **2–6**.

The IR spectra of complexes **2–6** display general characteristic features. The strong vibrational band observed over the frequency range 1605–1615 cm^{-1} is typically assigned to the coordinated carbonyl stretching frequency, ν(C=O) [47–49]. The broad band centered at 3462 cm^{-1} in **1a** and 3448 cm^{-1} in **1b** is assigned to the ν(O-H) stretching frequency of the coordinated aqua/methanol ligands.

3.2. Description of the Crystal Structures 2–6

Molecular plots and coordination figures of **2–6** complexes are depicted in Figures 1–5, and selected bond parameters are summarized in Table 2. Each Ho(III) center of the neutral and monomeric complex **2–6** are ligated by six oxygen donor atoms of three β-diketonato ligands anions (btfa$^-$) for **2–4** or (ntfa$^-$) for **5** and **6**, respectively, in the chelating coordination mode. The coordination number (CN) 8 in **2** is completed by two N-donor atoms of one phen chelating ligand. The Ho-N/O bond lengths in **2** are in the range of 2.3051(2)–2.5549(2) Å. The CN = 8 in **3–6** of the HoO$_6$N$_2$ coordination sphere around Ho is achieved by the ligation of one 2,2′-bipy (**3** and **6**), di-^{t}Bu-bipy (**4**), and 5,5′-Me2-bipy (**5**) chelating ligands, respectively. The Ho-N/O bond lengths for **3** are in the range of 2.287(2)-2.527(3) Å, for **4**, from 2.292(3) to 2.541(4) Å, for **5**, from 2.293(4) to 2.518(4) Å, and for **6**, from 2.268(6) to 2.530(7) Å, respectively. The O-Ho-O bite angles of the β-diketonato groups fall in the range from 71.5(2) to 76.21(1)° in **2–6**, whereas the corresponding N-Ho-N bite angles of the chelating phen, bipy, di-^{t}bu-bipy, and 5,5′-Me$_2$bipy ligands in compounds **2–6** vary from 63.80(13) to 64.66(1)°.

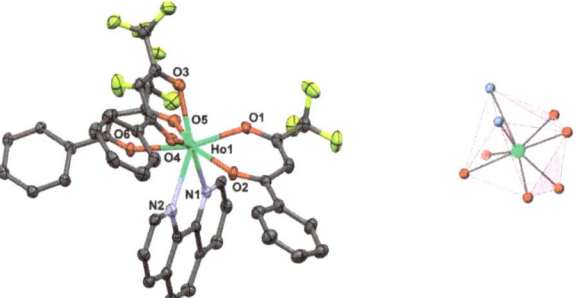

Figure 1. Left: partially labeled structure [Ho(btfa)$_3$(phen)] (**2**). Color code: turquoise = Ho, red = O, yellow = F, grey = C. Right: coordination polyhedron of Ho(III) ion in compound **2**.

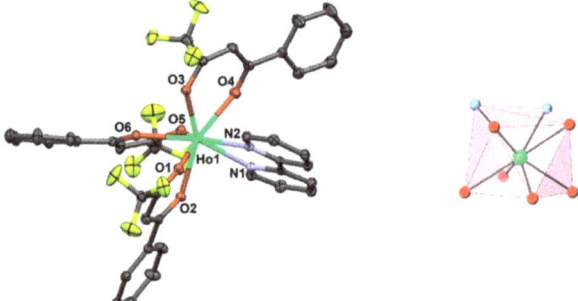

Figure 2. Left: partially labeled structure [Ho(btfa)$_3$(bipy)] (**3**). Color code: turquoise = Ho, red = O, yellow = F, blue = N, grey = C. Right: coordination polyhedron of Ho(III) ion in compound **3**.

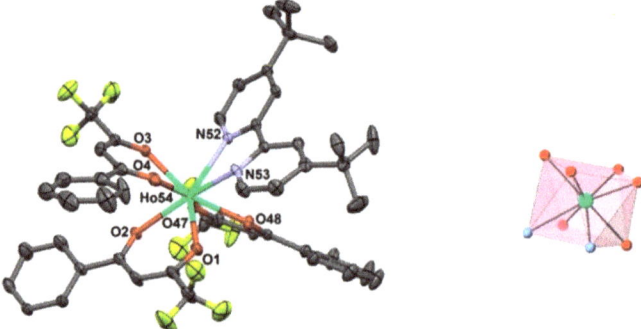

Figure 3. Left: partially labeled structure [Ho(btfa)$_3$(di-^{t}bubipy)] (**4**). Color code: turquoise = Ho, red = O, yellow = F, blue = N, grey = C. Right: coordination polyhedron of Ho(III) ion in compound **4**.

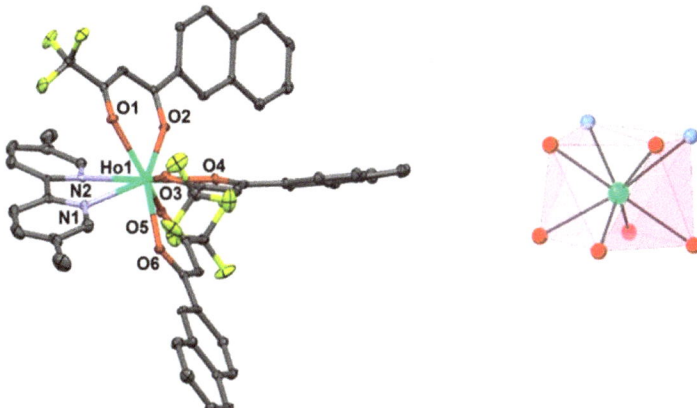

Figure 4. Left: partially labeled structure [Ho(ntfa)$_3$(5,5′-Me$_2$-bipy)] (**5**). Color code: turquoise = Ho, red = O, yellow = F, blue = N, grey = C. Right: coordination polyhedron of Ho(III) ion in compound **5**.

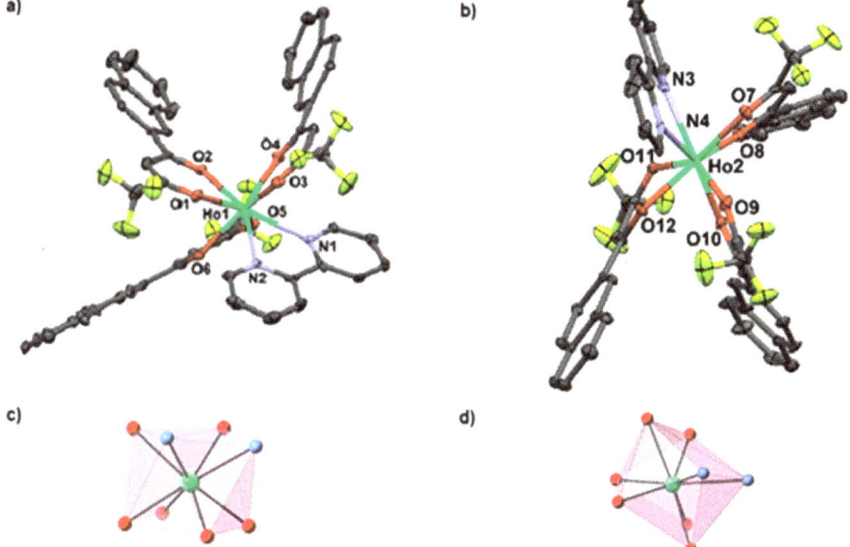

Figure 5. (**a**) Partially labeled structure of [Ho(ntfa)$_3$(bipy)] (**6**): (**a**) Ho atom 1, (**b**) Ho atom 2, (**c**) coordination polyhedron of Ho atom1 and (**d**) coordination polyhedron of Ho atom 2. Color code: turquoise = Ho, red = O, yellow = F, blue = N, grey = C.

Table 2. Selected bond distances (Å) and bite angles (°) for compounds **2–6**.

Compound 2		Compound 3		Compound 4	
Ho1-O1	2.3111(2)	Ho1-O1	2.315(2)	Ho54-O1	2.292(3)
Ho1-O2	2.3051(2)	Ho1-O2	2.343(2)	Ho54-O2	2.331(3)
Ho1-O3	2.3064(2)	Ho1-O3	2.297(2)	Ho54-O3	2.323(3)
Ho1-O4	2.3139(2)	Ho1-O4	2.326(2)	Ho54-O4	2.320(3)
Ho1-O5	2.3647(2)	Ho1-O5	2.287(2)	Ho54-O47	2.305(3)
Ho1-O6	2.3225(2)	Ho1-O6	2.330(2)	Ho54-O48	2.323(3)
Ho1-N1	2.5477(2)	Ho1-N1	2.524(2)	Ho54-N52	2.535(4)
Ho1-N2	2.5549(2)	Ho1-N2	2.527(3)	Ho54-N53	2.541(4)
O1-Ho1-O2	72.59(1)	O1-Ho1-O2	72.58(7)	O1-Ho54-O2	73.50(11)
O3-Ho1-O5	72.74(1)	O3-Ho1-O4	72.69(3)	O3-Ho54-O4	73.70(13)
O4-Ho1-O6	76.21(1)	O5-Ho1-O6	73.61(8)	O47-Ho54-O48	72.77(11)
N1-Ho1-N2	64.66(1)	N1-Ho1-N2	63.89(8)	N52-Ho54-N53	63.80(13)
Compound 5		Compound 6			
Ho1-O1	2.336(4)	Ho1-O1	2.273(6)	Ho2-O7	2.342(6)
Ho1-O2	2.293(4)	Ho1-O2	2.323(6)	Ho2-O8	2.285(6)
Ho1-O3	2.321(4)	Ho1-O3	2.362(6)	Ho2-O9	2.283(6)
Ho1-O4	2.311(4)	Ho1-O4	2.268(6)	Ho2-O10	2.319(6)
Ho1-O5	2.328(4)	Ho1-O5	2.320(6)	Ho2-O11	2.337(6)
Ho1-O6	2.313(4)	Ho1-O6	2.318(6)	Ho2-O12	2.341(6)
Ho1-N1	2.515(5)	Ho1-N1	2.495(7)	Ho2-N3	2.511(7)
Ho1-N2	2.518(4)	Ho1-N2	2.530(7)	Ho2-N4	2.516(7)
O1-Ho1-O2	71.77(13)	O1-Ho1-O2	73.8(2)	O7-Ho2-O8	72.1(2)
O3-Ho1-O4	72.95(12)	O3-Ho1-O4	71.5(2)	O9-Ho2-O10	73.5(2)
O5-Ho1-O6	72.37(13)	O5-Ho1-O6	72.7(2)	O11-Ho2-O12	72.0(2)
N1-Ho1-N2	64.38(16)	N1-Ho1-N2	64.5(3)	N3-Ho2-N4	64.6(2)

Various non-covalent interactions (ring···ring, C-H(X)···ring [72], hydrogen bonds) are summarized in Tables S1–S5 for compounds **2–6**, respectively.

In order to analyze the degree of distortion of the coordination polyhedra for compounds **2–6** from their ideal polyhedron geometry, calculations using the continuous shape measures theory with the SHAPE software were performed [73,74]. The HoO$_6$N$_2$ coordination polyhedron of **2–6** shows an intermediate distortion between various ideal eight-vertex polyhedra geometries. These are a square antiprism (SPAR-8), triangular dodecahedron (TDD-8), and biaugmented trigonal prism (BTPR-8) with continuous shape values of 1.382, 1.236, and 1.779 for **2**; 0.553, 2.351, and 2.051 for **3**; 0.417, 2.515, and 2.254 for **4**; and 0.497, 2.210, and 1.958 for **5**.

The corresponding calculations of the degree of distortion of the HoO$_6$N$_2$ coordination polyhedra of compound **6** [(Ho(ntfa)$_3$(bipy)] reveals an intermediate distortion between various coordination polyhedra geometries. These are a square antiprism (SPAR-8), triangular dodecahedron (TDD-8), and biaugmented trigonal prism (BTPR-8) with continuous shape measures values of 0.412, 2.348, and 2.33 for Ho(1)O$_6$N$_2$ and 0.3944, 2.165, and 2.115 for Ho(2)O$_6$N$_2$.

3.3. Photoluminescence of the Complexes

The luminescence spectra of compounds **2–6** were measured in the solid state at room temperature. The excitation spectra recorded at the emission wavelength (λ_{em}) of 661 nm reveal a broad, intense band around 367 nm for **2–4** and 380 nm for **5** and **6**. This broad band corresponds to the $\pi \to \pi^*$ transition from the ligands. The luminescence emission spectra of the samples were recorded upon the excitation wavelengths (λ_{ex}) of 367 nm for **2–4** and 380 nm for **5** and **6**. All spectra display a characteristic band at 661 nm ($^5F_5 \to {}^5I_8$) corresponding to the metal-centered emission and is assigned to the Ho^{3+} *f-f* transition from the $^5F_5$ excited state to the $^5I_8$ ground state. For this band, the Stark splitting of the degenerate 4*f* levels under the crystal field is perceived. In addition, compounds **2–4** showed a weak band at 545 nm, which can be assigned to an *f-f* transition from higher-energy states ($^5F_4$, $^5S_2$) to the ground state $^5I_8$ [75–77]. The triplet states of the ntfa and btfa ligands were calculated by Sato and Wada in Gd(III) complexes [78], taking into account the sensitization effect of the energy transfer from the singlet state of the ligand (S_1) to the lower-in-energy ligand triplet state (T_1) through the intersystem crossing. These calculations showed that the ntfa T_1 state falls around 19,600 cm^{-1} for ntfa and 21,400 cm^{-1} for btfa. Thus, we can suggest that the energy transfer from the T_1 of the ntfa ligand to the $^5F_4$ and $^5S_2$ (18,348 cm^{-1}) thermal state is inefficient because the two states are too close in energy, and as a result, the $^5F_4$, $^5S_2 \rightarrow ^5I_8$ transitions are not identified for compounds **5** and **6**, but they are seen for btfa complexes **2–4** [79]. Typical representative UV-Vis and luminescence emission spectra (Vis and NIR regions) are depicted in Figure 6 for **3** and Figure 7 for **6** as representatives of the two categories of **2–4** and **5** and **6** complexes, respectively (for luminescence spectra of **2**, **4**, and **5**, see Figures S12–S14).

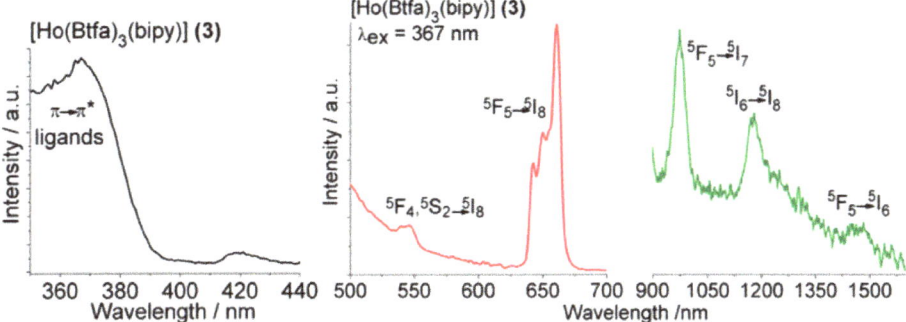

Figure 6. Spectra for complex **3**. Luminescence excitation (**black line**), emission in the visible range (**red line**), and in the NIR (**green line**) regions.

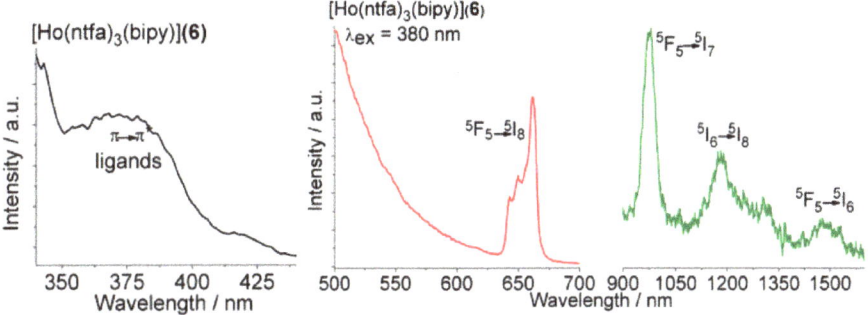

Figure 7. Spectra for complex **6**. Luminescence excitation (**black line**), emission in the visible (**red line**), and in the NIR (**green line**) regions.

Furthermore, the luminescence emissions of the compounds **2–6** were recorded in the NIR region from 900 to 1600 nm, where three weak bands were detected at 973, 1179, and 1474 nm. The first and most intense band is assigned to the $^5F_5 \rightarrow ^5I_7$ transition. The band located at 1179 nm accounts for the $^5I_6 \rightarrow ^5I_8$ transition; the very weak band at 1474 nm corresponds to the $^5F_5 \rightarrow ^5I_6$ transition [80]. The results obtained here agree with other Ho(III) coordination compounds, where the study of the sensitization of Ho^{3+} luminescence by the energy transfer from chromophore ligands has been performed [81–85].

3.4. Magnetic Properties of the Complexes

3.4.1. Ac Magnetic Susceptibility Studies

In order to study the dynamic magnetic properties and the possible Single Molecular Magnet (SMM) behavior (slow relaxation of magnetization) of the synthesized compounds, ac magnetic susceptibility, measurements were recorded for solvent-free compounds **2–5**. Compounds **2–5** do not show a dependence on the in-phase and out-of-phase components in front of the temperature and frequency, neither in the minimum dc field (0 T) nor in the maximum applied dc magnetic field (0.1 T). Therefore, these compounds do not show slow relaxation of the magnetization and consequently will not show SMM's behavior.

3.4.2. Dc Magnetic Susceptibility Studies

Powder samples of complexes **2–5** were measured under applied magnetic fields of 0.3 T (300–2 K). The data are plotted as $\chi_M T$ products versus T in Figure 8. Magnetization dependence of the applied field at 2 K for compounds **2–5** was also recorded and is shown in Figure 9.

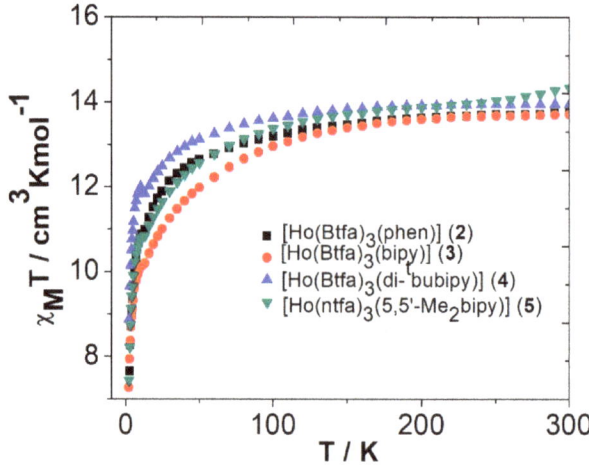

Figure 8. $\chi_M T$ vs. T plots for compounds **2–5**.

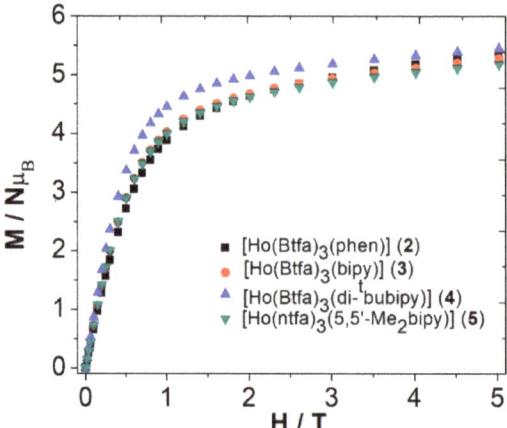

Figure 9. Field dependence of the magnetization plots at T = 2 K for compounds **2–5**.

The magnetic measurement on **2–5** reveals that the $\chi_M T$ values at 300 K are 13.8, 13.7, 13.9, and 14.3 cm^3 K mol^{-1}, respectively, which are in the range of the theoretical value for a magnetically uncoupled Ho(III) compound (14.07 cm^3·K·mol^{-1}) in the $^5I_8$ ground state ($g_J = 5/4$) [86]. On cooling the samples, $\chi_M T$ values remain constant up to 125 K. Below this temperature, $\chi_M T$ values decrease to finite values of 6.8, 7.3, 8.9, and 7.4 cm^3·K·mol^{-1} at 2 K for compounds **2–5**, respectively. The decrease in $\chi_M T$ values at low temperatures could be due to the depopulation of the sublevels generated for the spin–orbit coupling and the ligand-field effect (Stark sublevels).

Magnetization dependence on magnetic static applied field at T = 2 K for complexes **2–5** (Figure 9) reveals no saturation at high fields with similar values of 5.4, 5.3, 5.4, and 5.2 $N\mu_B$ at 5 T for **2–5**, respectively. The magnetization saturation point expected for a mononuclear Ho^{3+} complex should be $\approx$4 $N_A \mu_B$.

The $1/\chi_M$ versus T plots for **2–5** are shown in Figure 10. Between 2 K and 300 K, the $1/\chi_M$ versus T plots are linear for the four compounds and well described by the Curie–Weiss law $1/\chi_M = (T-\theta)/C$, where C = 13.9 cm^3·K·mol^{-1} and θ = −4.9 K for **2**, C = 13.9 cm^3·K·mol^{-1} and θ = −3.50 K for **3**, C = 14.0 cm^3·K·mol^{-1} and θ = −2.3 K for **4**, and C = 14.3 cm^3·K·mol^{-1} and θ = −4.6 K for **5**.

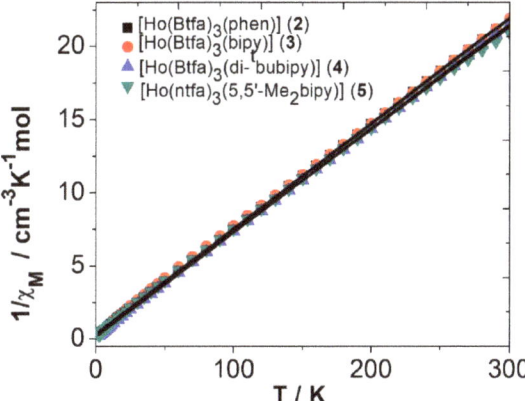

Figure 10. $1/\chi_M$ vs. T plots for compounds **2–5**. Solid lines represent the fitting using the Curie–Weiss law $1/\chi_M = (T - \theta)/C$.

4. Conclusions

A novel series of five mono-bipyridyl adducts of Ho^{3+}-trifluoro-phenyl (btfa$^-$) and -naphthalen-2-yl (ntfa$^-$) β-diketonato complexes [Ho(btfa)$_3$(phen)] (**2**), [Ho(btfa)$_3$(bipy)] (**3**), [Ho(btfa)$_3$(di-^{t}bubipy)] (**4**), [Ho(ntfa)$_3$(Me$_2$bipy)] (**5**), and [Ho(ntfa)$_3$(bipy)$_2$] (**6**) were synthesized from their precursors diaqua tris(β-diketonato) species. The compounds were structurally characterized, where coordination numbers CN = 8 were observed. The distortion of the coordination polyhedra of Ho^{3+} centers was analyzed with the SHAPE program. All the complexes display CN 8. In a fashion that is similar to their Ln^{3+} analog complexes (Ln = La, Pr, and Nd) derived from the same set of ligands [47–49]. The solid-state luminescence emission of the complexes revealed a strong, intense emission band at 661 nm in the visible and three other bands in NIR regions. The magnetic measurements of the complexes **2–5** revealed that the $\chi_M T$ values are within the range of 14.0 ± 0.3 cm^3·mol^{-1}·K at 300 K, which is predicted for a magnetically uncoupled Ho^{3+} compound (14.07 cm^3·mol^{-1}·K) in the $^5I_8$ ground state ($g_J = 5/4$) [86]. The luminescence emission and magnetic results reported here for the Ho^{3+} compounds demonstrate that these properties are not significantly affected by either the small changes in the geometrical shape of the Ho^{3+} complexes or their local symmetry. Additionally, results are almost independent of the nature of the ancillary bipyridyl ligands or the nature of the β-diketone coligands. Similar results were obtained with pyridyl adducts derived from the same coligands with Pr(III) and Nd(III) compounds [48,49].

Supplementary Materials: Non-covalent interactions (ring···ring, C-H(F)···ring, hydrogen bonds) are summarized in Tables S1–S5 for compounds **2–6**, respectively. PXRD pattern (Figure S1a,b, S2–S6), packing views (Figures S7–S11) for compounds **2–6**, excitation and emission spectra of compounds **2**, **4**, and **5**, recorded in the solid state at room temperature, are given in Figures S12–S14, respectively. CCDC deposition numbers: CCDC 2120112-2120116 contain the supplementary crystallographic data for **2–6**, respectively. These data can be obtained free of charge from The Cambridge Crystallographic Data Centre via www.ccdc.cam.ac.uk/data_request/cif.

Author Contributions: Conceptualization, F.A.M., R.V., and M.F.-B.; methodology, F.A.M., R.V. and M.F.-B.; software, F.A.M., R.C.F., M.F.-B. and R.V.; validation, F.A.M., R.V., R.C.F. and S.S.M.; formal analysis, F.B., S.S. and Á.T.; investigation, F.B., Á.T., R.C.F., M.F.-B., S.S. and R.C.F.; resources, F.A.M., R.V. and R.C.F.; data curation, F.A.M., F.B., R.C.F., R.V., M.F.-B., Á.T., S.S. and S.S.M.; writing original draft preparation, F.A.M., F.B., R.C.F., R.V., Á.T., M.F.-B., S.S. and S.S.M.; writing—review and editing, F.A.M., Á.T., R.V. and S.S.M.; visualization, F.A.M., Á.T., R.V., M.F.-B. and S.S.; supervision, F.A.M., R.V. and S.S.M.; project administration, F.A.M. and R.V.; funding acquisition, R.V. All authors have read and agreed to the published version of the manuscript.

Funding: R.V. acknowledges the financial support from MINECO Project PGC2018-094031-B-I00.

Institutional Review Board Statement: Not applicable.

Informed Consent Statement: Not applicable.

Data Availability Statement: Data is contained within the article or supplementary material.

Conflicts of Interest: The authors declare no conflict of interest.

Sample Availability: Samples of the compounds are not available.

References

1. Bünzli, J.-C.G.; McGill, I.I. Rare Earth Elements. In *Ullmann's Encyclopedia of Industrial Chemistry*; Wiley: Hoboken, NJ, USA, 2018; pp. 1–53.
2. Bünzli, J.-C.G. Lanthanide Photonics: Shaping the nano world. *Trends Chem.* **2019**, *1*, 751–762. [CrossRef]
3. Bünzli, J.-C.G. *Lanthanides, Kirk-Othmer Encyclopedia of Chemical Technology*; Wiley Online Library: New York, NY, USA, 2013; pp. 1–43.
4. Cotton, S. *Lanthanide and Actinide Chemistry*; John Wiley & Sons Ltd.: Chichester, UK, 2006.
5. Harrowfield, J.M.; Silber, H.B.; Paquette, S.J. *Metal Ions in Biological Systems*; Sigel, A., Sigel, H., Eds.; Marcel Dekker: New York, NY, USA, 2003.
6. Bünzli, J.-C.G. Lanthanide luminescence for biomedical analyses and imaging. *Chem. Rev.* **2010**, *110*, 2729–2755. [CrossRef]
7. Eliseeva, S.V.; Bünzli, J.-C.G. Lanthanide luminescence for functional materials and bio-sciences. *Chem. Soc. Rev.* **2010**, *39*, 189–227. [CrossRef] [PubMed]
8. Brayshaw, L.L.; Smith, R.-C.G.; Badaoui, M.; James, A.; Irving, J.A.; Price, R.S. Lanthanides compete with calcium for binding to cadherins and inhibit cadherin-mediated cell adhesion. *Metallomics* **2019**, *11*, 914–924. [CrossRef]
9. Allen, K.N.; Imperiali, B. Lanthanide-tagged proteins—An illuminating partnership. *Curr. Opin. Chem. Biol.* **2010**, *15*, 247–254. [CrossRef]
10. Pałasz, A.; Segovia, Y.; Skowronek, R.; Worthington, J.J. Molecular neurochemistry of the lanthanides. *Synapse* **2019**, *73*, e22119. [CrossRef] [PubMed]
11. Jastrza, R.; Nowak, M.; Skrobańska, M.; Tolińska, A.; Zabiszak, M.; Gabryel, M.; Marciniak, Ł.; Kaczmarek, M.T. DNA as a target for lanthanide (III) complexes influence. *Coord. Chem. Rev.* **2019**, *382*, 145–159. [CrossRef]
12. Campello, M.P.C.; Palma, E.; Correia, I.; Paulo, P.M.R.; Matos, A.; Rino, J.; Coimbra, J.; Pessoa, J.C.; Gambino, D.; Paulo, A.; et al. Lanthanide complexes with phenanthroline-based ligands: Insights into cell death mechanisms obtained by microscopy techniques. *Dalton Trans.* **2019**, *48*, 4611–4624. [CrossRef] [PubMed]
13. Qin, Q.P.; Wang, Z.F.; Tan, M.X.; Huang, X.L.; Zou, H.H.; Zou, B.Q.; Shi, B.B.; Zhang, S.H. Complexes of lanthanides(III) with mixed 2,2'-bipyridyl and 5,7-dibromo-8-quinolinoline chelating ligands as a new class of promising anti-cancer agents. *Metallomics* **2019**, *11*, 1005–1015. [CrossRef]
14. Dos Santos, C.M.G.; Harte, A.J.; Quinn, S.J.; Gunnlaugson, T. Recent developments in the field of supramolecular lanthanide luminescent sensors and self-assemblies. *Coord. Chem. Rev.* **2008**, *252*, 2512–2527. [CrossRef]
15. Staszak, K.; Wieszczycka, K.; Marturano, V.; Tylkowski, B. Lanthanides complexes—Chiral sensing of biomolecules. *Coord. Chem. Rev.* **2019**, *397*, 76–90. [CrossRef]

16. Eliseeva, S.V.; Bünzli, J.-C.G. Rare earths: Jewels for functional materials of the future. *New J. Chem.* **2011**, *35*, 1165–1176. [CrossRef]
17. Carlos, L.D.; Ferreira, R.A.S.; De Zea Bermudez, V.; Julian-Lopez, B.; Escribano, P. Progress on lanthanide-based organic–inorganic hybrid phosphors. *Chem. Soc. Rev.* **2011**, *40*, 536–549. [CrossRef]
18. Ward, M.D. Mechanisms of sensitization of lanthanide (III)-based luminescence in transition metal/lanthanide and anthracene/lanthanide dyads. *Coord. Chem. Rev.* **2010**, *254*, 2634–2642. [CrossRef]
19. Chen, F.-F.; Chen, Z.-Q.; Bian, Z.-Q.; Huang, C.-H. Sensitized luminescence from lanthanides in d–f bimetallic complexes. *Coord. Chem. Rev.* **2010**, *254*, 991–1010. [CrossRef]
20. Cui, Y.; Yue, Y.; Qian, G.; Chen, B. Luminescent functional metal–organic frameworks. *Chem. Rev.* **2012**, *112*, 1126–1162. [CrossRef] [PubMed]
21. Huang, H.; Gao, W.; Zhang, X.-M.; Zhou, A.-M.; Liu, J.-P. 3D LnIII-MOFs: Displaying slow magnetic relaxation and highly sensitive luminescence sensing of alkylamines. *CrystEngComm* **2019**, *21*, 694–702. [CrossRef]
22. Greenspon, A.S.; Marceaux, B.L.; Hu, E.L. Robust lanthanide emitters in polyelectrolyte thin films for photonic applications. *Nanotechnology* **2018**, *29*, 075302. [CrossRef] [PubMed]
23. Forrester, P.R.; Patthey, F.; Fernandes, E.; Sblendorio, D.P.; Brune, H.; Natterer, F.D. Quantum state manipulation of single atom magnets using the hyperfine interaction. *Phys. Rev. B* **2019**, *100*, 180405. [CrossRef]
24. Coldeway, D. Storing Data in a Single Atom Proved Possible by IBM Researchers. *TechCrunch*. 9 March 2017. Available online: https://techcrunch.com/2017/03/08/storing-data-in-a-single-atom-proved-possible-by-ibm-researchers/ (accessed on 10 March 2017).
25. Hoard, R.W.; Mance, S.C.; Leber, R.L.; Dalder, E.N.; Chaplin, M.R.; Blair, K.; Nelson, D.H.; Van Dyke, D.A. Field enhancement of a 12.5-T magnet using holmium poles. *IEEE Trans. Magn.* **1985**, *21*, 448–450. [CrossRef]
26. Wollin, T.A.; Denstedt, J.D. The holmium laser in urology. *J. Clin. Laser Med. Surg.* **1998**, *16*, 13. [CrossRef] [PubMed]
27. Lucas, J.; Lucas, P.; Le Mercier, T.; Rollat, A.; Davenport, W. Rare earth doped lasers and optical amplifiers. In *Rare Earths*; Elsevier: Amsterdam, The Netherlands, 2015; pp. 319–332. [CrossRef]
28. Placer, J.; Gelabert-Mas, A.; Vallmanya, F.; Manresa, J.M.; Menéndez, V.; Cortadellas, R.; Arango, O. Holmium laser enucleation of prostate: Outcome and complications of self-taught learning curve. *Urology* **2009**, *73*, 1042–1048. [CrossRef] [PubMed]
29. Da Rosa, P.P.F.; Kitagawa, Y.; Hasegawa, Y. Luminescent lanthanide complex with seven-coordination geometry. *Coord. Chem. Rev.* **2020**, *406*, 213153. [CrossRef]
30. Hasegawa, Y.; Kitagawa, Y.; Nakanishi, T. Effective photosensitized, electrosensitized, and mechanosensitized luminescence of lanthanide complexes. *NPG Asia Mater.* **2018**, *10*, 52–70. [CrossRef]
31. Wu, D.-F.; Liu, Z.; Ren, P.; Liu, X.-H.; Wang, N.; Cui, H.-J.Z.; Gao., L. A new family of dinuclear lanthanide complexes constructed from 8-hydroxyquinoline Schiff base and β-diketone: Magnetic properties and near-infrared luminescence. *Dalton Trans.* **2019**, *48*, 1392–1403. [CrossRef] [PubMed]
32. Carlos, L.D.; Ferreira, R.A.S.; De Zea Bermudez, V.; Ribeiro, S.J.L. Lanthanide-containing light-emitting organic–inorganic hybrids: A bet on the future. *Adv. Mater.* **2009**, *21*, 509–534. [CrossRef]
33. Lis, S.; Elbanowski, M.; Makowska, B.; Hnatejko, Z. Energy transfer in solution of lanthanide complexes. *J. Photochem. Photobiol. A Chem.* **2002**, *150*, 233–247. [CrossRef]
34. De Sa, G.F.; Malto, O.L.; De Mello Donega, C.; Simas, A.M.; Longo, R.L.; Santa-Cruz, P.A.; Da Silva, E.R., Jr. Spectroscopic properties and design of highly luminescent lanthanide coordination complexes. *Coord. Chem. Rev.* **2000**, *196*, 165–195. [CrossRef]
35. Bünzli, J.-C.G.; Piguet, C. Taking advantage of luminescent lanthanide ions. *Chem. Soc. Rev.* **2005**, *34*, 1048–1077. [CrossRef]
36. Su, C.Y.; Kang, B.S.; Liu, H.Q.; Wang, Q.G.; Chen, Z.N.; Lu, Z.L.; Tong, Y.X.; Mak, T.C.W. Luminescent lanthanide complexes with encapsulating polybenzimidazole tripodal ligands. *Inorg. Chem.* **1999**, *38*, 1374–1375. [CrossRef]
37. Alpha, B.; Lehn, J.M.; Mathis, G. Energy transfer luminescence of europium(III) and terbium(III) cryptates of macrobicyclic polypyridine ligands. *Angew. Chem. Int. Ed.* **1987**, *26*, 266–267. [CrossRef]
38. Ziessel, R.; Maestri, M.; Prodi, L.; Balzani, V.; Dorsselaer, A. Dinuclear europium (3+), terbium (3+) and gadolinium (3+) complexes of a branched hexaazacyclooctadecane ligand containing six 2,2′-bipyridine pendant units. *Inorg. Chem.* **1993**, *32*, 1237–1241. [CrossRef]
39. Armelao, L.; Quici, S.; Barigelletti, F.; Accorsi, G.; Bottaio, G.; Cavazzini, M.; Tondello, E. Design of luminescent lanthanide complexes: From molecules to highly efficient photo-emitting materials. *Coord. Chem. Rev.* **2010**, *254*, 487–505. [CrossRef]
40. Binnemans, K. Rare earth β-diketonates. In *Gschneider, Design of Luminescent Lanthanide Complexes: From Molecules to Highly Efficient Photo-Emitting Materials; Handbook on the Physics and Chemistry of Rare Earths*; Bünzli, J.-C.G., Pecharsky, V.K., Eds.; Elsevier: Amsterdam, The Netherlands, 2005; Volume 35, pp. 107–272.
41. Hasegawa, Y.; Nakagawa, T.; Kawai, T. Recent progress of luminescent metal complexes with photochromic units. *Coord. Chem. Rev.* **2010**, *254*, 2643–2651. [CrossRef]
42. Yu, J.; Zhang, H.; Fu, L.; Deng, R.; Zhou, L.; Li, H.; Liu, F.; Fu, H. Synthesis, structure and luminescent properties of a new praseodymium(III) complex with β-diketone. *Inorg. Chem. Commun.* **2003**, *6*, 852–854. [CrossRef]
43. Vicente, R.; Tubau, À.; Speed, S.; Mautner, F.A.; Bierbaumer, F.; Fischer, R.C.; Massoud, S.S. Slow magnetic relaxation and luminescence properties in neodymium(III)-4,4,4-trifluoro-1-(2-naphthyl)butane-1,3-dionato complexes incorporating bipyridyl ligands. *New J. Chem.* **2021**, *45*, 14713–14723. [CrossRef]

44. Hyre, A.S.; Doerrer, L.H. A structural and spectroscopic overview of molecular lanthanide complexes with fluorinated O-donor ligands. *Coord. Chem. Rev.* **2020**, *404*, 213098. [CrossRef]
45. Gao, H.-L.; Wang, N.-N.; Wang, W.-M.; Shen, H.-Y.; Zhou, X.-P.; Chang, Y.-X.; Zhang, R.X.; Cui, J.-Z. Fine-tuning the magnetocaloric effect and SMMs behaviors of coplanar RE4 complexes by β-diketonate coligands. *Inorg. Chem. Front.* **2017**, *4*, 860–870. [CrossRef]
46. Chang, Y.-X.; Gao, N.; Wang, M.-Y.; Wang, W.-T.; Fan, Z.-W.; Ren, D.-D.; Wu, Z.-L.; Wang, W.-M. Two phenoxo-O bridged dinuclear Dy(III) complexes exhibiting distinct slow magnetic relaxation induced by different β-diketonate ligands. *Inorg. Chim. Acta* **2020**, *505*, 119499. [CrossRef]
47. Mautner, F.A.; Bierbaumer, F.; Fischer, R.C.; Torvisco, A.; Vicente, R.; Font-Bardía, M.; Tubau, À.; Speed, S.; Massoud, S.S. Diverse coordination numbers and geometries in pyridyl adducts of lanthanide(III) complexes based on β-diketonate. *Inorganics* **2021**, *9*, 74. [CrossRef]
48. Mautner, F.A.; Bierbaumer, F.; Fischer, R.C.; Vicente, R.; Tubau, À.; Ferran, A.; Massoud, S.S. Structural characterization, magnetic and luminescent properties of praseodymium(III)-4,4,4-trifluoro-1-(2-naphthyl)butane-1,3-dionato(1-) complexes. *Crystals* **2021**, *11*, 179. [CrossRef]
49. Mautner, F.A.; Bierbaumer, F.; Gyurkac, M.; Fischer, R.C.; Torvisco, A.; Massoud, S.S.; Vicente, R. Synthesis and characterization of lanthanum(III) complexes containing 4,4,4-trifluoro-1-(2-naphthalen-yl)-butane-1,3-dionate. *Polyhedron* **2020**, *179*, 114384. [CrossRef]
50. Zhang, S.; Ke, H.; Shi, Q.; Zhang, J.; Yang, Q.; Wei, Q.; Xie, G.; Wang, W.; Yang, D.; Chen, S. Dysprosium(iii) complexes with a square-antiprism configuration featuring mononuclear single-molecule magnetic behaviours based on different β-diketonate ligands and auxiliary ligands. *Dalton Trans.* **2016**, *45*, 5310–5320. [CrossRef]
51. Li, D.-P.; Zhang, X.-P.; Wang, T.-W.; Ma, B.-B.; Li, C.-H.; Li, Y.-Z.; You, X.-Z.; You, X.-Z. Distinct magnetic dynamic behavior for two polymorphs of the same Dy(III) complex. *Chem. Commun.* **2011**, *47*, 6867–6869. [CrossRef] [PubMed]
52. Yu, J.; Deng, R.; Sun, L.; Li, Z.; Zhang, H. Photophysical properties of a series of high luminescent europium complexes with fluorinated ligands. *J. Lumin.* **2011**, *131*, 328–335. [CrossRef]
53. Fernandes, J.A.; Ferreira, R.A.S.; Pillinger, M.; Carlos, L.D.; Jepsen, J.; Hazell, A.; Ribeiro-Claro, P.; Goncalves, I.S. Investigation of europium(III) and gadolinium(III) complexes with naphthoyltrifluoroacetone and bidentate heterocyclic amines. *J. Lumin.* **2005**, *113*, 50–63. [CrossRef]
54. Thompson, L.C.; Atchison, F.W.; Young, V.G. Isomerism in the adduct of tris(4, 4,4,-trifluoro-1-(2-naphthyl)-1,3-butanedionato) europium(III) with dipyridyl. *J. Alloys Compd.* **1998**, *275*, 765–768. [CrossRef]
55. Trieu, T.-N.; Dinh, T.-H.; Nguyen, H.-H.; Abram, U.; Nguyen, M.-H. Novel lanthanoide(III) ternary complexes with naphtoyltrifluoroacetone: A synthetic and spectroscopic study. *Z. Anorg. Allg. Chem.* **2015**, *641*, 1934–1940. [CrossRef]
56. Taydakov, I.V.; Akkuzina, A.; Avetisov, R.I.; Khomyakov, A.V.; Saifutarov, R.R.; Avetissov, I.C. Effective electroluminescent materials for OLED applications based on lanthanide 1,3-diketonates bearing pyrazole moiety. *J. Lumin.* **2016**, *177*, 31–39. [CrossRef]
57. Maggini, I.; Traboulsi, H.; Yoosaf, K.; Mohanraj, J.; Wouters, J.; Pietraszkiewicz, O.; Pietraszkiewicz, M.; Armaroli, N.; Bonifazi, D. Electrostatically-driven assembly of MWCNTs with a europium complex. *Chem. Commun.* **2011**, *47*, 1625–1627. [CrossRef] [PubMed]
58. Lunstroot, L.; Nockemann, P.; Van Hecke, K.; Van Meervelt, L.; Gorller-Walrand, C.; Binnemans, K.; Driesen, K. Visible and near-infrared emission by samarium (III)-containing ionic liquid mixtures. *Inorg. Chem.* **2009**, *48*, 3018–3026. [CrossRef] [PubMed]
59. Bruno, S.M.; Ferreira, R.A.S.; Paz, F.A.A.; Carlos, L.D.; Pillinger, M.; Ribeiro-Claro, P.; Goncalves, I.S. Structural and photoluminescence studies of a europium(III) tetrakis(β-diketonate) complex with tetrabutylammonium, imidazolium, pyridinium and silica-supported imidazolium counterions. *Inorg. Chem.* **2009**, *48*, 4882–4895. [CrossRef] [PubMed]
60. Tu, H.-R.; Sun, W.-B.; Li, H.-F.; Chen, P.; Tian, Y.-M.; Zhang, W.-Y.; Zhang, Y.-Q.; Yan, P.-F. Complementation and joint contribution of appropriate intramolecular interaction and local ion symmetry to improve magnetic relaxation in a series of dinuclear Dy2 single-molecule magnets. *Inorg. Chem. Front.* **2017**, *4*, 499–508. [CrossRef]
61. Martin-Ramos, P.; Coya, C.; Alvarez, A.L.; Ramos-Silva, M.; Zaldo, C.; Paixao, J.A.; Chamorro-Posada, P.; Martin-Gil, J. Charge transport and sensitized 1.5 μm electroluminescence properties of full solution-processed NIR-OLED based on novel Er(III) fluorinated β-diketonate ternary complex. *J. Phys. Chem. C* **2013**, *117*, 10020–10030. [CrossRef]
62. Dasari, S.; Singh, S.; Sivakumar, S.; Patra, A.K. Dual-sensitized luminescent europium(III) and terbium(III) complexes as bioimaging and light-responsive therapeutic agents. *Chem. Eur. J.* **2016**, *22*, 7387–17396. [CrossRef]
63. Bruno, S.M.; Ananias, D.; Paz, F.A.A.; Pillinger, M.; Valente, A.A.; Carlos, L.D.; Goncalves, I.S. Crystal structure and temperature-dependent luminescence of a heterotetranuclear sodium–europium(III) β-diketonate complex. *Dalton Trans.* **2015**, *44*, 488–492. [CrossRef]
64. Fernandes, J.A.; Braga, S.S.; Pillinger, M.; Ferreira, R.A.S.; Carlos, L.D.; Hazell, A.; Ribeiro-Claro, P.; Goncalves, I.S. β-Cyclodextrin inclusion of europium(III) tris(β-diketonate)-bipyridine. *Polyhedron* **2006**, *25*, 1471–1476. [CrossRef]
65. Shen, F.; Hu, J.; Xie, M.; Wang, S.; Huang, X. Synthesis and structural investigation of lanthanide organometallics involving cyclopentadienyl and 2-napthoyltrifluoroacetonato chelate ligands Synthesis and structural investigation of lanthanide organometallics involving cyclopentadienyl and 2-naptoyl-trifluoroacetonato chelate ligands. *J. Organomet. Chem.* **1995**, *485*, C6–C9.

66. Shen, F.; Hu, J.; Xie, M.; Wang, S.; Huang, X. Synthesis and structural study of cyclopentadienyl lanthanide derivatives containing the 2-naphthoyltrifluoro-acetonato ligand. *Polyhedron* **1996**, *15*, 1151–1155. [CrossRef]
67. Bruker. *APEX, SAINT v. 8.37A*; Bruker AXS, Inc.: Madison, WI, USA, 2015.
68. Sheldrick, G.M. *SADABS v. 2*; University of Goettingen: Goettingen, Germany, 2001.
69. Sheldrick, G.M. A Short history of SHELX. *Acta Crystallogr. A* **2008**, *64*, 112–122. [CrossRef]
70. Sheldrick, G.M. Crystal structure refinement with SHELXL. *Acta Crystallogr. C Struct. Chem.* **2015**, *71*, 3–8. [CrossRef]
71. Macrae, C.F.; Edington, P.R.; McCabe, P.; Pidcock, E.; Shields, G.P.; Taylor, R.; Towler, T.; Van de Streek, J.J. Mercury: Visualization and analysis of crystal structures. *Appl. Cryst.* **2006**, *39*, 453–457. [CrossRef]
72. Spek, A.L. *PLATON, a Multipurpose Crystallographic Tool*; Utrecht University: Utrecht, The Netherlands, 1999.
73. Alvarez, S.; Alemany, P.; Casanova, D.; Cirera, J.; Llunell, M.; Avnir, D. Shape maps and polyhedral interconversion paths in transition metal chemistry. *Chem. Soc. Rev.* **2005**, *249*, 1693–1708. [CrossRef]
74. Cirera, J.; Alvarez, S. Stereospinomers of pentacoordinate iron porphyrin complexes: The case of the [Fe(porphyrinato)(CN)]$^-$ anions. *Dalton Trans.* **2013**, *42*, 7002–7008. [CrossRef]
75. Boyer, J.C.; Vetrone, F.; Capobianco, J.A.; Speghini, A.; Zambelli, M.; Bettinelli, M. Investigation of the upconversion processes in nanocrystalline $Gd_3Ga_5O_{12}$:Ho^{3+}. *J. Lumin.* **2004**, *106*, 263–268. [CrossRef]
76. Zhang, T.; Wang, J.; Jiang, J.; Pan, R.; Zhang, B. Microstructure and photoluminescence properties of Ho-doped $(Ba,Sr)TiO_3$ thin films. *Thin Solid Films* **2007**, *515*, 7721–7725. [CrossRef]
77. Lim, C.S.; Aleksandrovsky, A.; Molokeev, M.; Oreshonkov, A.; Atuchin, V. The modulated structure and frequency upconversion properties of $CaLa_2(MoO_4)_4$: Ho^{3+}/Yb^{3+} phosphors prepared by microwave synthesis. *Phys. Chem. Chem. Phys.* **2015**, *17*, 19278–19287. [CrossRef]
78. Susumu, S.; Masanobu, W. Relations between intramolecular energy transfer efficiencies and triplet state energies in rare earth β-diketone chelates. *Bull. Chem. Soc. Jpn.* **1970**, *43*, 1955–1962.
79. Latva, M.; Takalo, H.; Mukkala, V.-M.; Matachescu, C.; Rodríguez-Ubis, J.C.; Kankare, J. Luminescent Lanthanoid Calixarene Complexes and Materials. *J. Lumin.* **1997**, *75*, 149–169. [CrossRef]
80. Quici, S.; Cavazzini, M.; Marzanni, G.; Accorsi, G.; Armaroli, N.; Ventura, B.; Barigelletti, F. Visible and near-infrared intense luminescence from water-soluble lanthanide [Tb(III), Eu(III), Sm(III), Dy(III), Pr(III), Ho(III), Yb(III), Nd(III), Er(III)] complexes. *Inorg. Chem.* **2005**, *44*, 529–537. [CrossRef] [PubMed]
81. Komissar, D.A.; Metlin, M.T.; Ambrozevich, S.A.; Taydakov, I.V.; Tobokhova, A.S.; Varaksina, E.A.; Selyukov, A.S. Luminescence properties of pyrazolic 1,3-diketone Ho^{3+} complex with 1,10-phenanthroline. *Spectrochim. Acta Part A Mol. Biomol. Spectrosc.* **2019**, *222*, 117229–117238. [CrossRef]
82. Dang, S.; Yu, J.; Yu, J.; Wang, X.; Sun, L.; Feng, J.; Fan, W.; Zhang, H. Novel holmium (Ho) and praseodymium (Pr) ternary complexes with fluorinated-ligand and 4,5-diazafluoren-9-one. *Mater. Lett.* **2011**, *65*, 1642–1644. [CrossRef]
83. Dang, S.; Sun, L.-N.; Song, S.-Y.; Zhang, H.-J.; Zheng, G.-L.; Bi, Y.-F.; Guo, H.-D.; Guo, Z.-Y.; Feng, J. Syntheses, crystal structures and near-infrared luminescent properties of holmium (Ho) and praseodymium (Pr) ternary complexes. *Inorg. Chem. Commun.* **2008**, *11*, 531–534. [CrossRef]
84. Coban, M.B.; Amjad, A.; Aygun, M.; Kara, H. Sensitization of HoIII and SmIII luminescence by efficient energy transfer from antenna ligands: Magnetic, visible and NIR photoluminescence properties of GdIII, HoIII and SmIII coordination polymers. *Inorg. Chim. Acta* **2017**, *455*, 25–33.
85. Ahmed, Z. Iftikhar, K. Sensitization of visible and NIR emitting lanthanide(III) ions in noncentrosymmetric complexes of hexafluoroacetylacetone and unsubstituted monodentate pyrazole. *Phys. Chem. A* **2013**, *117*, 11183–11201. [CrossRef] [PubMed]
86. Atwood, D.A. (Ed.) *The Rare Earth Elements: Fundamentals and Applications*; John Wiley & Sons Ltd.: Chichester, UK, 2005.

Article

A New Pt(II) Complex with Anionic *s*-Triazine Based *NNO*-Donor Ligand: Synthesis, X-ray Structure, Hirshfeld Analysis and DFT Studies

Mezna Saleh Altowyan [1], Saied M. Soliman [2,*], Jamal Lasri [3,*], Naser E. Eltayeb [3], Matti Haukka [4], Assem Barakat [5,*] and Ayman El-Faham [2]

[1] Department of Chemistry, College of Science, Princess Nourah bint Abdulrahman University, P.O. Box 84428, Riyadh 11671, Saudi Arabia; msaltowyan@pnu.edu.sa
[2] Department of Chemistry, Faculty of Science, Alexandria University, P.O. Box 426, Ibrahimia, Alexandria 21321, Egypt; ayman.elfaham@alexu.edu.eg
[3] Department of Chemistry, Rabigh College of Science and Arts, King Abdulaziz University, Jeddah 21589, Saudi Arabia; nasertaha90@gmail.com
[4] Department of Chemistry, University of Jyväskylä, P.O. Box 35, FI-40014 Jyväskylä, Finland; matti.o.haukka@jyu.fi
[5] Department of Chemistry, College of Science, King Saud University, P.O. Box 2455, Riyadh 11451, Saudi Arabia
* Correspondence: saeed.soliman@alexu.edu.eg (S.M.S.); jlasri@kau.edu.sa (J.L.); ambarakat@ksu.edu.sa (A.B.)

Abstract: The reaction of PtCl$_2$ with *s*-triazine-type ligand (**HTriaz**) (1:1) in acetone under heating afforded a new **[Pt(Triaz)Cl]** complex. Single-crystal X-ray diffraction analysis showed that the ligand (**HTriaz**) is an *NNO* tridentate chelate via two N-atoms from the *s*-triazine and hydrazone moieties and one oxygen from the deprotonated phenolic OH. The coordination environment of the Pt(II) is completed by one Cl^{-1} ion *trans* to the Pt-N$_{(hydrazone)}$. Hirshfeld surface analysis showed that the most dominant interactions are the H···H, H···C and O···H intermolecular contacts. These interactions contributed by 60.9, 11.2 and 8.3% from the whole fingerprint area, respectively. Other minor contributions from the Cl···H, C···N, N···H and C···C contacts were also detected. Among these interactions, the most significant contacts are the O···H, H···C and H···H interactions. The amounts of the electron transfer from the ligand groups to Pt(II) metal center were predicted using NBO calculations. Additionally, the electronic spectra were assigned based on the TD-DFT calculations.

Keywords: Pt(II) complex; *s*-triazine; Hirshfeld; NBO; TD-DFT; X-ray

1. Introduction

s-triazine and their metal complexes have gained much attention for their properties and potential applications in many fields [1]. In the last decade, *s*-triazine and their complexes have been explored in the pharmaceutical field, catalytic process including Heck and Suzuki-Miyaura cross-coupling reactions, olefin polymerization, hydrogen transfer reactions, decarbonylation of ketones, asymmetric allylic alkylation, and some derivatives have been designed to develop photoelectronic materials [1]. Several ligands have been synthesized based on the *s*-triazine as a core structure and have been explored in coordination chemistry [1]. Mukherjee et al. constructed a complicated coordinated molecule by coordination-driven self-assembly of homometallic Pd/Pt-based *s*-triazine ligand as interlocked molecular cages [2]. Motloch et al. reported the synthesis of the Pt(II)/Pd(II) complex with *s*-triazine-type ligands for the purpose of hydrogen bonded/metal-coordination hybrid [3]. Another representative example was designed, synthesized and characterized by He et al. via self-assembly of supramolecular coordination complexes using platinum salt with two different types of pyridyl-derivatized ligands [4]. The photophysical properties of these supramolecular coordination complexes showed potential metal ion-responsive materials [4]. In the same field of photophysical study, a host–guest coordination cage has been assembled, and demonstrated a primary ultrafast excited dynamic process including excited-state energy and charge transfer. This tailored architecture was designed by the Han research group [5]. This fascinating s-triazine ligand has attracted great attention due to its several applications [6–13]. Mao et al. designed and synthesized two trigeminal star-like platinum complexes which stabilized hTel G4 with high selectivity and affinity, targeting telomerase inhibitors [14]. Additionally, some Pd(II)-s-triazine complexes have been constructed and assessed against breast cancer cell lines (MCF7 and MDA-MB-231) and have exhibited good potentials [15,16]. The design of new s-triazine-based ligands and their coordination modes with different metal centers is still a challenge [17–19]. Recently, Barakat et al. designed, synthesized and characterized a new hydrazono-s-triazine-based ligand and later explored the coordination chemistry of this ligand with a palladium(II) center. This study revealed that palladium coordinated via the s-triazine-type ligand as an *NNO*-donor [20]. Additionally, reaction of PdCl$_2$ with 4,4′-(6-(3,5-dimethyl-1*H*-pyrazol-1-yl)-1,3,5-triazine-2,4-diyl)dimorpholine (**MPT**) and *N*-methyl-*N*-phenyl-4,6-di(1*H*-pyrazol-1-yl)-1,3,5-triazin-2-amine (**BPT**) ligands afforded the corresponding [Pd(MPT)Cl$_2$] and [Pd(BPT)Cl]ClO$_4$ tetracoordinated Pd(II) complexes. In these Pd(II) complexes, the s-triazine ligands worked as bidentate and tridentate chelates, respectively [16]. Both complexes were found to have improved anticancer activities against MDA-MB-231 and MCF-7 cell lines compared to the corresponding free ligands. On the other hand, the reaction of PdCl$_2$ with 2,4-*bis*(3,5-dimethyl-1*H*-pyrazol-1-yl)-6-methoxy-1,3,5-triazine proceeded with partial hydrolysis of the ligand to 6-(3,5-dimethyl-1*H*-pyrazol-1-yl)-1,3,5-triazine-2,4(1*H*,3*H*)-dione (**HPT**) and the square planar complex [Pd(PT)Cl(H$_2$O)]*H$_2$O was obtained [15]. In addition, the Pd(II) complex was found to have almost equal activities against MDA-MB-231 and MCF-7 cell lines. Interestingly, the reaction of the same ligand with PtCl$_2$ proceeded with complete hydrolysis of the ligand as indicated by the formation of [Pt(3,5-dimethyl-1*H*-pyrazole)$_2$Cl$_2$] [15].

During our study, we have explored the utility of the hydrazono-s-triazine-based ligand towards metalation with the divalent platinum ion to synthesize a new Pt(II) complex based on s-triazine hydrazone ligand (Figure 1). Its 3D molecular and supramolecular structures were elucidated by single-crystal X-ray diffraction and Hirshfeld analyses. The chemical insights of the Pt(II) complex have also been demonstrated.

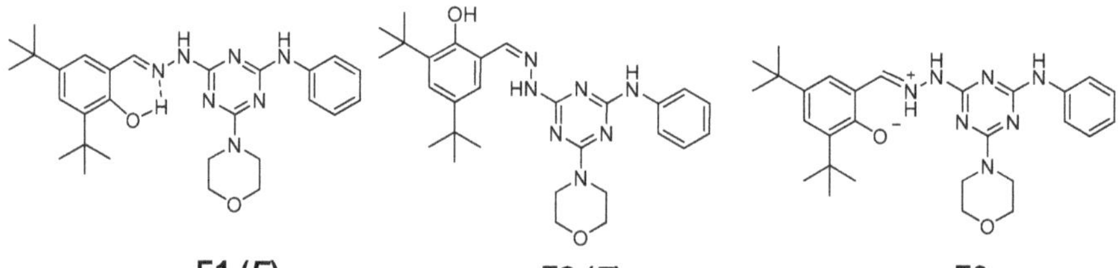

F1 (*E*)   **F2 (*Z*)**   **F3**

Figure 1. Structure of s-triazine hydrazone ligand (**HTriaz**).

2. Results and Discussion

2.1. [Pt(Triaz)Cl] Complex Synthesis and Chracterization

The Pt(II) complex **[Pt(Triaz)Cl]** was synthesized by reaction of (**HTriaz**) ligand with platinum (II) chloride (1:1) in acetone under heating (Scheme 1). The new Pt(II) complex was characterized by FT-IR, UV–Vis, single-crystal X-ray diffraction and CHNPt analyses. The reported structure by single-crystal X-ray diffraction agreed very well with the elemental analysis results. Additionally, the FT-IR spectra of **[Pt(Triaz)Cl]** exhibited vibrational characteristics of the functional groups, e.g., NH (3428 cm^{-1}), aromatic C–H (3120 cm^{-1}), aliphatic C–H (2957 and 2866 cm^{-1}), C=N/C=C (1630 cm^{-1}).

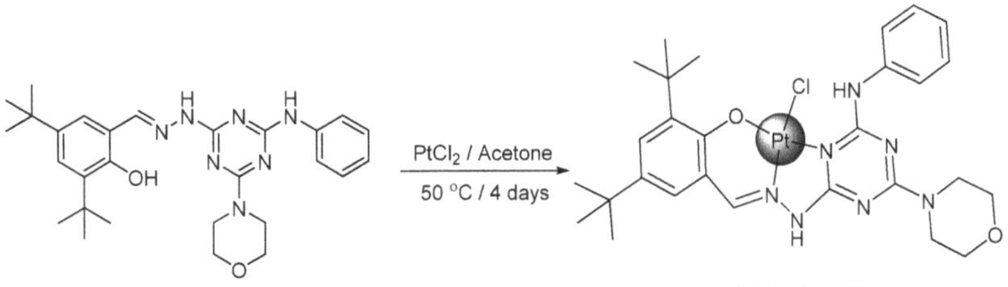

Scheme 1. Synthesis of [Pt(Triaz)Cl] complex.

2.2. Crystal Structure Description

The X-ray structure of **[Pt(Triaz)Cl]** including atom numbering and thermal ellipsoids drawn at 50% probability level is shown in Figure 2 (upper part). The **[Pt(Triaz)Cl]** complex crystallized in *I2/a* space group (Table S1; Supplementary data). The asymmetric unit comprised one **[Pt(Triaz)Cl]** complex unit and one acetone as a crystal solvent. The ligand (**Triaz^{-1}**) is a *NNO* tridentate ligand. The donor atoms of this ligand are two nitrogen atoms from the *s*-triazine and the hydrazone fragments in addition to the phenolic oxygen atom. The coordination environment of the Pt(II) is completed by one Cl^{-1} *trans* to the Pt-N$_{(hydrazone)}$. The Pt to donor atoms (N4, N7, O2 and Cl1) distances are 2.055(4), 1.945(4), 1.991(4) and 2.331(1) Å, respectively. The angle between the *trans*-bonds O2-Pt1-N4 and N7-Pt1-Cl1 are 173.35(16) and 177.01(13) Å, respectively (Table 1). The results are in good agreement with the X-ray structure of the structurally related **[Pd(Triaz)Cl]** complex [20].

Table 1. [Pt(Triaz)Cl] complex bond lengths [Å] and angles [°].

Atoms	Distance	Atoms	Distance
Pt1-N7	1.945(4)	Pt1-N4	2.055(4)
Pt1-O2	1.991(4)	Pt1-Cl1	2.3308(13)
Atoms	Angle	Atoms	Angle
N7-Pt1-O2	93.09(16)	N7-Pt1-Cl1	177.01(13)
N7-Pt1-N4	80.42(18)	O2-Pt1-Cl1	83.91(11)
O2-Pt1-N4	173.35(16)	N4-Pt1-Cl1	102.58(13)

On the other hand, the angles between the *cis*-bonds are in the range of 83.91(11)–102.58(13)°, indicating a distorted square planar coordination environment around the Pt(II). The structure of this complex showed one intramolecular N-H···O H-bond between the N–H group from the organic ligand as a H-bond donor and the carbonyl oxygen atom from the acetone molecule as H-bond acceptor. The hydrogen-acceptor and donor-acceptor distances are 2.028 and 2.777(7) Å, respectively, while the N6-H6···O3 angle is 141.6°. A view of packing along *ac*-plane is shown in the lower part of Figure 2.

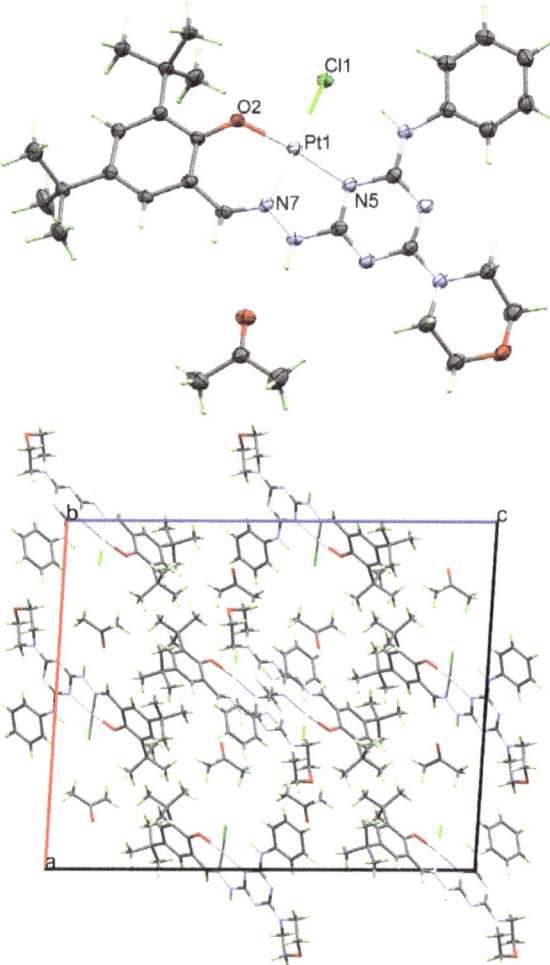

Figure 2. X-ray structure (**upper**) and packing view along *ac*-plane (**lower**) for [**Pt(Triaz)Cl**] complex.

2.3. Analysis of Molecular Packing

Hirshfeld surfaces mapped over d_{norm}, shape index (SI) and curvedness for the studied complex are shown in Figure 3, while the different contacts and their contribution percentages in the molecular packing are present in Figure 4.

As can be seen from Figure 4, the most dominant interactions are the H···H, H···C and O···H intermolecular contacts. These interactions contributed 60.9, 11.2, and 8.3% of the whole fingerprint area while the corresponding values for the **Pd(II)** complex are 60.6, 11.6, and 8.1, respectively. Other minor contributions from the Cl···H, C···N, N···H and C···C contacts were also detected. Generally, the most significant contacts are the O···H and H···C interactions. The latter belongs to the C-H···π interactions. In the corresponding **Pd(II)** complex, the O···H, H···H and H···C interactions are the most important. These intermolecular contacts appeared as red spots in d_{norm} and characterized by spikes in the fingerprint plots as shown in Figure 5. The O···H interactions appeared as one spike in the upper left part of the fingerprint plot due to the N–H···O (1.934 Å) and C–H···O (2.416 Å) interactions between the carbonyl group as hydrogen bond acceptor and the surface as hydrogen bond donor. On the other hand, the C–H···π interactions are characterized by two spikes with interaction distances ranges from 2.630 Å (H4A···C15) to 2.785 Å (H19B···C16). In the corresponding **Pd(II)** complex, the O···H and H···C interactions are 1.839 and 2.608 Å, respectively which are slightly shorter than the corresponding values of the **[Pt(Triaz)Cl]** complex. In the former, all H···H interactions have long interaction distances while in the latter, most H···H interactions also have long interaction distances, except for the H11···H2B contact, which appeared as a red spot in the d_{norm}. The H11···H2B contact distance is 2.003 Å. A summary of all contacts with shorter distances than the vdW radii sum of the interacting elements is listed in Table 2.

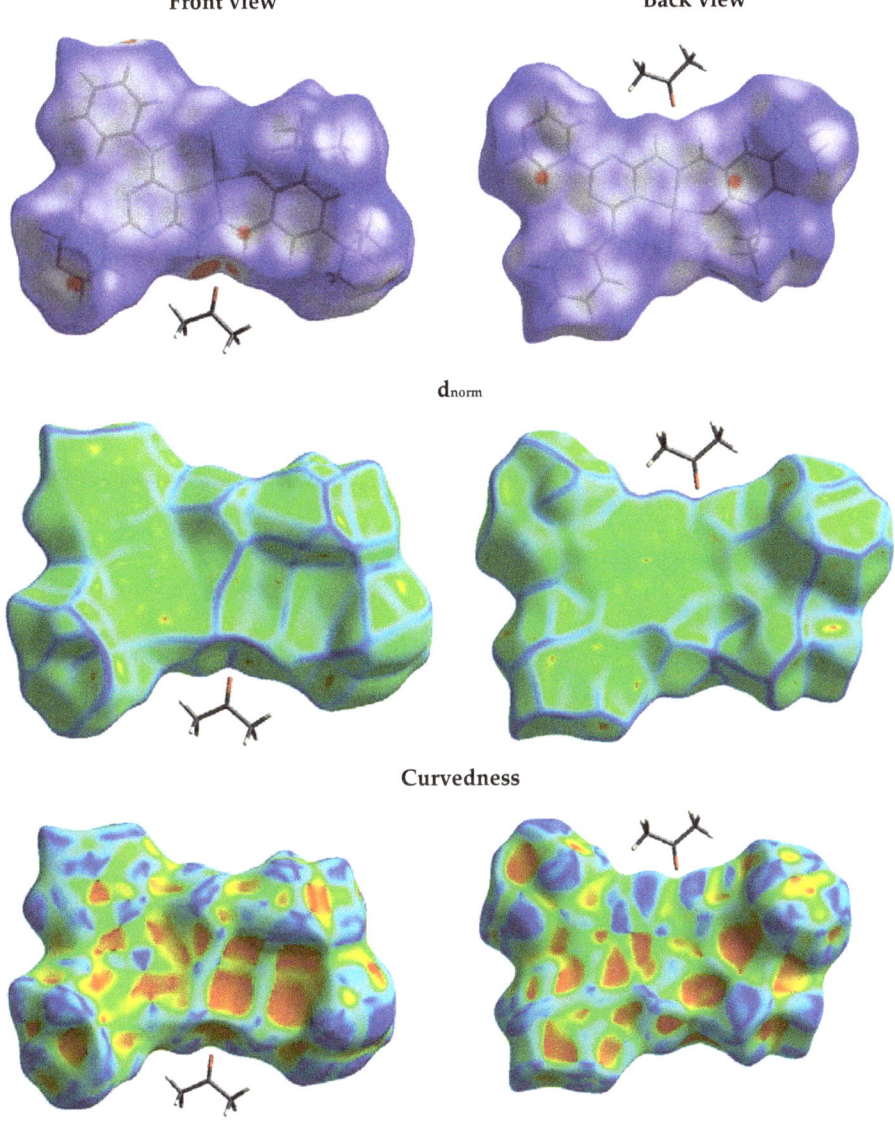

Figure 3. Hirshfeld surfaces of **[Pt(Triaz)Cl]**.

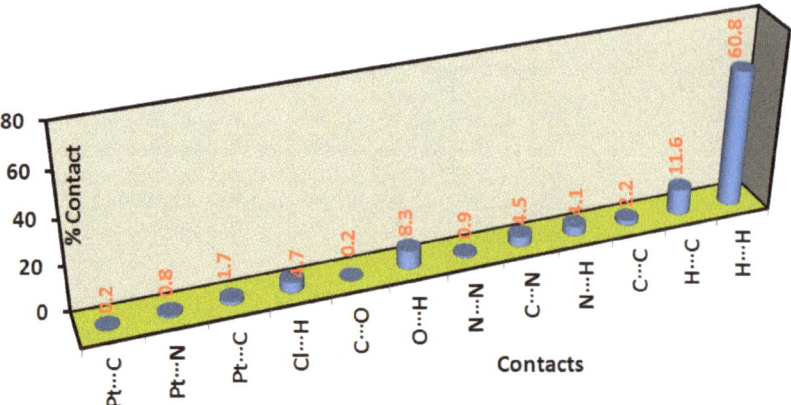

Figure 4. Percentages of intermolecular contacts in [Pt(Triaz)Cl].

Table 2. Short interactions and their contact distances in [Pt(Triaz)Cl].

Contact	Distance	Contact	Distance
O3···H1	2.416	H4A···C15	2.630
O3···H6	1.934	H4A···C28	2.777
H19B···C16	2.785	H11···C2	2.689
H20B···C14	2.656	H11···H2B	2.003

2.4. DFT Studies

The optimized structures of [Pt(Triaz)Cl] and two possible geometrical isomers (**F1** (*E*) and **F2** (*Z*); Figure 1) of the free ligand are shown in Figure 6. The total energies of the ligand isomers are −1622.4327 and −1622.4126 a.u. for **F1** and **F2**, respectively. Hence, **F1** is the more stable than **F2** by 12.6019 kcal/mol. This result agreed with our previous studies [21]. The extra stability of **F1** could be attributed to the presence of intramolecular O–H···N hydrogen bond between the hydrazone nitrogen atom and the OH proton with hydrogen-acceptor and donor-acceptor distances of 1.729 and 2.608 Å, respectively. Another possible isomer in which the labile proton is bonded to the Schiff base nitrogen atom leading to a zwitterion species is abbreviated in Figure 1 as **F3**. The structure of **F3** was optimized using the same level of theory. Interestingly, the geometry optimization ended to the same optimized structure of **F1** indicating that the form **F1** is more favored than the NH zwitter ionic form **F3**. Additionally, the proton affinity of **Triaz**$^-$ was calculated based on the enthalpy change (ΔH) of the reaction **Triaz**$^-$ + **H**$^+$ → **HTriaz** to be 353.06 kcal/mol. On the other hand, the Pt(II) affinity **Triaz**$^-$ was calculated to be 589.111 kcal/mol. In this regard, one could conclude that the higher affinity of **Triaz**$^-$ to the Pt(II) could be attributed to the chelate effect where the coordination between the Pt(II) ion and the tridentate **Triaz**$^-$ ligand lead to the formation of two chelate rings which could be the driving force for the deprotonation of the **HTriaz** and breaking the intramolecular O–H···N hydrogen bonding interaction of **F1**.

On the other hand, the optimized structure of the [Pt(Triaz)Cl] complex agreed very well with the experimental X-ray structure (Table S2, Supplementary data). In addition, good correlations were obtained between the calculated and experimental geometric parameters. The correlation coefficients for bond distances and angles are 0.9979 and 0.9758, respectively (Figure 7). The ligand and its Pt(II) complex are polar compounds where the calculated dipole moments are 7.933 and 2.289 Debye, respectively. It is clear that complexation of the ligand with Pt(II) decreased the polarity of the system.

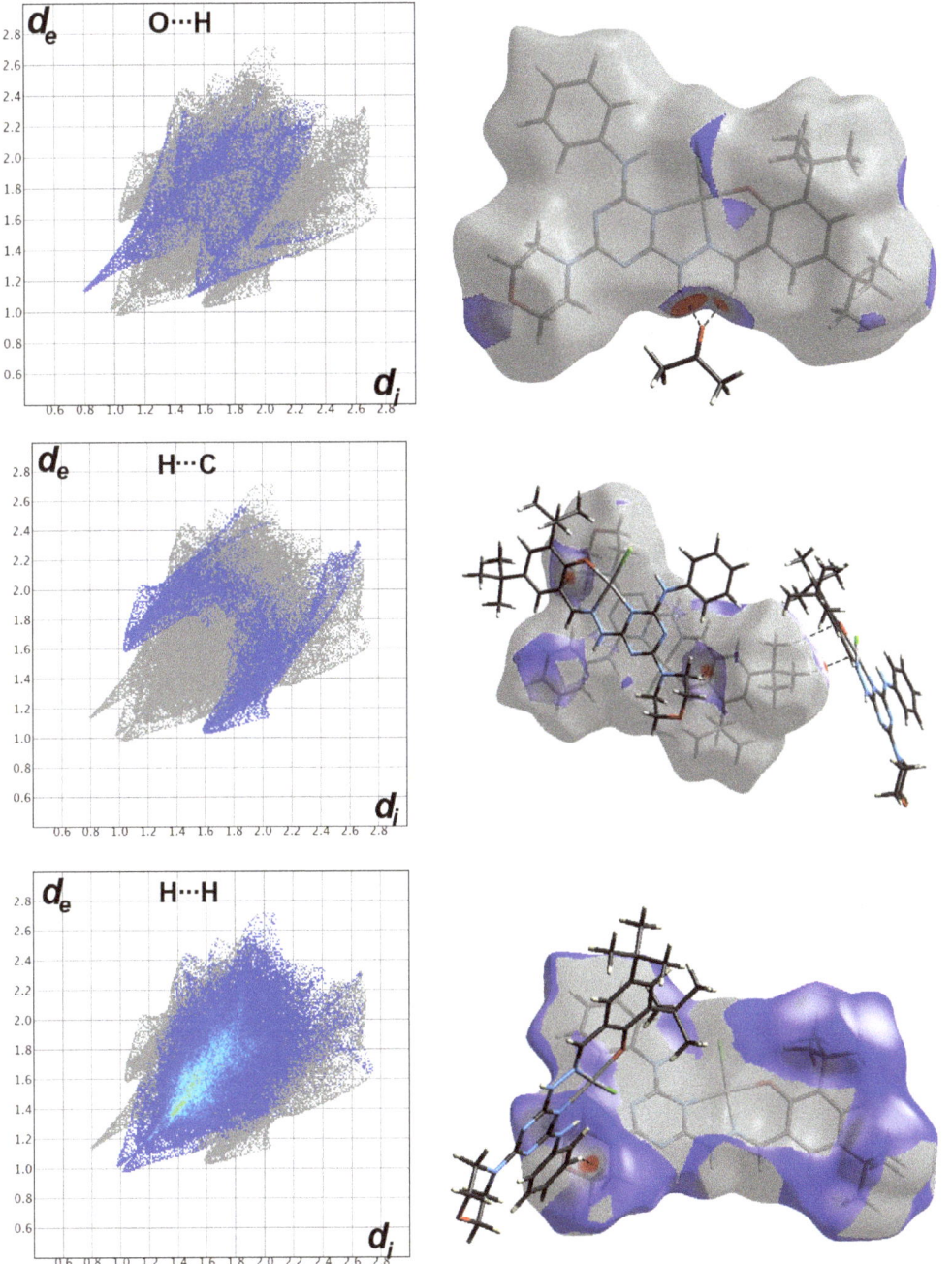

Figure 5. d$_{norm}$ maps (**right**) and fingerprint plots (**left**) of the O···H, H···C and H···H contacts in [Pt(Triaz)Cl].

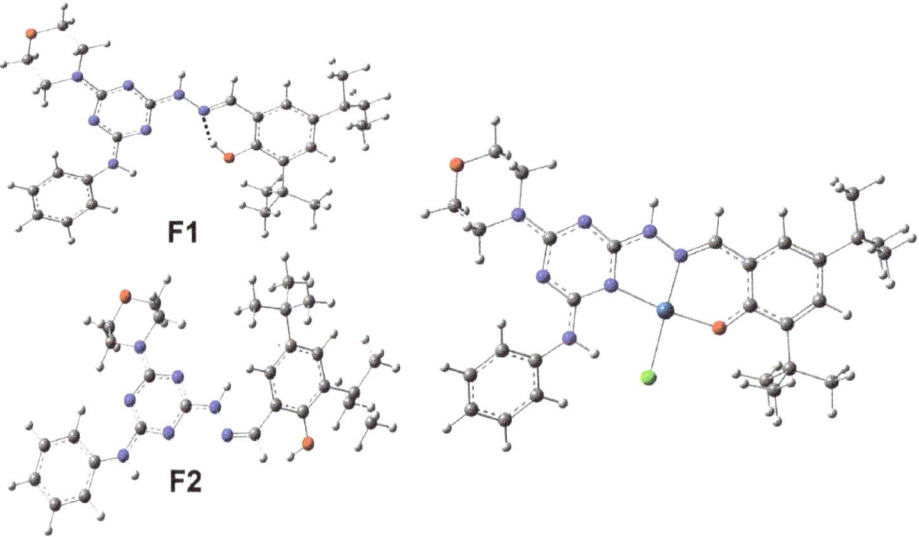

Figure 6. The optimized geometries of the two isomers (**F1** and **F2**) of the ligand **HTriaz** (**left**) and [Pt(Triaz)Cl] complex (**right**).

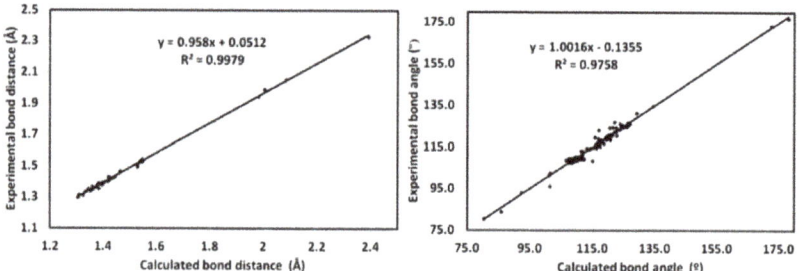

Figure 7. Correlations between the calculated and experimental bond distances (**left**) and angles (**right**).

The interaction between Pt(II) as a Lewis acid and ligand as a Lewis base affect the net charge at both fragments. The calculated charges at Pt, Cl, and the anionic ligand are depicted in Table 3. The charge at the Pt(II) is changed to +0.5 instead of +2.0 due to the large electron density transferred from the ligand groups. The amount of negative electron density transferred from the ligand groups are 0.56 and 0.95 e for the Cl^{-1} and $Triaz^{-1}$, respectively.

Table 3. The calculated charge at Pt, Cl and the anionic ligand.

Atom/Group	Optimized	X-ray
Pt	0.4998	0.4857
Cl	−0.4410	−0.4402
Triaz	−0.0588	−0.0455

2.5. UV–Vis Spectra

The experimental and calculated UV–Vis spectra of the studied Pt(II) complex in ethanol as solvent are presented in Figure 8. The longest wavelength band was observed experimentally at 427 nm. The TD-DFT calculations predicted this band at 409 nm with oscillator strength of 0.1646. This electronic transition was assigned to HOMO→LUMO (93%) excitation. In addition, the TD-DFT calculations predicted intense absorptions at 322 nm (exp. 338 nm) and 305 nm (exp. 320 nm) with oscillator strengths of 0.2102 and 0.2196, respectively. These electronic transition bands were assigned to H−1→LUMO (83%) and HOMO→L+2 (84%), respectively. Experimentally, the region below 300 nm showed an intense absorption at 261 nm, which is calculated at 266 nm (f = 0.3628). This band was assigned to H−1→L+2 (89%) excitation. Presentation of molecular orbitals (MOs) included in these electronic transitions are shown in Figure 9. Theoretically, an absorption band and a shoulder were predicted at 247 nm (f = 0.1883) and 226 nm (f = 0.1040), respectively. The former was assigned to the mixed H−3→L+2 (56%) and HOMO→L+5 (11%) transitions while the latter was assigned for H−3→L+3(26%) and HOMO→L+6 (17%)/L+7 (35%) transitions.

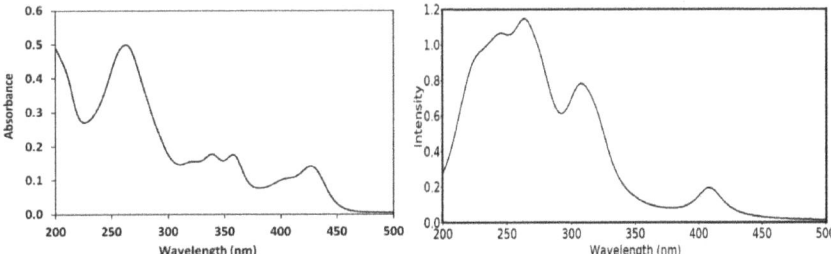

Figure 8. The experimental (**left**) and calculated (**right**) UV–Vis spectra of the studied [Pt(Triaz)Cl] complex.

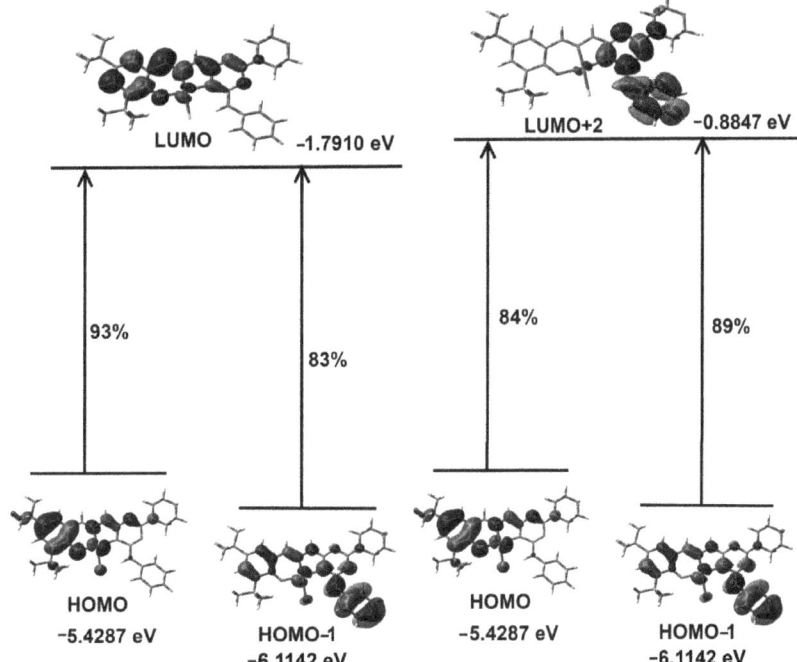

Figure 9. The MOs included in the electronic transitions of the studied [Pt(Triaz)Cl] complex.

3. Materials and Methods

3.1. Materials and Methods

Chemicals were purchased from Sigma–Aldrich (Chemie GmbH, 82024 Taufkirchen, Germany). The CHN analyses were determined using Perkin–Elmer 2400 instrument (PerkinElmer, Inc.940 Winter Street, Waltham, MA, USA). Pt content was determined using a Shimadzu atomic absorption spectrophotometer (AA-7000 series, Shimadzu, Ltd., Japan). FT-IR spectrum was assessed on a Perkin–Elmer 1000 FT-IR spectrometer, Waltham, MA, USA (Figure S1). The UV–Vis electronic spectrum of the Pt(II) complex at 3.0×10^{-4} mol L^{-1} in absolute ethanol as solvent was carried out using a UV–Vis spectrophotometer (Perkin–Elmer Lambda 35, Waltham, MA, USA) in 1 cm cell in the spectral range of 200–500 nm. Mass spectrum was recorded on JMS-600 H JEOL spectrometer (JEOL Ltd., Tokyo, Japan). ^{1}H and ^{13}C NMR spectra of **[Pt(Triaz)Cl]** were recorded on DMSO-d_6 using a JEOL 500 MHz spectrometer (JEOL Ltd., Tokyo, Japan) at room temperature.

3.2. Synthesis of the Ligand (HTriaz)

The ligand (**HTriaz**) has been prepared using our published method [20,22] and the NMR spectral data agreed with the reported data [20].

3.3. Synthesis of [Pt(Triaz)Cl] Complex

The (**HTriaz**) ligand (60.0 mg, 0.119 mmol) was dissolved in 30 mL of acetone then PtCl$_2$ (31.6 mg, 0.119 mmol) was added. The reaction mixture was heated at 50 °C for 4 days. Then, the resulting solution mixture was filtered, and the filtrate was left for slow evaporation at room temperature to afford the final product **[Pt(Triaz)Cl]** as reddish-brown block crystals. Yield; C$_{31}$H$_{42}$ClN$_7$O$_3$Pt 79%; Anal. Calcd. for: C, 47.06; H, 5.35; N, 12.39; Pt, 24.65. Found: C, 47.24; H, 5.29; N, 12.20; Pt, 24.46. FT-IR (KBr) cm^{-1}: 3428 (NH), 3263, 3120, 2957, 2866 (C-H), 1540 and 1630 (C=N and C=C) (Figure S1; Supplementary data); ^{1}H NMR (500 MHz, DMSO-d_6, ppm): δ 1.24 (s, 9H, 3CH$_3$), 1.37 (s, 9H, 3CH$_3$), 3.64 (t, 4H, J = 4.0 Hz, 2CH$_2$ (morpholine ring), 3.71 (t, 4H, J = 3.6 Hz, 2CH$_2$ (morpholine ring), 7.14 (t, 1H, J = 6.8 Hz, C$_6$H$_5$), 7.30 (d, 1H, J = 2.0 Hz, C$_6$H$_5$), 7.30–7.35 (m, 3H, C$_6$H$_5$ and C$_6$H$_2$), 7.63-7.57 (m, 3H, C$_6$H$_5$ and C$_6$H$_2$ and CH=N), 8.46 (s, 1H, NH), 10.91 (s, 1H, NH) (Figure S2; Supplementary data). ^{13}C NMR (126 MHz, DMSO-d_6) δ 194.92, 178.15, 166.74, 164.72, 143.14, 141.50, 140.00, 138.81, 133.36, 131.26, 130.51, 128.42, 125.73, 125.09, 124.80, 123.88, 123.08, 122.93, 116.34, 116.17, 111.63, 74.32, 74.26, 62.38, 54.28, 49.14, 47.13, 36.54 (Figure S3; Supplementary data).

3.4. X-ray Structure Determinations

The details of the crystal structure determination are found in Table S1 and all technical experiments are provided in the supplementary materials [23–27].

3.5. Hirshfeld and DFT Calculations

Crystal Explorer 17.5 [28] was used to perform the analysis of molecular packing. Details of DFT and TD-DFT calculations [29–34] as well as proton affinity [35] are given in supplementary data.

4. Conclusions

A novel Pt(II) complex **[Pt(Triaz)Cl]** with tridentate *NNO*-donor ligand-based *s*-triazine scaffold was achieved. The chemical structure of **[Pt(Triaz)Cl]** was confirmed by CHNPt analyses and single-crystal X-ray diffraction. The Pt(II) coordination environment is distorted square planar. The structure of this complex showed one intramolecular N–H···O hydrogen bond between the N–H group from the organic ligand as a hydrogen bond donor and the carbonyl oxygen atom from the acetone molecule as a hydrogen bond acceptor. The supramolecular structure of the studied Pt(II) complex is analyzed using Hirshfeld calculations. Additionally, the calculated UV–Vis spectra were assigned based on the results of the TD-DFT calculations. The natural charges were calculated, and the results indicated that the amount of the electron transfer from the Cl^{-1} and $Triaz^{-1}$ is 0.56 and 0.95 e, respectively.

Supplementary Materials: The following supporting information can be downloaded online. Table S1: Crystal data and structure refinement for [Pt(Triaz)Cl]; Table S2. The calculated geometric parameters of [Pt(Triaz)Cl]; Figure S1: FT-IR spectra of the studied Pt(II) complex; Figure S2: ^{1}H NMR spectra of the studied Pt(II) complex; Figure S3: ^{13}C NMR spectra of the studied Pt(II) complex.

Author Contributions: Conceptualization, J.L., A.B. and A.E.-F.; methodology, M.S.A., J.L., N.E.E. and A.B.; software, S.M.S., M.H.; M.S.A., J.L., N.E.E. and A.B.; formal analysis, J.L., N.E.E. and M.H.; investigation, M.S.A., J.L. and N.E.E.; resources, J.L. and A.B.; writing—original draft preparation, S.M.S. and A.B.; writing—review and editing, J.L., and A.E.-F.; supervision, A.E.-F.; funding acquisition, M.S.A. All authors have read and agreed to the published version of the manuscript.

Funding: Princess Nourah bint Abdulrahman University Researchers Supporting Project number (PNURSP2022R86), Princess Nourah bint Abdulrahman University, Riyadh, Saudi Arabia.

Institutional Review Board Statement: Not applicable.

Informed Consent Statement: Not applicable.

Data Availability Statement: Not applicable.

Acknowledgments: Princess Nourah bint Abdulrahman University Researchers Supporting Project number (PNURSP2022R86), Princess Nourah bint Abdulrahman University, Riyadh, Saudi Arabia.

Conflicts of Interest: The authors declare no conflict of interest.

Sample Availability: Samples of the compound **[Pt(Triaz)Cl]** is available from the authors.

References

1. Barakat, A.; El-Faham, A.; Haukka, M.; Al-Majid, A.M.; Soliman, S.M. s-Triazine pincer ligands: Synthesis of their metal complexes, coordination behavior, and applications. *Appl. Organomet. Chem.* **2021**, *35*, e6317. [CrossRef]
2. Kumar, A.; Mukherjee, P.S. Multicomponent Self-Assembly of PdII/PtII Interlocked Molecular Cages: Cage-to-Cage Conversion and Self-Sorting in Aqueous Medium. *Chem. Eur. J.* **2020**, *26*, 4842–4849. [CrossRef] [PubMed]
3. Motloch, P.; Hunter, C.A. Stimuli-Responsive Self-Sorting Hybrid Hydrogen-Bonded/Metal-Coordinated Cage. *Chem. Eur. J.* **2021**, *27*, 3302–3305. [CrossRef] [PubMed]
4. He, Z.; Li, M.; Que, W.; Stang, P.J. Self-assembly of metal-ion-responsive supramolecular coordination complexes and their photophysical properties. *Dalton Trans.* **2017**, *46*, 3120–3124. [CrossRef] [PubMed]
5. Zhang, R.L.; Yang, Y.; Yang, S.Q.; Han, K.L. Unveiling excited state energy transfer and charge transfer in a host/guest coordination cage. *Phys. Chem. Chem. Phys.* **2018**, *20*, 2205–2210. [CrossRef] [PubMed]
6. Fu, H.L.K.; Leung, S.Y.L.; Yam, V.W.W. A rational molecular design of triazine-containing alkynylplatinum (II) terpyridine complexes and the formation of helical ribbons via Pt···Pt, π–π stacking and hydrophobic–hydrophobic interactions. *Chem. Commun.* **2017**, *53*, 11349–11352. [CrossRef]
7. Liu, N.; Lin, T.; Wu, M.; Luo, H.K.; Huang, S.L.; Hor, T.A. Suite of Organoplatinum (II) triangular metallaprism: Aggregation-induced emission and coordination sequence induced emission tuning. *J. Am. Chem. Soc.* **2019**, *141*, 9448–9452. [CrossRef]
8. Marzo, T.; Cirri, D.; Ciofi, L.; Gabbiani, C.; Feis, A.; Di Pasquale, N.; Stefanini, M.; Biver, T.; Messori, L. Synthesis, characterization and DNA interactions of [Pt$_3$(TPymT)Cl$_3$], the trinuclear platinum (II) complex of the TPymT ligand. *J. Inorg. Biochem.* **2018**, *183*, 101–106. [CrossRef]
9. Ismail, A.M.; El Sayed, S.A.; Butler, I.S.; Mostafa, S.I. New Palladium (II), Platinum (II) and Silver (I) complexes of 2-amino-4, 6-dithio-1,3,5-triazine; synthesis, characterization and DNA binding properties. *J. Mol. Struct.* **2020**, *1200*, 127088. [CrossRef]
10. Yetim, N.K.; Sarı, N. Novel dendrimers containing redox mediator: Enzyme immobilization and applications. *J. Mol. Struct.* **2019**, *1191*, 158–164. [CrossRef]
11. Paul, L.E.; Therrien, B.; Furrer, J. The complex-in-a-complex cation [Pt (acac) 2-(p-cym)$_6$Ru$_6$(tpt)$_2$(dhnq)$_3$]$^{6+}$: Its stability towards biological ligands. *Inorg. Chim. Acta* **2018**, *469*, 1–10. [CrossRef]
12. Asman, P.W. Kinetics and mechanistic study of polynuclear platinum (II) polypyridyl complexes; A paradigm shift in search of new anticancer agents. *Inorg. Chim. Acta* **2018**, *469*, 341–352. [CrossRef]
13. Zhang, W.; Wang, J.; Xu, Y.; Li, W.; Shen, W. Fine tuning phosphorescent properties of platinum complexes via different N-heterocyclic-based C-N-N ligands. *J. Organomet. Chem.* **2017**, *836*, 26–33. [CrossRef]
14. Zheng, X.H.; Mu, G.; Zhong, Y.F.; Zhang, T.P.; Cao, Q.; Ji, L.N.; Zhao, Y.; Mao, Z.W. Trigeminal star-like platinum complexes induce cancer cell senescence through quadruplex-mediated telomere dysfunction. *Chem. Comm.* **2016**, *52*, 14101–14104. [CrossRef]

15. Lasri, J.; Haukka, M.; Al-Rasheed, H.H.; Abutaha, N.; El-Faham, A.; Soliman, S.M. Synthesis, structure and in vitro anticancer activity of Pd (II) complex of pyrazolyl-s-triazine ligand; A new example of metal-mediated hydrolysis of s-triazine pincer ligand. *Crystals* **2021**, *11*, 119. [CrossRef]
16. Lasri, J.; Al-Rasheed, H.H.; El-Faham, A.; Haukka, M.; Abutaha, N.; Soliman, S.M. Synthesis, structure and in vitro anticancer activity of Pd (II) complexes of mono-and bis-pyrazolyl-s-triazine ligands. *Polyhedron* **2020**, *187*, 114665. [CrossRef]
17. Soliman, S.M.; Albering, J.H.; Sholkamy, E.N.; El-Faham, A. Mono-and penta-nuclear self-assembled silver (I) complexes of pyrazolyl s-triazine ligand; synthesis, structure and antimicrobial studies. *Appl. Organomet. Chem.* **2020**, *34*, e5603. [CrossRef]
18. Soliman, S.M.; El-Faham, A. Synthesis and structure diversity of high coordination number Cd (II) complexes of large s-triazine bis-Schiff base pincer chelate. *Inorg. Chim. Acta* **2019**, *488*, 131–140. [CrossRef]
19. Soliman, S.M.; El-Faham, A.; Elsilk, S.E.; Farooq, M. Two heptacoordinated manganese (II) complexes of giant pentadentate s-triazine bis-Schiff base ligand: Synthesis, crystal structure, biological and DFT studies. *Inorg. Chim. Acta* **2018**, *479*, 275–285. [CrossRef]
20. Soliman, S.M.; Lasri, J.; Haukka, M.; Elmarghany, A.; Al-Majid, A.M.; El-Faham, A.; Barakat, A. Synthesis, X-ray structure, Hirshfeld analysis, and DFT studies of a new Pd (II) complex with an anionic s-triazine NNO donor ligand. *J. Mol. Struct.* **2020**, *1217*, 128463. [CrossRef]
21. Barakat, A.; El-Sendury, F.F.; Almarhoon, Z.; Al-Rasheed, H.H.; Badria, F.A.; Al-Majid, A.M.; Ghabbour, H.A.; El-Faham, A. Synthesis, X-ray crystal structures, and preliminary antiproliferative activities of new s-triazine-hydroxybenzylidene hydrazone derivatives. *J. Chem.* **2019**, *2019*, 1–10. [CrossRef]
22. Al-Rasheed, H.H.; Al Alshaikh, M.; Khaled, J.M.; Alharbi, N.S.; El-Faham, A. Ultrasonic irradiation: Synthesis, characterization, and preliminary antimicrobial activity of novel series of 4,6-disubstituted-1,3,5-triazine containing hydrazone derivatives. *J. Chem.* **2016**, *2016*, 3464758. [CrossRef]
23. Otwinowski, Z.; Minor, W. Processing of X-ray Diffraction Data Collected in Oscillation Mode. In *Methods in Enzymology, Volume 276, Macromolecular Crystallography, Part A*; Carter, C.W., Sweet, J., Eds.; Academic Press: New York, NY, USA, 1997; pp. 307–326.
24. Sheldrick, G.M. *SADABS—Bruker Nonius Scaling and Absorption Correction*; Bruker AXS, Inc.: Madison, WI, USA, 2012.
25. Sheldrick, G.M. Crystal Structure Refinement with SHELXL. *Acta Cryst. C* **2015**, *C71*, 3–8. [CrossRef] [PubMed]
26. Hübschle, C.B.; Sheldrick, G.M.; Dittrich, B. *ShelXle*: A Qt graphical user interface for *SHELXL*. *J. Appl. Cryst.* **2011**, *44*, 1281–1284. [CrossRef] [PubMed]
27. Rikagu Oxford Diffraction. *CrysAlisPro*; Agilent Technologies Inc.: Oxford, UK, 2020.
28. Turner, M.J.; McKinnon, J.J.; Wolff, S.K.; Grimwood, D.J.; Spackman, P.R.; Jayatilaka, D.; Spackman, M.A. Crystal Explorer17 (2017) University of Western Australia. Available online: https://crystalexplorer.scb.uwa.edu.au/ (accessed on 20 May 2021).
29. Frisch, M.J.; Trucks, G.W.; Schlegel, H.B.; Scuseria, G.E.; Robb, M.A.; Cheeseman, J.R.; Scalmani, G.; Barone, V.; Mennucci, B.; Petersson, G.A.; et al. *GAUSSIAN 09*; Revision A02; Gaussian Inc.: Wallingford, CT, USA, 2009.
30. Dennington, R., II; Keith, T.; Millam, J. *GaussView*; Version 4.1; Semichem Inc.: Shawnee Mission, KS, USA, 2007.
31. Reed, A.E.; Curtiss, L.A.; Weinhold, F. Intermolecular interactions from a natural bond orbital, donor-acceptor viewpoint. *Chem. Rev.* **1988**, *88*, 899–926. [CrossRef]
32. Marten, B.; Kim, K.; Cortis, C.; Friesner, R.A.; Murphy, R.B.; Ringnalda, M.N.; Sitkoff, D.; Honig, B. New model for calculation of solvation free energies: Correction of self-consistent reaction field continuum dielectric theory for short-range hydrogen-bonding effects. *J. Phys. Chem.* **1996**, *100*, 11775–11788. [CrossRef]
33. Tannor, D.J.; Marten, B.; Murphy, R.; Friesner, R.A.; Sitkoff, D.; Nicholls, A.; Ringnalda, M.; Goddard, W.A.; Honig, B. Accurate first principles calculation of molecular charge distributions and solvation energies from ab initio quantum mechanics and continuum dielectric theory. *J. Am. Chem. Soc.* **1994**, *116*, 11875–11882. [CrossRef]
34. Scalmani, G.; Frisch, M.J.; Mennucci, B.; Tomasi, J.; Cammi, R.; Barone, V. Geometries and properties of excited states in the gas phase and in solution: Theory and application of a time-dependent density functional theory polarizable continuum model. *J. Chem. Phys.* **2006**, *124*, 1–15. [CrossRef]
35. Moser, A.; Range, K.; York, D.M. Accurate proton affinity and gas-phase basicity values for molecules important in biocatalysis. *J. Phys. Chem. B* **2010**, *114*, 13911–13921. [CrossRef]

Article

Selective Formation of Intramolecular Hydrogen-Bonding Palladium(II) Complexes with Nucleosides Using Unsymmetrical Tridentate Ligands

Ryoji Mitsuhashi [1,*], Yuya Imai [2], Takayoshi Suzuki [3] and Yoshihito Hayashi [2]

1 Institute of Liberal Arts and Science, Kanazawa University, Kakuma, Kanazawa 920-1192, Ishikawa, Japan
2 Department of Chemistry, Kanazawa University, Kakuma, Kanazawa 920-1192, Ishikawa, Japan; helibnnaalclkvcu@stu.kanazawa-u.ac.jp (Y.I.); hayashi@se.kanazawa-u.ac.jp (Y.H.)
3 Research Institute for Interdisciplinary Science, Okayama University, 3-1-1 Tsushima-naka, Okayama 700-8530, Japan; suzuki@okayama-u.ac.jp
* Correspondence: mitsuhashi@staff.kanazawa-u.ac.jp

Abstract: Three palladium(II) complexes with amino-amidato-phenolato-type tridentate ligands were synthesized and characterized by ^{1}H NMR spectroscopy and X-ray crystallography. The strategic arrangement of a hydrogen-bond donor and acceptor adjacent to the substitution site of the Pd^{II} complex allowed the selective coordination of nucleosides. Among two pyrimidine-nucleosides, cytidine and 5-methyluridine, cytidine was successfully coordinated to the Pd^{II} complex while 5-methyluridne was not. On the other hand, both purine-nucleosides, adenosine and guanosine, were coordinated to the Pd^{II} complex. As purines have several coordination sites, adenosine afforded three kinds of coordination isomers expected from the three different donors. However, guanosine afforded a sole product according to the ligand design such that the formation of double intramolecular hydrogen-bond strongly induced the specific coordination by *N1*-position of guanine moiety. Furthermore, the preference of the nucleosides was evaluated by scrambling reactions. It was found that the preference of guanosine is nearly twice as high as adenosine and cytidine, owing to the three-point interaction of a coordination bond and two hydrogen bonds. These results show that the combination of a coordination and hydrogen bonds, which is reminiscent of the Watson–Crick base pairing, is an effective tool for the precise recognition of nucleosides.

Keywords: palladium(II) complex; hydrogen-bonding interactions; crystal structure; nucleoside

Citation: Mitsuhashi, R.; Imai, Y.; Suzuki, T.; Hayashi, Y. Selective Formation of Intramolecular Hydrogen-Bonding Palladium(II) Complexes with Nucleosides Using Unsymmetrical Tridentate Ligands. *Molecules* 2022, 27, 2098. https://doi.org/10.3390/molecules27072098

Academic Editor: Luigi A. Agrofoglio

Received: 25 February 2022
Accepted: 21 March 2022
Published: 24 March 2022

Publisher's Note: MDPI stays neutral with regard to jurisdictional claims in published maps and institutional affiliations.

Copyright: © 2022 by the authors. Licensee MDPI, Basel, Switzerland. This article is an open access article distributed under the terms and conditions of the Creative Commons Attribution (CC BY) license (https://creativecommons.org/licenses/by/4.0/).

1. Introduction

Hydrogen-bonding interactions have attracted much attention in the past and present as they are involved in many chemical and biological systems. Hydrogen bonds exhibit a directionality and reversibility upon bond formation. Although the bond energy is much smaller than that of a covalent bond, such unique properties enable construction of supramolecular structures of a molecule to control the chemical and physical properties [1–8]. Furthermore, it is also possible to precisely recognize a molecule by tuning the relative position of hydrogen-bond donors and acceptors in a molecule as represented by the formation of the base pair in DNA [9,10].

Coordination of biomolecules such as nucleobases to a transition metal ion often exhibits cytotoxic properties [11–14]. For example, a well-known mononuclear Pt^{II} complex, cisplatin *cis*-[PtCl$_2$(NH$_3$)$_2$], inhibits the replication of DNA by selective coordination by two continuous guanine residues of DNA. On the other hand, we previously reported synthesis and crystal structures of cobalt and manganese complexes with amino-amidato-phenolato-type tridentate ligands (Figure 1) [15,16]. In combination with a Pd^{II} ion, for which square planar geometry is expected, the amino-amidato-phenolato ligand affords a vacant coordination site for the recognition of nucleosides by arranging a hydrogen-bond donor and acceptor to accommodate a specific nucleoside. For this Pd^{II} complex fragment, selective coordination of a nucleobase is expected, owing to a labile coordination bond in Pd^{II} complex and the arrangement of hydrogen-bonding sites. In this study, we synthesized Pd^{II} complexes with two amino-amidato-phenolato-type ligands and evaluated the effects of hydrogen-bonding interaction on the selectivity of nucleosides upon coordination to the Pd^{II} complex fragment. For such tridentate ligands, it is expected to form a Pd–N coordination and two intramolecular hydrogen bonds with guanine and cytosine moieties. We demonstrated that this three-point interaction enables the precise recognition of nucleobases.

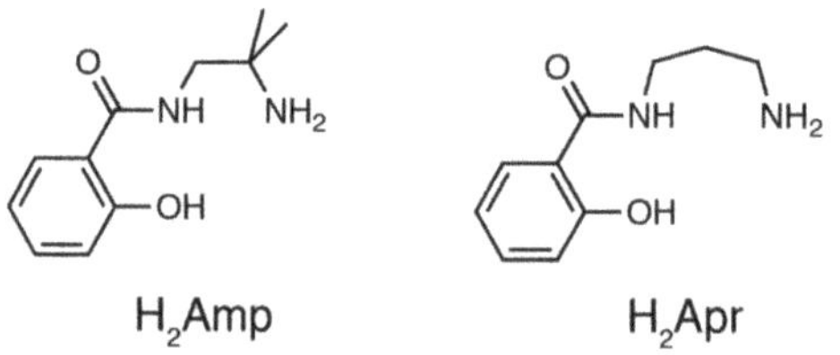

Figure 1. Chemical structures of ligand precursors H$_2$Amp and H$_2$Apr.

2. Results and Discussion

2.1. Preparation

Preparation of Pd^{II} Complexes

The ligand precursors H$_2$Amp, H$_2$Apr, and Pd^{II} starting material were prepared according to the previously reported procedures [15,17]. The Pd^{II} complex was synthesized by a reaction of [PdCl$_2$(CH$_3$CN)$_2$] and H$_2$Amp followed by an addition of triethylamine and N(CH$_3$)$_4$OH·5H$_2$O as a base in methanol. The product was crystallized by diffusing diethyl ether vapor into the methanol solution to afford yellow crystals of N(CH$_3$)$_4$[Pd(Amp)Cl]·H$_2$O (**1**). To accept the coordination by nucleosides, the presence of chloride ligand is not suitable because of its coordination ability. When the ^{1}H NMR spectrum of this product was measured in CD$_3$OD, it exhibited rather broad resonances (Figure 2a). This implies that the complex [Pd(Amp)Cl]$^-$ undergoes a rapid exchange reaction of Cl$^-$ ligand with a solvent methanol molecule. Thus, we exploited this reaction to isolate a chloride-free complex, [Pd(Amp)(CH$_3$CN)]·CH$_3$CN (**1'**), by using KO^{t}Bu instead of N(CH$_3$)$_4$OH·5H$_2$O. In this reaction, potassium salt of **1** was not obtained; instead, **1'** was selectively produced due to the elimination of KCl from the reaction mixture. **1'** is more suitable for the substitution reaction with a nucleoside because acetonitrile is a better leaving group.

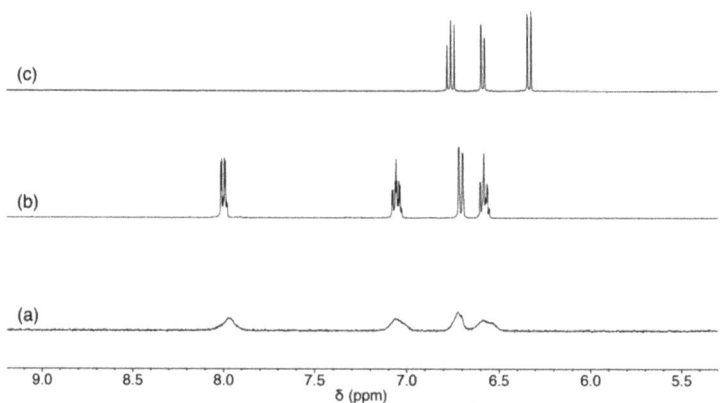

Figure 2. ^{1}H NMR spectra of (a) **1**, (b) **1'** in CD$_3$OD, and (c) **2** in CD$_3$CN.

An analogous Pd^{II} complex with Apr^{2-} was also prepared by a similar procedure with **1'**, and [Pd(Apr)(CH$_3$CN)] (**2**) was obtained. Although the ligand structure of Apr^{2-} is analogous to Amp^{2-}, **2** exhibited a different coordination mode of the tridentate ligands. The ^{1}H NMR spectrum in CD$_3$CN only showed three aryl-H resonances, which indicates that one of the protons on the aromatic group was removed by the base (Figure 2c). The elimination of the proton induced the formation of a Pd–C bond, and the resulting direct coordination of the ring carbon makes the acetonitrile coordination site rather hydrophobic. The direct coordination of the aromatic ring was also supported by the significant up-field shift of aryl-H resonances. The formation of **2** has parallels with orthometalation reaction in organometallic chemistry.

2.2. Crystal Strucre of Pd^{II} Complexes

2.2.1. Crystal Structures of **1** and **1'**

Single-crystals of **1** and **1'** were prepared by diffusing diethyl ether vapor to the reaction solution or slow evaporation of the solution, respectively. The crystallographic information is summarized in Table 1. In both crystals, the Pd^{II} ion takes general square planar coordination geometry (Figure 3). In both **1** and **1'**, Amp^{2-} ligand coordinates to a Pd^{II} center with a phenolato-O, amidato-N, and amino-N atoms as a tridentate ligand. In **1**, the remaining coordination site is occupied by a Cl$^-$ ion to afford an anionic Pd^{II} complex, [Pd(Amp)Cl]$^-$. In **1'**, on the other hand, the Pd^{II} center is coordinated by an acetonitrile molecule instead of Cl$^-$ ligand, although the reaction solvent was acetonitrile in both cases. This is because the Cl$^-$ ion was removed from the system during the reaction due to low solubility of KCl. The bonding parameters around the Pd^{II} ion are summarized in Table 2. The bond lengths around the Pd^{II} center are nearly identical between **1** and **1'** except those of the ancillary ligand Cl$^-$ and CH$_3$CN.

Table 1. Selected bond parameters for **1**.

Atom–Atom	Length/Å	Atom–Atom–Atom	Angle/°
Pd(1)–O(1)	1.9877 (13)	N(1)–Pd(1)–O(1)	93.70 (6)
Pd(1)–N(1)	1.9647 (14)	N(1)–Pd(1)–N(2)	83.87 (6)
Pd(1)–N(2)	2.0264 (16)	O(1)–Pd(1)–N(2)	177.15 (5)
Pd(1)–Cl(1)	2.3389 (4)	N(1)–Pd(1)–Cl(1)	176.50 (4)
		O(1)–Pd(1)–Cl(1)	89.79 (4)
		N(2)–Pd(1)–Cl(1)	92.64 (4)

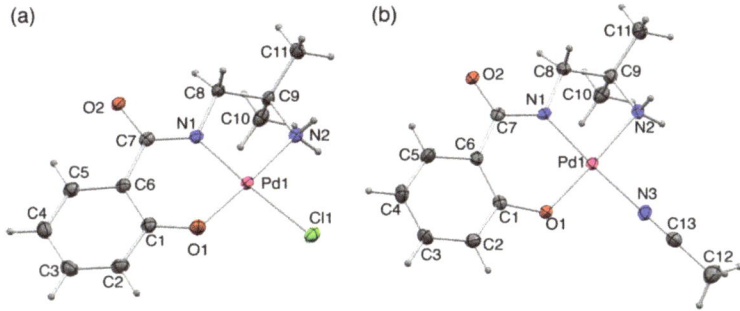

Figure 3. Perspective views of (**a**) [Pd(Amp)Cl]$^-$ in **1**, and (**b**) [Pd(Amp)(CH$_3$CN)] in **1'** (50% probability levels).

Table 2. Selected bond parameters for **1'**.

Atom–Atom	Length/Å	Atom–Atom–Atom	Angle/°
Pd(1)–O(1)	1.9771 (19)	N(1)–Pd(1)–O(1)	94.73 (8)
Pd(1)–N(1)	1.956 (2)	N(1)–Pd(1)–N(2)	82.77 (9)
Pd(1)–N(2)	2.025 (2)	O(1)–Pd(1)–N(2)	176.79 (8)
Pd(1)–N(3)	2.027 (2)	N(1)–Pd(1)–N(3)	176.99 (8)
		O(1)–Pd(1)–N(3)	88.02 (8)
		N(2)–Pd(1)–N(3)	94.44 (9)

In **1**, intermolecular hydrogen bonds are formed between neighboring complex anions and water molecules of crystallization to construct two-dimensional sheet structures along the *bc* plane. The two-dimensional sheet structures in the *bc* plane are also formed in **1'** by the intermolecular hydrogen bonds between the neighboring complexes. These results are indicative of the strong ability of Amp^{2-} to induce hydrogen-bonding interactions.

2.2.2. Crystal Structure of **2**

In the crystal of **2**, [Pd(Apr)(CH$_3$CN)] also takes a general square planar coordination geometry (Figure 4 and Table 3). The asymmetric unit consists of two independent [Pd(Apr)(CH$_3$CN)] molecules, and each molecule is stacked to form a centrosymmetric dimer structure by NH π interactions (the symmetric operation for Pd1 moiety is 1−x, 1−y, 2−z and for Pd2 moiety is 2−x, 2−y, 1−z). Despite the structural similarity with Amp^{2-}, Apr^{2-} ligand coordinates to a Pd^{II} center with phenyl-C, amidato-N, and amino-N atoms. The difference in the coordination modes can be attributed to the combination of chelate rings. For a linear tridentate ligand in octahedral complexes, the combination of six-membered and five-membered chelates is more favored to afford *meridional* coordination geometry, whereas the combination of two six-membered chelates favors *facial* coordination geometry [15,16]. The coordination of a linear-tridentate ligand in square planar complexes may correspond to the coordination mode of the *meridional* type in the octahedral complexes. The observation of aromatic carbon coordination in the formation of complex **2** coincides with the larger stability gain by the combination of five- and six-membered chelates in the octahedral complexes. When Apr^{2-} ligand coordinates in the same way to Amp^{2-}, the resulting formation of two six-membered rings corresponds to *facial* coordination in the octahedral complexes, but in square planar complexes, it is impossible. The average bond length between the Pd^{II} ion and the terminal amino-N atom, 2.13 Å, is significantly longer than those in **1** and **1'** (2.03 Å) because of a strong *trans* influence of phenyl-C donor [18]. The hydroxy group of the ligand forms an intramolecular hydrogen bond with the O atom of the carbonyl group.

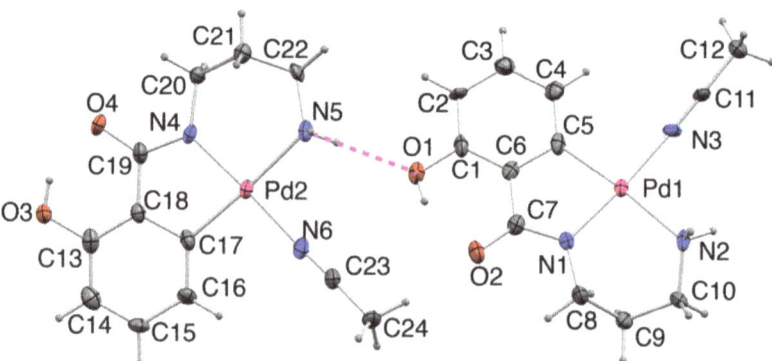

Figure 4. The asymmetric unit of **2** (50% probability levels). An intermolecular hydrogen bond is shown as magenta dashed line.

Table 3. Selected bond parameters for **2**.

Atom–Atom	Length/Å	Atom–Atom–Atom	Angle/°
Pd(1)–C(5)	1.982 (5)	N(1)–Pd(1)–C(5)	81.8 (2)
Pd(1)–N(1)	1.974 (5)	N(1)–Pd(1)–N(2)	92.2 (2)
Pd(1)–N(2)	2.144 (4)	C(5)–Pd(1)–N(2)	173.2 (3)
Pd(1)–N(3)	1.986 (5)	N(1)–Pd(1)–N(3)	175.65 (17)
Pd(2)–C(17)	1.981 (6)	C(5)–Pd(1)–N(3)	95.1 (2)
Pd(1)–N(4)	1.998 (4)	N(2)–Pd(1)–N(3)	90.72 (19)
Pd(1)–N(5)	2.121 (5)	N(4)–Pd(2)–C(17)	81.8 (2)
Pd(1)–N(6)	2.013 (5)	N(4)–Pd(2)–N(5)	94.33 (18)
		C(17)–Pd(2)–N(5)	173.84 (19)
		N(4)–Pd(2)–N(6)	176.3 (2)
		C(17)–Pd(2)–N(6)	95.3 (2)
		N(5)–Pd(2)–N(6)	88.41 (19)

2.3. Reaction of 1 with Nucleosides

Although **1′** is suitable for the ligand exchange, KCl was easily contaminated as impurity during the synthesis. Thus, we employed **1** for the substitution study with nucleosides as the substitution reaction of the chloride ligand is readily proceeded in methanol. Reactivity and selectivity of [Pd(Amp)Cl]$^-$ with four nucleosides, guanosine, adenosine, cytidine, and 5-methyluridine, were evaluated in methanol by ^{1}H NMR spectroscopy (Figure 5). The reaction was performed by mixing **1** and the nucleoside in methanol.

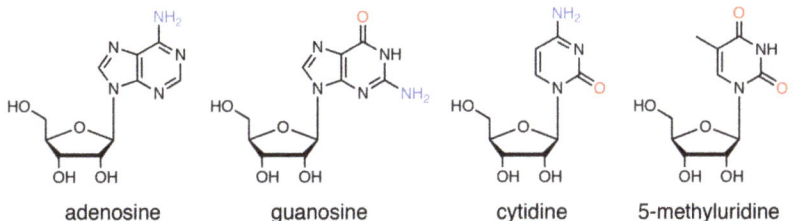

Figure 5. Chemical structures of the nucleosides employed in this study. Hydrogen-bond donors and acceptors are shown in blue and red color.

2.3.1. Reaction of 1 with Pyrimidine-Nucleosides

Two pyrimidine nucleosides, cytidine and 5-methyluridine, were evaluated. Cytidine was suitable for selective coordination to [Pd(Amp)] fragment because it has only one possible coordination site and both hydrogen-bond donor and acceptor sites were available within the proximity of the coordination site (2- and 4-position of the pyrimidine ring). The ^{1}H NMR spectrum indicated the formation of [Pd(Amp)(cytidine)]-adduct (**Pd-C**) as shown in Figure 6. The broad resonances observed in the spectrum of **1** became converged in clear multiplets in the presence of cytidine, implying that **Pd-C** is reasonably stable without further ligand exchange reaction. The splitting and geminal coupling of the signal corresponding to the coordinating amino group implies the presence of nonequivalent protons by the intramolecular hydrogen bond (4.6–4.3 ppm in Figure S4, Supplementary Materials). In the spectrum, only a trace amount of minor product was observed (<5%), which was attributed to a negligible coordination mode such as carbonyl-O coordination. The observation of a selective formation of **Pd-C** indicated that it has a specific binding mode by the coordination of N-donor site on pyrimidine that was supported by two pairs of donor-acceptor interactions (Figure 7). This three-point interaction is reminiscent of the Watson–Crick base pairing in DNA.

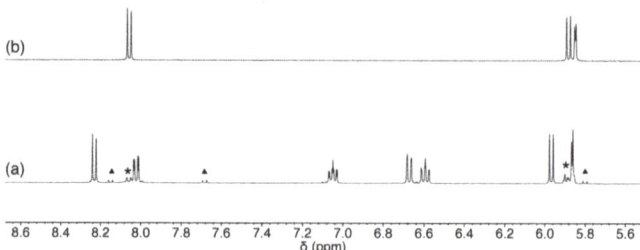

Figure 6. ^{1}H NMR spectra of (**a**) the reaction mixture of **1** and cytidine, (**b**) cytidine in CD$_3$OD. The resonances from a minor product and unreacted cytidine are marked with triangles and asterisks, respectively.

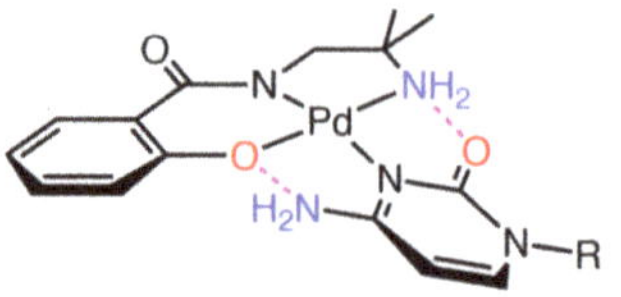

Figure 7. Chemical structure of **Pd-C** (R indicates the ribose moiety). Hydrogen-bond donor and acceptor are shown in blue and red, respectively. Dashed lines in magenta indicate intramolecular hydrogen bonds.

In the case of 5-methyluridine, on the other hand, no shift was observed for the resonances of both the nucleoside and [Pd(Amp)Cl]$^-$, indicating that 5-methyluridine was not able to coordinate to the Pd center (Figure 8). In the presence of 1 equiv. of triethylamine, a few sets of resonances corresponding to 5-methyluridine, including unreacted, were observed. From the integration ratio, over 80% of 5-methyluridine remained unreacted, although the formation of a minor amount of [Pd(Amp)(5-methyluridine)]-adduct was observed. The resonances corresponding to [Pd(Amp)]-fragment became sharp and the formation of a new species [Pd(Amp)(OCD$_3$)]$^-$ was suggested. These results showed that two hydrogen-bond acceptors in 5-methyluridine distract its coordination to the [Pd(Amp)]-fragment due to the electrostatic repulsive force between O atoms of Amp$^-$ ligand and 5-methyluridine. Thus, [Pd(Amp)]-fragment exhibited high selectivity on the coordination of pyrimidine-nucleosides, owing to the preference of hydrogen-bonding interactions between Amp$^-$ and the pyrimidine moieties.

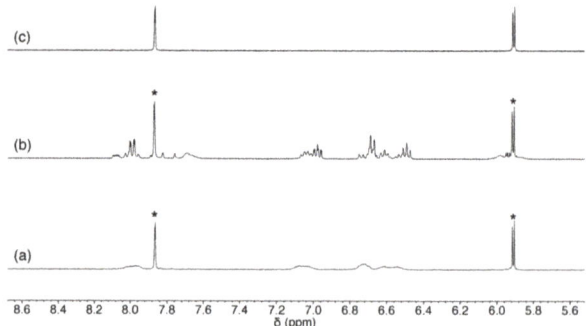

Figure 8. ^{1}H NMR spectra of (**a**) the reaction mixture of **1** and 5-methyluridine, (**b**) **1**, 5-methyluridine, and triethylamine, and (**c**) 5-methyluridine in CD$_3$OD. The resonances for unreacted 5-methyluridine are marked with asterisks.

2.3.2. Reaction of 1 with Purine-Nucleosides

Two purine-nucleosides, adenosine and guanosine, were evaluated as well as the pyrimidine-nucleosides. Due to the increasing possible coordination sites (N1-, N3-, and N7-position), purine-nucleosides might afford different products compared with the previous cases. Adenosine possesses a hydrogen-bond donor at 6-position and no hydrogen-bond acceptor exists at the coordination site. For [Pd(Amp)(adenosine)]-adduct (**Pd-A**), three different coordination modes are possible depending on which N atom is coordinated (Figure 9). The ^{1}H NMR spectrum indicated that the product is a mixture of **Pd-A** (Figure 10). The resonances according to the protons at adenine moiety imply that three species (**Pd-A1**, **Pd-A2**, and **Pd-A3**) are present in the reaction solution. The formation ratio of the products was determined to be **Pd-A1**:**Pd-A2**:**Pd-A3** = 6:3:2 from the integration of the resonances of the aryl-H of the adenine moiety. The N1-coordination mode is favorable because it is supported by a hydrogen bond (Figure 9a). Such intramolecular hydrogen-bond formation with adenine was reported in a Pd^{II} complex with coordination at the N9 position of adenine, although the N9 position is not available for coordination in our current system [19]. In contrast, no intramolecular hydrogen-bond interactions are expected for the other two modes (Figure 9b,c). Moreover, the N3 coordination must endure a large degree of steric hindrance between the ligand and ribose moiety. Thus, the major product **Pd-A1** is in the N1-coordination mode, and the two other sets of signals **Pd-A2** and **Pd-A3** correspond to the N7- and N3-coordination modes, respectively.

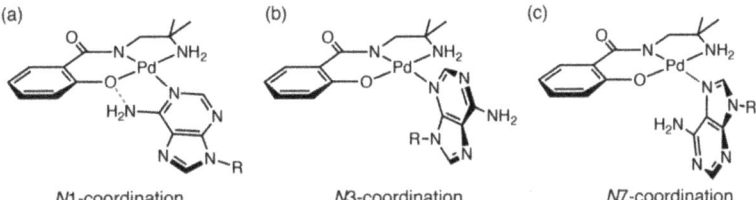

Figure 9. Three possible coordination modes for **Pd-A** (R indicates the ribose moiety). (**a**) N1-coordination mode, (**b**) N3-coordination mode, and (**c**) N7-coordination mode. Hydrogen bonds are shown in magenta.

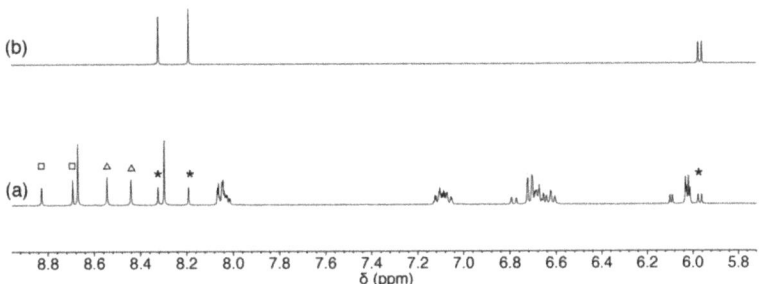

Figure 10. ^{1}H NMR spectra of (**a**) the reaction mixture of **1** and adenosine, and (**b**) adenosine in CD$_3$OD. The resonances of aryl-H atoms for two minor products and adenosine are marked in the spectrum (**Pd-A2**: open triangle; **Pd-A3**: open square; adenosine: asterisk).

Guanosine is structurally similar to cytidine in terms of the position of hydrogen-bond donors and acceptors (2- and 6-position, respectively). Although guanosine need to be deprotonated for the coordination at the N1 atom, formation of double intermolecular hydrogen bonds is expected for guanosine (Figure 11). The ^{1}H NMR spectrum of 1:1 mixture of **1** and guanosine in CD$_3$OD exhibited a sole set of resonances corresponding to [Pd(Amp)(guanosine)]-adduct (**Pd-G**), as shown in Figure 12. As observed in the case of cytidine, geminal coupling of the signals corresponding to the coordinating amino group was also observed, indicating *N*1 coordination of guanine moiety (5.3–5.0 ppm in Figure S7). It is to be noted that *N*7 is the most nucleophilic position in a neutral guanine moiety, and as a result, it is the primary binding site [20]. The *N*1 coordination in **Pd-G** without addition of base is indicative of the strong preference toward formation of intramolecular hydrogen bonds.

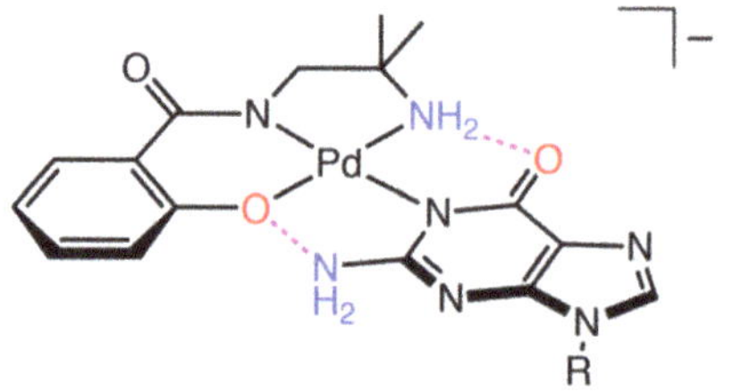

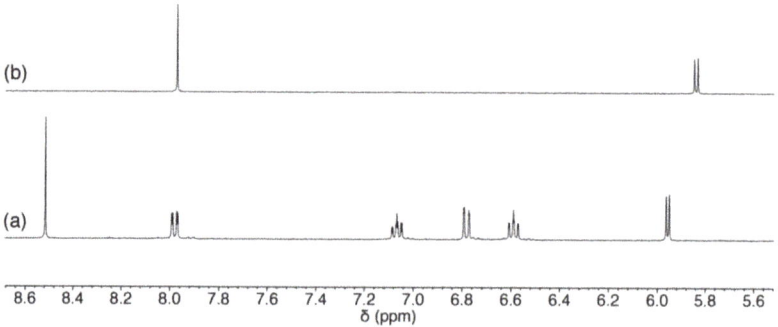

Figure 11. Chemical structure of **Pd-G** anion (R indicates the ribose moiety). Hydrogen-bond donor and acceptor are shown in blue and red, respectively. Dashed lines in magenta indicate intramolecular hydrogen bonds.

Figure 12. ^{1}H NMR spectra of (**a**) the reaction mixture of **1** and guanosine, and (**b**) guanosine in CD$_3$OD.

2.3.3. Scrambling Reaction of **1** with Two Nucleosides

To investigate the selectivity among nucleosides, complex **1** was reacted with two kinds of nucleosides at the same time. The Pd complex **1**, nucleoside-1, and nucleoside-2 were reacted in methanol at room temperature in a 1:1:1 molar ratio. 5-Methyluridine was excluded because it did not react with **1**. All sets of the combinations from two kinds of nucleosides were tested: adenosine vs. cytidine, adenosine vs. guanosine, and guanosine vs. cytidine (Figure 13).

In all cases, a mixture of a few Pd-nucleoside adducts was obtained. In the presence of adenosine and cytidine, **Pd-A1**, **Pd-A2**, and **Pd-C** were formed and formation of **Pd-A3** was negligible. From the formation ratio of **Pd-A1**:**Pd-A2**:**Pd-C** = 3:2:6, the coordination of cytidine is significantly more favorable than adenosine because of the formation of the double intramolecular hydrogen bonds. In the case of adenosine and guanosine, both **Pd-A1** and **Pd-A2** were observed, similar to the previous case. The formation ratio of **Pd-A1**:**Pd-A2**:**Pd-G** = 3:2:10 indicated that the coordination of guanosine is about twice as favorable as that of adenosine. These two experiments are indicative of the stabilization effect of double intramolecular hydrogen-bond formation on the coordination.

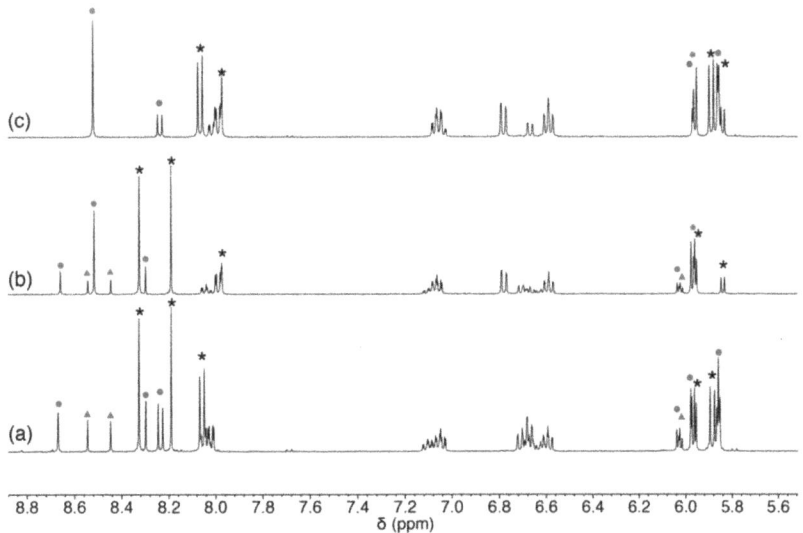

Figure 13. ^{1}H NMR spectra of **1** with two different nucleosides in CD$_3$OD. (**a**) Adenosine and cytidine, (**b**) adenosine and guanosine, and (**c**) guanosine and cytidine. The resonances attributed to the major and minor products for each nucleoside are indicated with solid circles and triangles, respectively (the resonances for Amp^{2-} ligand for each species are not marked). **Pd-A**, **Pd-G**, and **Pd-C** species are shown in blue, green, and red, respectively. Asterisks indicate unreacted nucleoside.

Finally, the coordination ability of guanosine and cytidine was compared. The NMR spectrum showed only two species, **Pd-G** and **Pd-C**, in the reaction. The formation ratio was found to be **Pd-G:Pd-C** = 2:1, which is reasonably consistent with the ratio expected from the previous two experiments, (**Pd–G/Pd-A**):(**Pd-C/Pd-A**) = 10:6. Thus, guanosine has the highest preference among the four nucleosides. Considering the number of hydrogen-bonding interactions of guanosine and cytidine with [Pd(Amp)]-moiety, the formation ratio should be comparable. As the anionic N1 position of the guanine moiety is more nucleophilic, however, the coordination of guanosine was more favorable than that of cytidine.

3. Materials and Methods

3.1. Measurements

Elemental analyses (C, H, and N) were performed at the Research Institute for Instrumental Analysis, Kanazawa University. ^{1}H NMR measurements were carried out at 22 °C on a JEOL 400SS spectrometer. Chemical shifts were referenced to the solvent residual peak [21].

3.2. Materials

All the chemicals were used as received without further purification. The ligand precursors N-(2-amino-2-methylpropyl)salicylamide (H$_2$Amp), N-3-aminopropylsalicylamide (H$_2$Apr), and the starting material of Pd complex, [PdCl$_2$(CH$_3$CN)$_2$], were synthesized according to the previously reported procedure [15,17].

3.3. Preparations

N(CH$_3$)$_4$[Pd(Amp)Cl]·H$_2$O (**1**). To an acetonitrile solution (20 mL) of [PdCl$_2$(CH$_3$CN)$_2$] (51.4 mg, 0.20 mmol) was added a solid H$_2$Amp (41.5 mg, 0.20 mmol) and triethylamine (28 µL). After stirring the mixture for 10 min, a solid (CH$_3$)$_4$NOH·5H$_2$O (36.2 mg, 0.20 mmol) was added followed by stirring at room temperature overnight. The reaction solution was concentrated to 2 mL by Ar gas bubbling. Diethyl ether solution was diffused to the solution to give orange, needle-shaped crystals. Yield: 65.8 mg, 53%. Anal. Calcd for N(CH$_3$)$_4$[Pd(Amp)Cl]·H$_2$O = C$_{15}$H$_{28}$ClN$_3$O$_3$Pd: C, 40.92; H, 6.41; N, 9.54%. Found: C, 40.64; H, 6.25; N, 9.54%. ^{1}H NMR (399 MHz, CD$_3$OD): δ 7.97 (broad, 1H, aryl-H), 7.05 (broad, 1H, aryl-H), 6.72 (broad, 1H, aryl-H), 6.58 (broad, 1H, aryl-H), 4.62 (s, 2H, NH$_2$), 3.41 (s, 2H, CH$_2$), 3.19 (s, 12H, CH$_3$), 1.40 (s, 6H, CH$_3$).

[Pd(Amp)(CH$_3$CN)]·CH$_3$CN (**1'**). To an acetonitrile solution (10 mL) of [PdCl$_2$(CH$_3$CN)$_2$] (25.9 mg, 0.10 mmol) was added a solid H$_2$Amp (20.8 mg, 0.1 mmol). After dissolving the ligand completely, a methanolic solution (5 mL) of KO^{t}Bu (22.3 mg, 0.20 mol) was added to the mixture, followed by stirring overnight. The reaction solution was concentrated to 3 mL by Ar gas bubbling. The residue was filtered, and the filtrate was crystallized by slow evaporation of the solvent. Yield: 22.4 mg, 57%. Anal. Calcd for [Pd(Amp)(CH$_3$CN)]·1.3H$_2$O = C$_{13}$H$_{19.6}$N$_3$O$_{3.3}$Pd: C, 41.40; H, 5.24; N, 11.14%. Found: C, 41.85; H, 5.49; N, 10.60%. ^{1}H NMR (399 MHz, CD$_3$OD): δ 8.00 (m, 1H, aryl-H), 7.06 (dd, J = 8.6, 6.8 Hz, 1H), 6.71 (d, J = 8.2 Hz, 1H), 6.58 (dd, J = 8.3, 6.8 Hz, 1H), 3.82 (s, 2H, CH$_2$), 2.52 (s, 3H, CH$_3$), 1.43 (d, J = 7.6 Hz, 6H).

[Pd(Apr)(CH$_3$CN)] (**2**). To an acetonitrile solution (8 mL) of [PdCl$_2$(CH$_3$CN)$_2$] (25.6 mg, 0.10 mmol) was added a solid H$_2$Apr (19.4 mg, 0.1 mmol) and triethylamine (14 μL). After dissolving the ligand completely, a methanolic solution (3 mL) of KO^{t}Bu (11.0 mg, 0.10 mol) was added to the mixture, followed by stirring overnight. The reaction solution was filtered to remove white precipitate. The filtrate was evaporated to dryness and dissolved in acetonitrile (2 mL). Diethyl ether vapor was diffused to the solution to give yellow microcrystals. Yield: 7.3 mg, 23%. This compound was extremely hygroscopic and unable to perform elemental analysis. ^{1}H NMR (399 MHz, CD$_3$CN): δ 6.76 (m, 1H, aryl-H), 6.59 (dd, J = 7.4, 0.7 Hz, 1H, aryl-H), 6.33 (dd, J = 8.2, 0.9 Hz, 1H, aryl-H), 3.22 (m, 2H, CH$_2$), 2.66 (m, 4H, CH$_2$, and NH$_2$), 1.57 (m, 2H, CH$_2$).

3.4. Crystallography

Crystallographic data are summarized in Table 4. Single-crystal X-ray diffraction data were obtained with a Rigaku XtaLAB AFC11 diffractometer with graphite-monochromated Mo Kα radiation (λ = 0.71073 Å). A single crystal was mounted with a glass capillary and flash-cooled with a cold N$_2$ gas stream. Data were processed using the CrysAlisPro software packages. The structure was solved by intrinsic phasing methods using the SHELXT [22] software packages and refined on F^2 (with all independent reflections) using the SHELXL [23] software packages. The non-hydrogen atoms were refined anisotropically. The hydrogen atoms except for OH and CH$_3$ groups were located at the calculated positions and refined isotropically using the riding models. The hydrogen atoms for OH and CH$_3$ groups were located by difference Fourier maps and allowed to rotate. The O–H atoms for water molecules were located at the position suitable for hydrogen bonds and refined freely with a restrained distance. In the case of **2**, only tiny single-crystals were obtained. As it is difficult to apply a sufficient absorption correction to a tiny crystal based on the actual size, the R_{int} value for **2** is apparently large. However, the quality of the crystal itself was reasonable. The anisotropic refinement for the acetonitrile ligands was restrained to obtain reasonable parameters. The Cambridge Crystallographic Data Centre (CCDC) deposition numbers are included in Table 4.

Table 4. Crystallographic data and refinement parameters of **1'** and **2**.

Complex	1	1'	2
Empirical formula	$C_{15}H_{28}ClN_3O_3Pd$	$C_{15}H_{20}N_4O_2Pd$	$C_{12}H_{15}N_3O_2Pd$
Formula weight	440.25	394.75	339.67
Crystal system	Monoclinic	Monoclinic	Triclinic
Crystal dimensions/mm	0.19 × 0.12 × 0.06	0.21 × 0.15 × 0.09	0.11 × 0.07 × 0.04
Space group	$P2_1/c$	$P2_1/c$	$P-1$
a/Å	11.8363 (3)	13.3692 (5)	8.7885 (4)
b/Å	8.7436 (2)	10.9488 (4)	10.8746 (5)
c/Å	18.3451 (5)	11.1827 (4)	13.2005 (5)
α/°			96.688 (4)
β/°	104.732 (3)	94.084 (3)	90.840 (3)
γ/°			104.073 (4)
V/Å³	1836.15 (8)	1632.73 (10)	1214.19 (9)
Z	4	4	4
T/K	100 (2)	100 (2)	100 (2)
ρ_{calcd}/g·cm^{-3}	1.593	1.606	1.858
μ/mm^{-1}	1.173	1.149	1.526
F(000)	904	800	680
$2\theta_{max}$/°	55	55	55
No. of reflections measured	17,469	11,616	17,415
No. of independent reflections	4207 (R_{int} = 0.0244)	3737 (R_{int} = 0.0390)	5543 (R_{int} = 0.1488)
Data/restraints/parameters	4207/2/222	3737/0/203	5543/18/329
R_1 [1] [$I > 2.00\ \sigma(I)$]	0.0217	0.0320	0.0555
wR_2 [2] (all reflections)	0.0558	0.0803	0.1348
Goodness of fit indicator	1.076	1.088	0.906
Highest peak, deepest hole/e Å^{-3}	0.445, −0.439	0.627, −1.156	1.646, −1.379
CCDC deposition number	2153005	2153006	2153007

[1] $R_1 = \Sigma||Fo| - |Fc||/\Sigma|Fo|$, [2] $wR_2 = [\Sigma(w(Fo^2 - Fc^2)^2)/\Sigma w(Fo^2)^2]^{1/2}$.

*3.5. Reaction of **1** with Nucleosides*

The reactions were performed by mixing a methanol solution of **1** with a solid nucleoside. Although some nucleosides are hardly soluble to typical organic solvents such as methanol and dimethylformamide, they dissolved well in methanol in the presence of **1**. A typical procedure for reactions of **1** with nucleosides is as follows.

A 5 mL methanolic solution of **1** (20 µmol) was added to a solid nucleoside (20 µmol) followed by stirring overnight. Half of the reaction mixture was taken and evaporated to dryness by Ar gas bubbling. The resulting yellow residue was dried in vacuo for 1 h. The residue was dissolved in ca. 0.7 mL of CD$_3$OD, followed by ^{1}H NMR measurement.

4. Conclusions

Palladium(II) complexes containing unsymmetric tridentate ligand, Amp^{2-}, and Apr^{2-}, were synthesized and crystallographically characterized. Although Amp^{2-} exhibited a typical O–N–N coordination mode, Apr^{2-} formed a Pd–C bond, resulting an observation of the C–N–N coordination mode. The Pd complex with Amp^{2-}, **1**, was employed for evaluating the effect of stereoselective intramolecular hydrogen-bonding interaction of four nucleosides (adenosine, guanosine, cytidine, and 5-methyluridine). Three of the four nucleosides resulted in coordination to the [Pd(Amp)]-fragment. In the case of 5-methyluridine, in which no intramolecular hydrogen bond is expected upon coordination, [Pd(Amp)(5-methyluridine)]-adduct was not formed. By contrast, the other three nucleosides, which form intramolecular hydrogen bonds upon coordination, were coordinated to the [Pd(Amp)]-fragment. These results clearly show that hydrogen-bonding interactions play an important role upon coordination to the [Pd(Amp)]-fragment. Preference of the nucleosides was further evaluated by reacting **1** with two different nucleosides, and guanosine showed the highest selectivity among the series.

Supplementary Materials: The following supporting information can be downloaded at https://www.mdpi.com/article/10.3390/molecules27072098/s1. Table S1: Hydrogen-bonding distances and angles; Figure S1: ^{1}H NMR spectrum of **1** in CD$_3$OD; Figure S2: ^{1}H NMR spectrum of **1′** in CD$_3$OD; Figure S3: ^{1}H NMR spectrum of **2** in CD$_3$CN; Figure S4: ^{1}H NMR spectra of (a) reaction mixture of **1** and cytidine, (b) cytidine in CD$_3$OD; Figure S5: ^{1}H NMR spectra of (a) reaction mixture of **1** and 5-methyluridine, (b) **1**, 5-methyluridine, and triethylamine, and (c) 5-methyluridine; cytidine in CD$_3$OD; Figure S6: ^{1}H NMR spectra of (a) reaction mixture of **1** and adenosine, (b) adenosine in CD$_3$OD; Figure S7: ^{1}H NMR spectra of (a) reaction mixture of **1** and guanosine, (b) guanosine in CD$_3$OD.

Author Contributions: Conceptualization, R.M. and T.S.; methodology, R.M.; validation, R.M.; investigation, R.M. and Y.I.; resources, R.M., T.S. and Y.H.; writing—original draft, R.M.; writing—review and editing, R.M., T.S. and Y.H.; visualization, R.M.; project administration, R.M. All authors have read and agreed to the published version of the manuscript.

Funding: This work was partly funded by Sakigake Project 2020 of Kanazawa University. This work was partly supported by a Grant-in-aid for Scientific Research No. 19K15525 from MEXT, Japan.

Data Availability Statement: The crystallographic data are available from the Cambridge Crystallographic Data Centre (CCDC). Other data not presented in Supplementary Materials are available on request from the corresponding author.

Acknowledgments: Single-crystal X-ray analysis was conducted at the Institute of Molecular Science, supported by Nanotechnology Platform (Molecule and Material Synthesis) of the MEXT, Japan.

Conflicts of Interest: The authors declare no conflict of interest.

Sample Availability: Not applicable.

References

1. Desiraju, G.R.; Steiner, T. *The Weak Hydrogen Bond in Structural Chemistry and Biology*; Oxford University Press: Oxford, UK, 2001. [CrossRef]
2. Gilli, G.; Gilli, P. *The Nature of Hydrogen Bond: Outline of a Comprehensive Hydrogen Bond Theory*; Oxford University Press: Oxford, UK, 2009. [CrossRef]
3. Burrows, A.D. Crystal Engineering Using Multiple Hydrogen Bonds. In *Supramolecular Assembly via Hydrogen Bond I Structure and Bondings*; Mingos, D.M.P., Ed.; Springer: Amsterdam, The Netherlands, 2004; Volume 108, pp. 55–96. [CrossRef]
4. Braga, D.; Maini, L.; Polito, M.; Grepioni, F. Crystal Engineering Using Multiple Hydrogen Bonds. In *Supramolecular Assembly via Hydrogen Bond II Hydrogen Bonding Interactions Between Ions: A Powerful Took in Molecular Crystal Engineering*; Mingos, D.M.P., Ed.; Springer: Amsterdam, The Netherlands, 2004; Volume 111, pp. 1–32. [CrossRef]
5. Mitsuhashi, R.; Suzuki, T.; Hosoya, S.; Mikuriya, M. Hydrogen-Bonded Supramolecular Structures of Cobalt(III) Complexes with Unsymmetrical Bidentate Ligands: Mer/fac Interconversion Induced by Hydrogen-Bonding Interactions. *Cryst. Growth Des.* **2017**, *17*, 207–213. [CrossRef]
6. Mitsuhashi, R.; Pedersen, K.S.; Ueda, T.; Suzuki, T.; Bendix, J.; Mikuriya, M. Field-induced single-molecule magnet behavior in ideal trigonal antiprismatic cobalt(II) complexes: Precise geometrical control by a hydrogen-bonded rigid metalloligand. *Chem. Commun.* **2018**, *54*, 8869–8872. [CrossRef]
7. Mitsuhashi, R.; Hosoya, S.; Suzuki, T.; Sunatsuki, Y.; Sakiyama, H.; Mikuriya, M. Hydrogen-bonding interactions and magnetic relaxation dynamics in tetracoordinated cobalt(II) single-ion magnets. *Dalton Trans.* **2019**, *48*, 395–399. [CrossRef] [PubMed]
8. Mitsuhashi, R.; Hosoya, S.; Suzuki, T.; Sunatsuki, Y.; Sakiyama, H.; Mikuriya, M. Zero-field slow relaxation of magnetization in cobalt(II) single-ion magnets: Suppression of quantum tunneling of magnetization by tailoring the intermolecular magnetic coupling. *RSC Adv.* **2020**, *10*, 43472–43479. [CrossRef]
9. Watson, J.D.; Crick, F.H.C. Molecular Structure of Nucleic Acids: A Structure for Deoxyribose Nucleic Acid. *Nature* **1953**, *171*, 737–738. [CrossRef] [PubMed]
10. Nikolova, E.N.; Kim, E.; Wise, A.A.; O'Brien, P.J.; Andricioaei, I.; Al-Hashimi, H.M. Transient Hoogsteen base pairs in canonical duplex DNA. *Nature* **2011**, *470*, 498–502. [CrossRef] [PubMed]
11. Rosenberg, B.; Camp, L.V.; Krigas, T. Inhibition of Cell Division in *Escherichia coli* by Electrolysis Products from a Platinum Electrode. *Nature* **1969**, *205*, 698–699. [CrossRef]
12. Rosenberg, B.; Camp, L.V.; Trosko, J.E.; Mansour, V.H. Platinum Compounds: A New Class of Potent Antitumour Agents. *Nature* **1969**, *222*, 385–386. [CrossRef] [PubMed]
13. Lovejoy, K.S.; Todd, R.C.; Zhang, S.; McCormick, M.S.; D'Aquino, J.A.; Reardon, J.T.; Sancar, A.; Giacomini, K.M.; Lippard, S.J. cis-Diammine(pyridine)chloroplatinum(II), a monofunctional platinum(II) antitumor agent: Uptake, structure, function, and prospects. *Proc. Natl. Acad. Sci. USA* **2008**, *105*, 8902–8907. [CrossRef] [PubMed]

14. Todd, R.C.; Lippard, S.J. Structure of duplex DNA containing the cisplatin 1,2-{Pt(NH$_3$)$_2$}$^{2+}$-d(GpG) cross-link at 1.77Å resolution. *J. Inorg. Biochem.* **2010**, *104*, 902–908. [CrossRef] [PubMed]
15. Mitsuhashi, R.; Suzuki, T.; Sunatsuki, Y.; Kojima, M. Hydrogen-bonding interactions, geometrical selectivity and spectroscopic properties of cobalt(III) complexes with unsymmetrical tridentate amine-amidato-phenolato type ligands. *Inorg. Chim. Acta* **2013**, *399*, 131–137. [CrossRef]
16. Mitsuhashi, R.; Ogawa, R.; Ishikawa, R.; Suzuki, T.; Sunatsuki, Y.; Kawata, S. Preparation, structures and properties of manganese complexes containing amine–(amido or amidato)–phenolato type ligands. *Inorg. Chim. Acta* **2016**, *447*, 113–120. [CrossRef]
17. Schlosser, M. (Ed.) *Organometallics in Synthesis: A Manual*, 2nd ed.; Wiley: Hoboken, NJ, USA, 2004; 1126p.
18. Hofmann, A.; Dahlenburg, L.; van Eldik, R. Cyclometalated analogues of platinum terpyridine complexes: Kinetic study of the strong sigma-donor cis and trans effects of carbon in the presence of a pi-acceptor ligand backbone. *Inorg. Chem.* **2003**, *42*, 6528–6538. [CrossRef] [PubMed]
19. Odoko, M.; Okabe, N. (Adeninato-N^9)[N-(2-aminoethyl)salicylideneiminato]palladium(II) 3.5-hydrate. *Acta Crystallogr. Sect. E Struct. Rep. Online* **2005**, *61*, m2670–m2673. [CrossRef]
20. Sherman, S.E.; Lippard, S.J. Structural aspects of platinum anticancer drug interactions with DNA. *Chem. Rev.* **1987**, *87*, 1153–1181. [CrossRef]
21. Gottlieb, H.E.; Kotlyar, V.; Nudelman, A. NMR chemical shifts of common laboratory solvents as trace impurities. *J. Org. Chem.* **1997**, *62*, 7512–7515. [CrossRef] [PubMed]
22. Sheldrick, G.M. SHELXT—Integrated space-group and crystal-structure determination. *Acta Crystallogr. Sect. A Found. Adv.* **2015**, *71*, 3–8. [CrossRef]
23. Sheldrick, G.M. Crystal structure refinement with SHELXL. *Acta Crystallogr. Sect. C Struct. Chem.* **2015**, *71*, 3–8. [CrossRef]

Article

Synthesis, Crystal Structure and Magnetic Properties of a Trinuclear Copper(II) Complex Based on P-Cresol-Substituted Bis(α-Nitronyl Nitroxide) Biradical

Sabrina Grenda *, Maxime Beau and Dominique Luneau *

Laboratoire des Multimatériaux et Interfaces (UMR 5615), Université Claude Bernard Lyon 1, 69100 Villeurbanne, France; beau-maxime@outlook.fr
* Correspondence: sabrina.grenda@univ-lyon1.fr (S.G.); dominique.luneau@univ-lyon1.fr (D.L.)

Abstract: Trinuclear copper(II) complex [Cu$^{II}_3$(NIT$_2$PhO)$_2$Cl$_4$] was synthesized with p-cresol-substituted bis(α-nitronyl nitroxide) biradical: 4-methyl-2,6-bis(1-oxyl-3-oxido-4,4,5,5-tetramethyl-2-imidazolin-2-yl)phenol (NIT$_2$PhOH). The crystal structure of this heterospin complex was determined using single-crystal X-ray diffraction analysis and exhibits four unusual seven-membered metallocycles formed from the coordination of oxygen atoms of the N-O groups and of bridging phenoxo (μ-PhO$^-$) moieties with copper(II) ions. The crystal structure analysis reveals an incipient agostic interaction between a square planar copper center and a hydrogen-carbon bond from one methyl group carried on the coordinated nitronyl-nitroxide radical. The intramolecular Cu⋯H-C interaction involves a six-membered metallocycle and may stabilize the copper center in square planar coordination mode. From the magnetic susceptibility measurements, the complex, which totals seven S = 1/2 spin carriers, has almost a ground state spin S = 1/2 at room temperature ascribed to strong antiferromagnetic interaction between the nitronyl nitroxide moieties and the copper(II) centers and in between the copper(II) centers through the bridging phenoxo oxygen atom.

Keywords: nitronyl nitroxide biradical; molecular magnetism; copper complex

1. Introduction

The coordination chemistry of nitronyl nitroxide free radicals has played a major role in the development of molecule-based magnetic materials following the so-called metal-radical approach pioneered thirty years ago [1]. Several well-known and good reasons make nitronyl nitroxide free radicals attractive organic spin carriers for magnetic coordination compounds. One is the quite easy synthesis of the nitronyl nitroxide (NN) moiety, which can be incorporated into many chemical groups in the α-position. This is usually done by the Ullman procedure via the reaction in methanol of 2,3-bis(hydroxylamino)-2,3-dimethylbutane with an appropriate aldehyde [2]. In addition, a palladium-catalyzed cross-coupling reaction of (nitronyl nitroxide-2-ido)(triphenylphosphine)gold(I) complex [Ph$_3$P-Au-NN] with aryl halides has also been developed to directly graft the NN moiety onto aromatic rings [3]. Therefore, an almost unlimited number of open-shell molecules may be synthesized. Thus, nitronyl nitroxide have been grafted on alcanes [4], pyridine [5–10], imidazole [11], triazole [12], bipyridine [13], phenol [14–17], pyrene [18], azulene [19], phthalocyanine [20], porphyrin [21], cyclotriphosphazene [22], or even on graphene [23]. Most importantly, nitronyl nitroxide has one of the best stabilities among the free radicals, and this is retained in most of its derivative so that they can generally be handled under mild but normal conditions, even when coordinated with most metal ions. Although the NO group is a weak Lewis base, coordination is effective by using electron-withdrawing ancillary ligands on the metal center, such as hexafluoroacetylacetone, or by incorporating the NO group into a chelate so that such extra ligand can be removed from the coordination sphere [11]. The nitronyl nitroxide stability in coordination compounds could be tempered after the discovery of valence tautomerism in some 2D coordination polymers with manganese(II), where nitronyl nitroxide radical is reduced on cooling [24–26]. However, the process reverses on reheating. To end, the two NO groups of the nitronyl nitroxide (NN) moiety on which the unpaired electron is equally delocalized [27] make it a bridging-ligands but open-shell molecule. This makes it easy to build heterospin systems in which direct magnetic interactions operate between the metal and radical spin carriers. This has provided many magnetic systems [28] among which molecular magnets [29–31], single-molecule magnets [31–36], or single-chain magnets [37–41]. However, that is not all. Research on metal-nitronyl nitroxide systems has led to the discovery of two other phenomena out of the classical magnetism relying on the interplay of magnetic interaction. These are molecular-spin transitions in some copper(II) complexes [42–44] and valence tautomerism in some manganese(II) coordination polymers [24–26]. Metal-nitronyl nitroxide coordination chemistry is, therefore, a versatile tool for different types of bistable systems [45].

As for most coordination chemistry, such a development would not have been possible without the crystallography tool, mainly single crystal X-ray diffraction. Crystal structure has indeed been essential to watch the molecular arrangement and to further understand the magnetic properties through the analysis of the structural features, as illustrated in the present paper.

Among metal-nitronyl nitroxide coordination compounds, those with copper(II) are singular regarding the magneto-structural relationships. Indeed, because of the lonly copper(II) magnetic orbital, the copper-radical magnetic interaction is very sensitive and dependent on the coordination mode. It can be either antiferromagnetic or ferromagnetic, depending on whether the coordination mode favors or not the overlap of the magnetic orbitals of the copper(II) with those of the radical [1]. This may even be temperature dependent and give rise to molecular spin transition. [42–44].

This versatile but richness of magnetic behaviors prompt us to investigate complexation of copper(II) by bis-nitronyl nitroxide diradical 4-methyl-2,6-bis(1-oxyl-3-oxido-4,4,5,5-tetramethyl-2-imidazolin-2-yl)phenol [14] (NIT$_2$PhOH, Scheme 1). Herein, we report the reaction with copper(II) chloride to afford a heterospin neutral complex [Cu^{II}$_3$(NIT$_2$PhO)$_2$Cl$_4$] where this diradical (NIT$_2$PhOH) acts both as a bridging and chelating ligand. The complex has been isolated in the crystalline solid state, and the temperature dependence of the magnetic susceptibility has been analyzed based on the structure characterized by single-crystal X-ray diffraction.

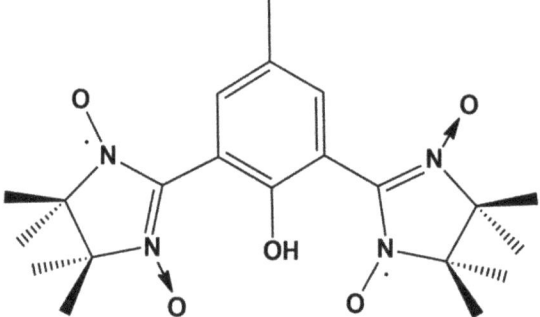

Scheme 1. Chemical structure of the biradical NIT$_2$PhOH.

2. Results

2.1. Synthesis of the Inorganic Complex

Trinuclear copper complex [Cu^{II}$_3$(NIT$_2$PhO)$_2$Cl$_4$] (NIT$_2$PhOH = 4-methyl-2,6-bis(1-oxyl-3-oxido-4,4,5,5-tetramethyl-2-imidazolin-2-yl)phenol) was synthesized as brown squared crystals from a mixture in methanol at room temperature of copper(II) chloride and NIT$_2$PhOH in a 2:1 molar ratio.

2.2. Crystal Structure

[Cu$_3$(NIT$_2$PhO)$_2$Cl$_4$] crystallizes in the monoclinic P2$_1$/n space group. Crystallographic data are reported in Table 1. The asymmetric unit exemplified in Figure 1 comprises all the compounds. It is made of three crystallographically independent copper(II) ions coordinated with two deprotonated bis-nitronyl-nitroxide ligands, so-called hereafter diradicals A and B involving their phenoxide oxygen atoms and the nitronyl nitroxide moieties (NN). The copper ions have their coordination sphere completed by four chloride ions among which two are bridging. This makes the complex electrically neutral. A simplified representation of the coordination mode of the two diradicals is also shown in Scheme 2.

Table 1. Crystallographic data for compounds of the complex [Cu^{II}$_3$(NIT$_2$PhO)$_2$Cl$_4$].

Formula	C$_{42}$H$_{58}$Cl$_4$Cu$_3$N$_8$O$_{10}$
M (g mol^{-1})	1167.38
T (K)	293(2)
λ (Å)	0.71073
Crystal system	Monoclinic
Space group	P2$_1$/n (#14)
a (Å)	19.8080(9)
b (Å)	11.4226(7)
c (Å)	22.0120(15)
α (deg)	90
β (deg)	99.450(5)
γ (deg)	90
V (Å^3)	4912.8(5)
Z	4
Dc (g cm^{-3})	1.578
F(000)	2404
θ range (deg.)	2.565 to 29.287
Limiting indices	$-27 \leq h \leq 26, -15 \leq k \leq 14, -28 \leq l \leq 30$
Refl. collected	31578
Rint (%)	6.66
Num. param.	622
GOF on F^2	1.036
R [a], ωR [b] (all data)	0.0648, 0.1643
Δρ$_{max}$/Δρ$_{min}$ (e.Å^3)	0.6/−0.7

[a] $R = \Sigma||Fo|-|Fc||/\Sigma|Fo|$, [b] $\omega R = [\Sigma(\omega(Fo^2-Fc^2)^2)/\Sigma(\omega(Fo^2)^2)]^{1/2}$ with $\omega = 1/[(\sigma^2 Fo^2) + (aP)^2 + bP]$ and $P = (\max(Fo^2) + 2Fc^2)/3$.

In diradical A and B, as generally observed, the phenoxide oxygen atoms are bridging for Cu2 and Cu1 (Diradical A) and for Cu1 and Cu3 (Diradical B). In both cases, nitronyl nitroxide (NN) moieties complete the coordination sphere of the copper(II) centers but depending on Diradical A or Diradical B, these NN moieties expose two types of coordination modes. In Diratical A, one of the NN moiety is chelating for Cu2, while the second NN moiety is chelating for Cu1, so that Diradical A may be viewed as bridging the Cu2 and Cu1 through the phenoxide oxygen atoms and through the NN moieties. In Diradical B, both NN moieties play differently and are solely chelated to Cu3. In all cases, the NN ligand displays the unusual seven-membered metallocycle reported elsewhere in complexes of closely related phenol substituted nitronyl nitroxide [15–17,46,47].

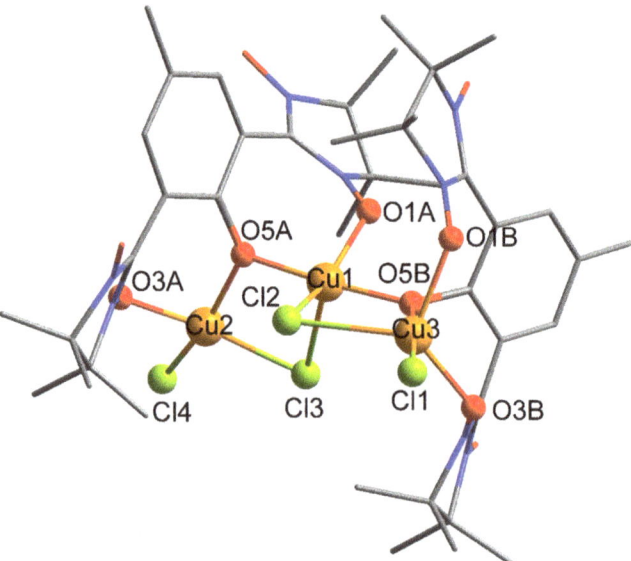

Figure 1. Crystal structure of the asymmetric unit of the complex [Cu$^{II}_3$(NIT$_2$PhO)$_2$Cl$_4$]. Atoms coordinated to copper are only labeled. Hydrogen atoms are omitted for clarity. Atoms are colored as follows: Cu orange; O red; Cl green; N blue; C grey.

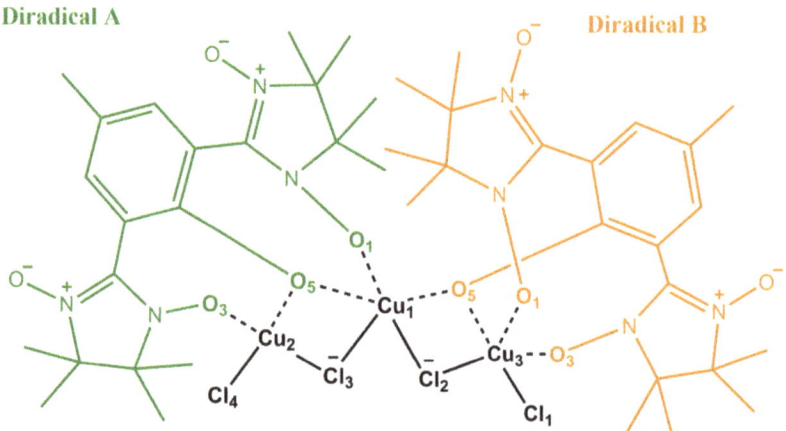

Scheme 2. Schematization of the coordination mode of the two diradicals.

Cu1 and Cu3 have a coordination of five, but Cu2 has an uncommon coordinance of four. Bond lengths and angles constitutive of the environment of the three copper(II) are listed in Table 2.

Table 2. Selected interatomic distances (Å) and angles (°).

Cu 1			
Cu1-O5A	1.934(3)	O5B-Cu1-O5A	175.99(14)
Cu1-O5B	1.937(3)	O1A-Cu-Cl2	148.64(11)
Cu1-O1A	1.947(3)	Cu1-O1A-N1A	126.4(3)
Cu1-Cl2	2.2596(14)	Cu1-Cl3-Cu2	82.91(5)
Cu1-Cl3	2.5869(15)	Cu1-O5B-Cu3	117.11(17)
Cu1—Cu2	3.211(1)	Cu1-O5A-Cu2	111.13(16)
Cu1—Cu3	3.281(1)		
Cu 2			
Cu2-O3A	1.918(4)	O5A-Cu2-Cl4	161.35(13)
Cu2-O5A	1.960(3)	O3A-Cu2-Cl3	158.53(13)
Cu2-Cl4	2.1520(18)	Cu2-O3A-N3A	112.7(4)
Cu2-Cl3	2.2492(17)		
Cu 3			
Cu3-O5B	1.909(3)	O5B-Cu3-Cl1	177.92(12)
Cu3-O3B	1.967(4)	O3B-Cu3-Cl2	127.3(11)
Cu3-O1B	2.110(4)	Cu3-O5B-Cu1	117.11(17)
Cu3-Cl1	2.1585(16)	Cu3-Cl2-Cu1	83.60(5)
Cu3-Cl2	2.6440(17)	Cu3-O1B-N1B	125.8(3)
		Cu3-O3B-N3B	120.1(3)
Nitronyl Nitroxide Moieties			
O1A-N1A *	1.295(5)	O1B-N1B *	1.300(5)
O2A-NA2	1.263(6)	O2B-N2B	1.270(5)
O3A-N3A *	1.310(5)	O3B-N3B *	1.297(5)
O4A-N4A	1.263(5)	O4A-N4A	1.254(5)

* denotes coordinated N-O groups.

Cu1 (Figure 2, Table 2) assumes an intermediary environment in between square planar pyramidal and trigonal bipyramidal with an Addison parameter of τ_5 = 0.46 [48]. The shortest bond length range 1.934–1.947 Å is found for Cu1-O bonds with oxygen atoms of one nitronyl nitroxyde radical of diradical A and the phenoxide oxygen atoms of bisradical A and B. The longest bond lengths 2.2596–2.5870 Å are found for Cu1-Cl bonds. As the length of the Cu1-Cl3 biding (2.59 Å) is the longest, we assume that the geometric environment might be a distorted square planar pyramidal with O5B, Cl2, O5A, O1A forming the base and Cl3 occupying the apical position, in agreement with an elongation Jahn Teller distortion as expected.

The length of the apical Cu1-Cl3 bonding (2.59 Å) is shorter than typical Cu-Cl bonds lengths found in the literature. For apical bridging, chloride values between 2.709 Å and 2.785 Å are found [49,50]. The equatorial Cu1-Cl2 (2.26 Å) bond length is in the range of typical Cu-Cl bonds found in the literature with values between 2.264 Å and 2.300 Å for a bridging chloride ion [50,51]. For Cu1-O5A (1.93 Å) and Cu1-O5B (1.94 Å), bond lengths are consistent with a μ-hydroxo character of the bridge with values between 1.923 Å and 1.97 Å, as reported in the literature [52,53]. Finally, the equatorial Cu1-O1A (1.93Å) distance is in the range for Cu-O$_{NO}$ bonding where values between 1.95 Å and 2.06 Å are reported [47].

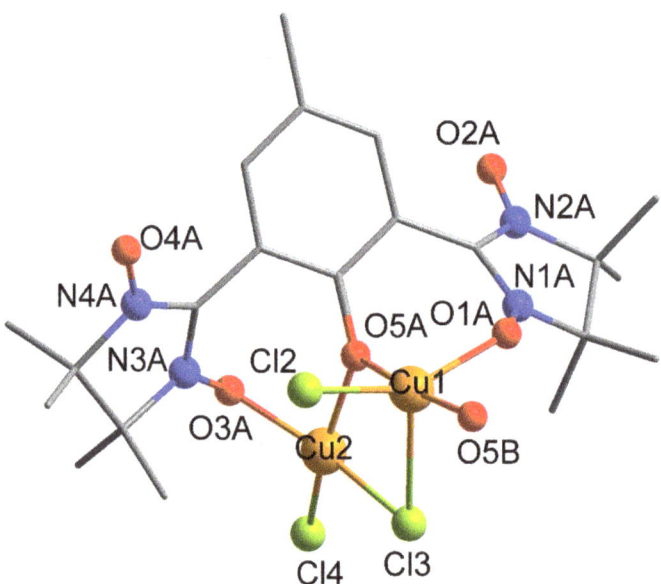

Figure 2. View focusing on the coordination mode involving copper(II) ions Cu1 and Cu2 with nitronyl nitroxide A. Atoms coordinated to copper and N-O groups are only labeled. Hydrogen atoms are omitted for clarity. Atoms are colored as follows: Cu orange; O red; Cl green; N blue; C grey.

Cu2 (Figures 2 and 3, Table 2) assumes a highly distorted square planar geometry with a Houser–Okuniewski parameter $\tau_4 = 0.28$ [54–56] and seems to be stabilized in its axial position by an intramolecular interaction with a hydrogen (H18A) from the methyl group (C18) carried on the NN-Radical moieties. This intramolecular interaction, together with the oxygen atom of nitroxide group (O3A), makes a chelate. This affords a six-membered metallacycle and brings the methyl carbon (C18) close to Cu2, as shown in Figure 3. These structural features suggest an incipient Cu(II) ··· H-C agostic interaction [57,58]. The Cu2···H18A and Cu2···C18 interatomic distances are respectively 2.6656(7) Å and 3.1112(6) Å with an angle M···H18A-C18 of 125.076(3)°, which fall in the range of structural critters for an agostic interactions parameters[59,60].

Cu3 (Figure 4, Table 2) assumes a distorted trigonal bipyramidal geometry with an Addison parameter of $\tau_5 = 0.84$ [48]. The shortest bond length range 1.909–2.110 Å is found for Cu3-O bonds with oxygen atoms of the two nitronyl nitroxyde radical of diradical B and its phenoxido oxygen atoms. The longest bond lengths 2.1585–2.6439 Å are found for Cu1-Cl bonds. As the bond length of axial Cu3-Cl1 biding (2.16 Å) and axial Cu3-O5B biding (1.909 Å) are shorter than equatorial Cu3-Cl2 (2.64 Å), Cu3-O1B (2.110 Å), and Cu3-O3B (1.966 Å), the environment of copper Cu3 can be described by an unusual compression Jahn–Teller distortion. Indeed, the equatorial Cu3-Cl2 (2.64 Å), Cu3-O1B (2.110 Å), and Cu3-O3B (1.966 Å) are much longer than the typical bond length, as described before, and axial Cu3-Cl1 biding (2.16 Å) is highly shorter than typical Cu-Cl length for a terminal chloride atom where a value of 2.267 Å is reported [50].

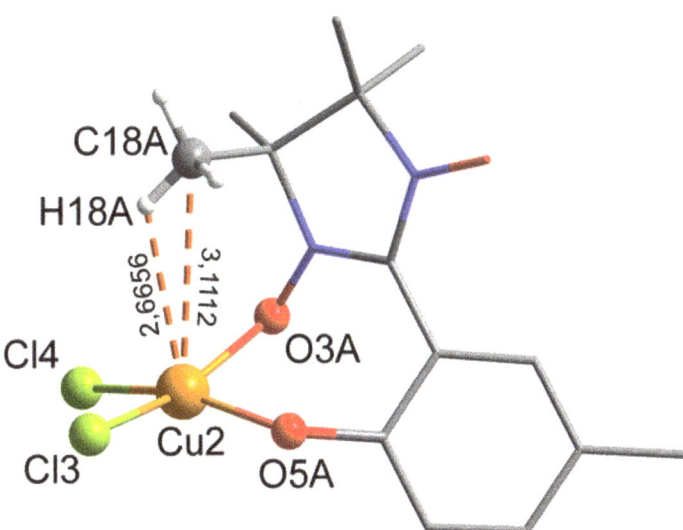

Figure 3. Intermolecular H···Cu2 interaction, represented by a dotted orange line, between the hydrogen atom carried by C18 and copper Cu2. Only atoms coordinated to Cu2 and the involved hydrogen are shown for clarity. Atoms are colored as follows: Cu orange; O red; Cl green; N blue; C grey.

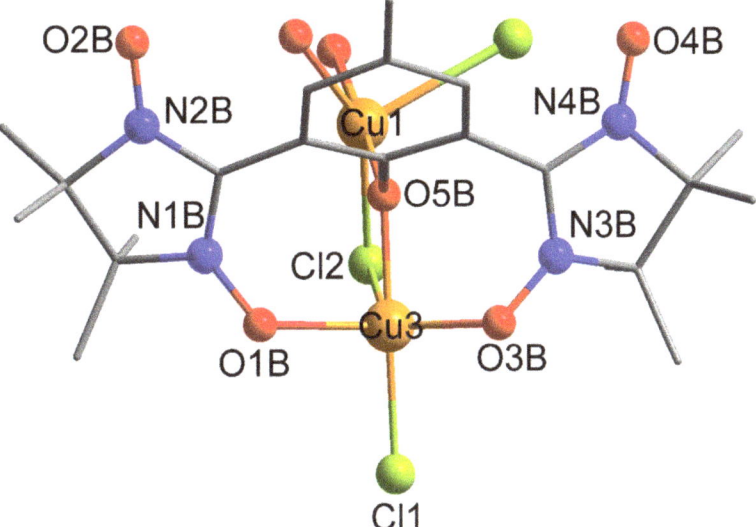

Figure 4. View focusing on the coordination mode involving copper(II) ion Cu3 with nitronyl nitroxide B. Atoms coordinated to copper and N-O groups are only labeled. Hydrogen atoms are omitted for clarity. Atoms are colored as follows: Cu orange; O red; Cl green; N blue; C grey.

Bond lengths of the N−O group within the four NN moieties are comprised in the range 1.254(6)–1.300(5) Å (Table 2) with small elongation (0.03 Å) of the coordinated ones and are characteristic of free and coordinated nitroxide radicals [4–23]. The O-N-C-N-O moieties are strictly planar (Table 3), as expected due to electron delocalization. These structural features demonstrate the persistence of the four nitronyl nitroxide radicals in [Cu$_3$Cl$_4$(NIT$_2$PhO)$_2$]. The dihedral angles between the O-N-C-N-O least-square plane and with attached phenyl ring (φ) are quite close for the four NN moieties in contrast with the free radical for which 38°, 64°, and 72° were found for φ [14].

Table 3. Deviation of O-N-C-N-O atoms in nitronyl nitroxide moieties from their least-square planes and dihedral angles with attached phenyl ring (φ) and Cu-O-N (δ) least-square planes.

Selected Planes	Atomic Deviation (Å) from Least-Square Planes					φ (°)	δ (°)
Diradical A	O1, O3	N1, N3	C1, C14	N2, N4	O2, O4		
O1-N1-C1-N2-O2	0.065(3)	−0.105(4)	−0.075(4)	0.031(4)	0.044(4)	50.6(2)	55.08
O3-N3-C14-N4-O4	−0.029(4)	0.031(4)	0.039(4)	−0.001(4)	−0.029(4)	45.2(1)	82.92
Diradical B	O1, O3	N1, N3	C1, C14	N2, N4	O2, O4		
O1-N1-C1-N2-O2	0.030(4)	−0.052(4)	−0.025(5)	0.028(4)	0.004(4)	42.6(2)	57.30
O3-N3-C14-N4-O4	0.022(4)	−0.022(4)	−0.031(4)	−0.004(4)	0.025(4)	46.4(1)	72.96

2.3. Magnetic Behaviour

The temperature dependence of the product of the magnetic susceptibility with the temperature ($\chi_M T$) of [Cu$^{II}_3$(NIT$_2$PhO)$_2$Cl$_4$] is shown in Figure 5. At 370 K, $\chi_M T$ is 0.57 emu·K·mol^{-1}, then, upon cooling, it decreases almost continuously down to 0.40 emu·K·mol^{-1} at 75 K. From there, it remains almost constant down to 30 K and then, it decreases abruptly to reach 0.14 emu·K·mol^{-1} at 3 K. The high temperature $\chi_M T$ value is very much lower than the expected value (~2.8 emu·K·mol^{-1}) for the seven magnetically independent spins S = 1/2, taking into account the three copper(II) ions and the four nitronyl nitroxide radicals. This, together with the continuous decreasing upon cooling down to a plateau with a $\chi_M T$ value corresponding to one resultant spin S = 1/2, is indicative of strong antiferromagnetic interactions operating in this hetero-spin complex.

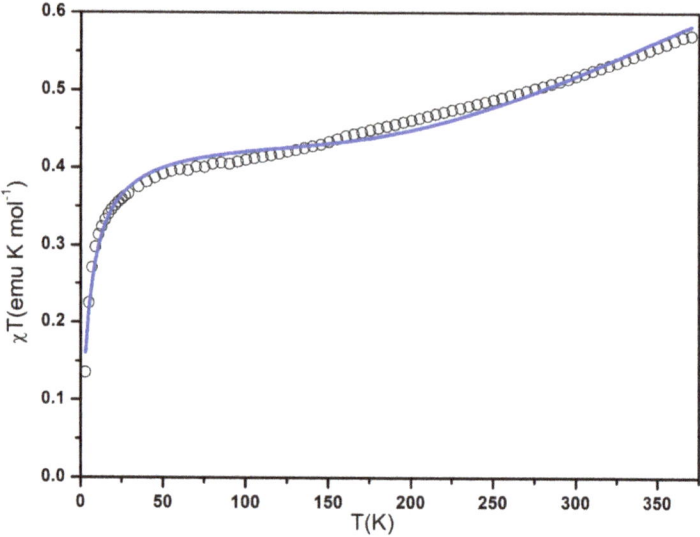

Figure 5. Temperature dependence of the product of magnetic susceptibility with temperature ($\chi_M T$) for [Cu$^{II}_3$(NIT$_2$PhO)$_2$Cl$_4$].

The understanding of the magnetic behavior at the molecule level requires considering a complex and multiple exchange interactions network, as schematized in Figure 6: *(i) Interaction in between the nitronyl nitroxide (NN) moieties through the phenyl ring (J_{NN-NN}).* From the study of the pure diradical (NIT$_2$PhOH), it was estimated to be weak and ferromagnetic (2J/k = 12 K) [14]. *(ii) Interaction between the NN moieties and next nearer Cu(II) through the phenyl rings via the bridging phenoxo oxygen atoms (J'_{NN-Cu}).* Herein, this exchange interaction may be significant but weaker than the exchange interaction between copper centers through the bridging phenoxide oxygen atom and chloride atoms and also compared with direct interactions between copper centers and N-oxide moieties. [16,17,58]. *(iii) Cu(II)···Cu(II) super-exchange interaction through the phenoxide oxygen atoms and chloro bridges (J_{Cu-Cu}).* According to the study of the crystal structure, we can determine the nature of the magnetic orbitals of metal centers. For Cu1 and copper Cu2, it was assumed to be respectively an elongated square planar pyramid and a square plane. In these two cases, the $dx^2 - y^2$ orbitals define the magnetic orbitals. For Cu3, the environment is a compressed trigonal bipyramid, which defines dz^2 orbital as the magnetic orbital. Regarding these results, the magnetic exchange is essentially governed by the $3dx^2 - y^2$(copper)–2p(oxygen) and the $3dz^2$(copper)–2p(oxygen) antibonding overlaps. Considering the Cu-Cl-Cu bond angles and Cu-Cl bond lengths, we do not expect the chloro bridges to mediate significant interaction (Tables 2 and 3). Magneto-structural correlations of μ-hydroxo-bridged copper(II) binuclear complexes have shown that the exchange interaction depends on Cu-Cu distance, Cu-O-Cu angle [59–61]. Here, the copper centers are separated by 3.212 Å for Cu1-Cu2 and 3.286 Å for Cu1-Cu3 and considering the quite large Cu-O-Cu angle of the phenoxido bridge (Table 2), this should give strong antiferromagnetic interaction [62]. The exchange interaction parameter (J) may be estimated to be greater than 700 cm^{-1} from the Hatfield's correlation 2J = 74.53Φ−7270 cm^{-1} and 2J = −4508d + 13018, taking the Cu-O-Cu angle (Φ) or Cu-Cu distance (d), respectively [63–65]. These interactions may be lower here because of the coordinated chloride ions. Indeed, from a previous study of unsymmetrical μ-hydroxo copper(II) complexes it has been reported that the nature of the exogenous bound ligands with varying electronegativity influences the exchange interactions within copper centers. It was found that bonding chloride ions greatly lower the antiferromagnetic couplings because they decrease the electron density on the coppers [66]. *(iv) Direct interaction between the NN moieties and Cu(II) spin carriers (J_{NN-Cu}).* According to the crystal structure, for the three copper centers, nitroxide radicals are equatorially coordinated to the square plane (Cu2), square planar pyramid (Cu1), and trigonal bipyramid (Cu3). For Cu1 and Cu2, this is to favor antiferromagnetic interaction [1]. Moreover, the Cu-O-N angles (126.4°–112.7°) and the small dihedral (δ) angle between the Cu-O-N and O-N-C-N-O least-square planes (Table 3) cause substantial overlap between the π* orbital of the nitroxide radical and the $3dx^2 - y^2$ orbitals of copper Cu1 and Cu2, leading to large antiferromagnetic couplings (−500 cm^{-1}) [1,62,66–68]. For Cu3, coordination of the nitroxide radical in the equatorial plane is not expected to cause any overlap with the dz^2 magnetic orbital of the trigonal bipyramid. This rules out any significant magnetic interaction.

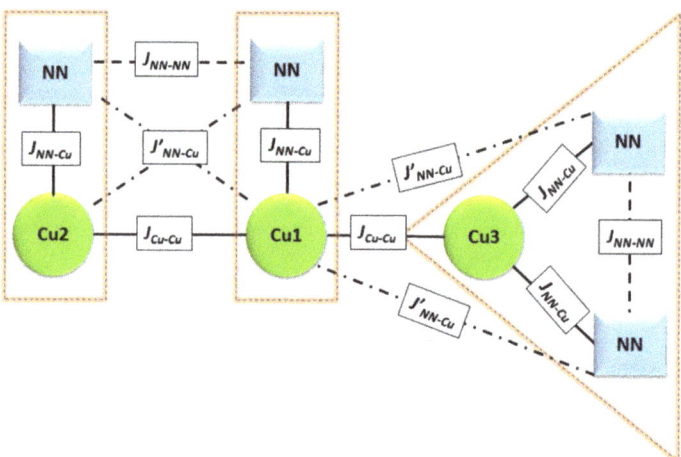

Figure 6. Schematization of the magnetic exchange couplings in $[Cu^{II}_3(NIT_2PhO)_2Cl_4]$.

With this scheme of possible interactions in mind, we tried to fit the temperature dependence of the product of the magnetic susceptibility with temperature ($\chi_M T$) using the PHI program [69]. All tentatives taking into account seven spin S = 1/2, using a different set of above interactions, were unsuccessful. The best fit was obtained only when considering a three spin system and three coupling constants: $J_{12} = -265(6)$ cm^{-1}; $J_{13} = -208(6)$ cm^{-1}; $J_{23} = -30(1)$ cm^{-1}; g1 = 2.32(1); g2 = g3 = 2; and zJ' = $-3.07(8)$ cm^{-1}. This is in agreement with the low $\chi_M T$ values reaching high temperature (370k), and it means that already in this temperature region, there are only three effective spins S = 1/2. One could think the nitronyl nitroxide radicals have been killed, and they are the three copper(II) ions only. This has to be ruled out because, as we have seen, the structural features demonstrate without any doubt the persistence of the four nitronyl nitroxide moieties. Moreover, the complex is neutral and this also rules out that the radicals could have been reduced or oxidized. Therefore, the unique possibility is that four spins S = 1/2 became silent because of antiferromagnetic couplings larger than the thermal energy. In agreement with the previous discussion of possible interactions, this is attributed to the antiferromagnetic coupling of each of the copper Cu1 and Cu2 with their attached nitronyl nitroxide from biradical A. As discussed above, the structural features are expected to result in large antiferromagnetic interaction (~500 cm^{-1}). This makes the four spins S = 1/2 comprising Cu1 and Cu2 and Diradical A silent. We may think this primes on the Cu—Cu interactions (J_{Cu-Cu}: Figure 6), which may be lower due to electron withdrawing effect of coordinated chloride ions, as discussed above. The resulting three spin S = 1/2 system is thus attributed to Cu3 interacting with the two nitroxide of biradical B. As we discuss above, Cu3 is in bipyramid trigonal coordination geometry, and this is not expected to cause direct antiferromagnetic interaction with the dz^2 magnetic orbital. In that case, the interaction of the two nitronyl nitroxide moieties of Diradical B with Cu3 is indirect and proceeds via the phenoxido oxygen atom (J'_{NN-Cu}: Figure 6). This is ascribed to J_{12} and J_{13}, with g1 holding for Cu3. Then, J_{23} is ascribed to the interaction between the nitronyl nitroxide radical via the phenyl ring, as this is not expected to be large (J_{NN-NN}: Figure 6). The interaction (J_{23}) is moderate but antiferromagnetic in contrast with the ferromagnetic one found for the free radical [14]. This is attributed to the difference in the dihedral angle between the O-N-C-N-O and phenyl least-square planes combined with the change in spin density distribution consecutive to coordination.

3. Materials and Methods

3.1. Materials

All chemicals and solvents were purchased as analytical grade and were used without further purification. 4-methyl-2,6-bis(1-oxyl-3-oxido-4,4,5,5-tetramethyl-2-imidazolin-2-yl)phenol (NIT$_2$PhOH) was synthesized following a reported procedure [14].

3.2. Synthesis of Complex $Cu_3(NIT_2Ph)_2Cl_4$

10 mL of a methanol solution of CuCl$_2$ (64 mg, 0.48 mmol), which was previously dried in a desiccator, was added to a 10 mL of a methanol solution of (NIT$_2$PhOH) (100 mg, 0.24 mmol). The dark brown solution was left for crystallization by slow evaporation. Dark brown crystals, which appeared after three weeks, were isolated by filtration and then washed with ethanol. The complex is insoluble in most usual solvents, which preclude such measurements as UV-vis and molar conductivity, as reported elsewhere [66]. Yield: 43.9 mg (0.04mmol, 33% in term of NIT$_2$PhOH ligand). Elemental analysis (%): C, 42.86; H, 4.98; Cu, 16.29; N, 9.65; Calculated for C$_{42}$H$_{58}$Cl$_4$Cu$_3$N$_8$O$_{10}$ (%): C, 43.21; H, 5.01; Cl, 12.15; Cu, 16.33; N, 9.60; O, 13.70; IR spectrum (υ/cm^{-1}) at 293(2) K: 2987 w, 2938 w, 1431 m, 1338 m, 1312 s, 1214 s, 1171 s, 1145 m, 1057 m, 940 m, 870 m, 798 m, 737 m, 598 m, 547 m, 444 s, 424 s.

3.3. Single-Crystal X-ray Diffraction

Single-crystal diffraction data were collected on a Xcalibur Gemini diffractometer with graphite monochromated Mo Kα radiation (λ = 0.71073 Å), using the related analysis software [70]. Absorption correction has not been performed because it caused a significant decrease in data quality: Increase of Rint value and decrease of the rate of completeness. The structures were solved using the SHELXT program [71] and refined by full-matrix least-square methods on F^2 with the 2018 version of SHELXL program [72] on OLEX2 software [61,73,74]. All non-hydrogen atoms were refined with anisotropic displacement parameters. Hydrogen atoms belonging to carbon atoms were placed geometrically in their idealized positions and refined using a riding model. Crystallographic data are presented in Table 1. Selected bond lengths and bond angles are collected in Tables 2 and 3. Crystallographic data for the structures have been deposited with the Cambridge Crystallographic Data Centre as supplementary publication nos: CCDC 2132162. Copies of the data can be obtained free of charge on application to CCDC, 12 Union Road, Cambridge CB2 1EZ, UK (fax, +44-(0)1223-336033; or e-mail, deposit@ccdc.cam.ac.uk).

3.4. Magnetic Measurements

Magnetic susceptibility data (2–300 K) were collected on powder samples using a SQUID magnetometer (Quantum Design model MPMS-XL) in a 1T applied magnetic field. A magnetization isotherm (2 K) was measured between 0–5 T. All data were corrected for the contribution of the sample holder and diamagnetism of the samples estimated from Pascal's constants [61,75,76].

4. Conclusions

This paper reports the synthesis of a neutral trinuclear copper(II) complex [Cu^{II}$_3$(NIT$_2$-PhO)$_2$Cl$_4$] affords by coordination with biradical, 4-methyl-2,6-bis(1-oxyl-3-oxido-4,4,5,5-tetramethyl-2-imidazolin-2-yl)phenol (NIT$_2$PhOH) completed by cloride. The crystal structure of this heterospin complex reveals a complicated arrangement in which two copper(II) ions, Cu1 and Cu2, are coordinated to each of the two nitronyl nitroxide moieties on one diradical (A). The third copper(II) ion is coordinated to both nitronyl nitroxide moieties of a second diradical (B). The three coppers are also bridged by the phenoxido oxygen atoms of deprotonated diradicals together with some of the cloride ions. The crystal structure analysis reveals an incipient agostic interaction between a square planar copper center and a hydrogen atom from one methyl group carried on the coordinated nitronyl-nitroxide radical. From the magnetic susceptibility measurements, this seven S = 1/2 spin carrier complex behaves as a three spin resulting systems S = 1/2 system. This is ascribed to strong antiferromagnetic direct interaction between the nitronyl nitroxide moieties of diradical A and their coordinated copper(II) ion, Cu1 and Cu2. This makes these four spin S = 1/2 silent, even at high temperature (370K), evidencing only the coupling within Cu3 and diradical B. This is a novel example of how much coordination nitronyl nitroxyde can generate strong magnetic interaction.

Author Contributions: Conceptualization, D.L.; methodology, D.L.; software, S.G. and D.L.; validation, D.L.; formal analysis, S.G. and D.L.; investigation, M.B. and S.G.; resources, D.L.; writing—original draft preparation, S.G.; writing—review and editing, S.G. and D.L.; visualization, S.G.; supervision, D.L.; project administration, D.L.; funding acquisition, D.L. All authors have read and agreed to the published version of the manuscript.

Funding: This research was funded by La Région Auvergne-Rhône-Alpes, grant number 19-008051-01&02-40890.

Institutional Review Board Statement: Not applicable.

Informed Consent Statement: Not applicable.

Data Availability Statement: Data are available from the authors D.L. and S.G.

Acknowledgments: D.L. thank Université Claude Bernard Lyon 1, CNRS and Laboratoire des Multimatériaux et Interfaces (UMR 5615) for laboratory facilities.

Conflicts of Interest: The authors declare no conflict of interest.

Sample Availability: Samples of the compound are available from the authors.

References

1. Caneschi, A.; Gatteschi, D.; Sessoli, R.; Rey, P. Toward molecular magnets: The metal-radical approach. *Acc. Chem. Res.* **1989**, *22*, 392–398. [CrossRef]
2. Ullman, E.F.; Osiecki, J.H.; Boocock, D.G.B.; Darcy, R. Stable free radicals. X. Nitronyl nitroxide monoradicals and biradicals as possible small molecule spin labels. *J. Am. Chem. Soc.* **1972**, *94*, 7049–7059. [CrossRef]
3. Tanimoto, R.; Suzuki, S.; Kozaki, M.; Okada, K. Nitronyl Nitroxide as a Coupling Partner: Pd-Mediated Cross-coupling of (Nitronyl nitroxide-2-ido)(triphenylphosphine)gold(I) with Aryl Halides. *Chem. Lett.* **2014**, *43*, 678–680. [CrossRef]
4. Caneschi, A.; Gatteschi, D.; Hoffman, S.K.; Laugier, J.; Rey, P.; Sessoli, R. Crystal and molecular structure, magnetic properties and EPR spectra of a trinuclear copper(II) complex with bridging nitronyl nitroxides. *Inorg. Chem.* **1988**, *27*, 2390–2392. [CrossRef]
5. Caneschi, A.; Ferraro, F.; Gatteschi, D.; Rey, P.; Sessoli, R. Structure and magnetic properties of a chain compound formed by copper(II) and a tridentate nitronyl nitroxide radical. *Inorg. Chem.* **1991**, *30*, 3162–3166. [CrossRef]
6. Caneschi, A.; Gatteschi, D.; Sessoli, R.; Rey, P. Structure and magnetic properties of a ring of four spins formed by manganese(II) and a pyridine substituted nitronyl nitroxide. *Inorg. Chim. Acta* **1991**, *184*, 67–71. [CrossRef]
7. Fegy, K.; Sanz, N.; Luneau, D.; Belorizky, E.; Rey, P. Proximate Nitroxide Ligands in the Coordination Spheres of Manganese(II) and Nickel(II) Ions. Precursors for High-Dimensional Molecular Magnetic Materials. *Inorg. Chem.* **1998**, *37*, 4518–4523. [CrossRef]
8. Romanov, V.; Tretyakov, E.; Selivanova, G.; Li, J.; Bagryanskaya, I.; Makarov, A.; Luneau, D. Synthesis and Structure of Fluorinated (Benzo[d]imidazol-2-yl)methanols: Bench Compounds for Diverse Applications. *Crystals* **2020**, *10*, 786. [CrossRef]
9. Romanov, V.; Bagryanskaya, I.; Gritsan, N.; Gorbunov, D.; Vlasenko, Y.; Yusubov, M.; Zaytseva, E.; Luneau, D.; Tretyakov, E. Assembly of Imidazolyl-Substituted Nitronyl Nitroxides into Ferromagnetically Coupled Chains. *Crystals* **2019**, *9*, 219. [CrossRef]
10. Mikuriya, M.; Tanaka, K.; Handa, M.; Hiromitsu, I.; Yoshioka, D.; Luneau, D. Adduct complexes of ruthenium(II,III) propionate dimer with pyridyl nitroxides. *Polyhedron* **2005**, *24*, 2658–2664. [CrossRef]
11. Luneau, D.; Rey, P. Magnetism of metal-nitroxide compounds involving bis-chelating imidazole and benzimidazole substituted nitronyl nitroxide free radicals. *Coord. Chem. Rev.* **2005**, *249*, 2591–2611. [CrossRef]
12. Sutter, J.-P.; Kahn, M.L.; Golhen, S.; Ouahab, L.; Kahn, E.O. Synthesis and Magnetic Behavior of Rare-Earth Complexes with N,O-Chelating Nitronyl Nitroxide Triazole Ligands: Example of a [GdIII{Organic Radical}2] Compound with an $S=9/2$ Ground State. *Chem. Eur. J.* **1998**, *4*, 571–576. [CrossRef]
13. Luneau, D.; Stroh, C.; Cano, J.; Ziessel, E.R. Synthesis, Structure, and Magnetism of a 1D Compound Engineered from a Biradical [5,5′-Bis(3″-oxide-1″-oxyl-4″,4″,5″,5″-tetramethylimidazolin-2″-yl)-2,2′-bipyridine] and Mn^{II} (hfac)$_2$. *Inorg. Chem.* **2005**, *44*, 633–637. [CrossRef]
14. Hase, S.; Shiomi, D.; Sato, K.; Takui, T. Phenol-substituted nitronyl nitroxide biradicals with a triplet (S = 1) ground state. *J. Mater. Chem.* **2001**, *11*, 756–760. [CrossRef]

15. Petrov, P.A.; Tret"yakov, E.V.; Romanenko, G.; Ovcharenko, V.I.; Sagdeev, R.Z. Seven-membered metallocycle in the CuIIcomplex with deprotonated 2-(2-hydroxy-3-nitrophenyl)-4,4,5,5-tetramethyl-4,5-dihydro-1H-imidazole-1-oxyl 3-oxide. *Bull. Acad. Sci. USSR Div. Chem. Sci.* **2004**, *53*, 109–113. [CrossRef]
16. Liu, R.; Liu, L.; Fang, D.; Xu, J.; Zhao, S.; Xu, W. Synthesis, Structure, and Magnetic Properties of Two Novel Dinuclear Complexes involving Lanthanide-phenoxo Anion Radical. *J. Inorg. Gen. Chem.* **2015**, *641*, 728–731. [CrossRef]
17. Spinu, C.A.; Pichon, C.; Ionita, G.; Mocanu, T.; Calancea, S.; Raduca, M.; Sutter, J.-P.; Hillebrand, M.; Andruh, M. Synthesis, crystal structure, magnetic, spectroscopic, and theoretical investigations of two new nitronyl-nitroxide complexes. *J. Coord. Chem.* **2021**, *74*, 279–293. [CrossRef]
18. Cassaro, R.; Friedman, J.; Lahti, P.M. Copper(II) coordination compounds with sterically constraining pyrenyl nitronyl nitroxide and imino nitroxide. *Polyhedron* **2016**, *117*, 7–13. [CrossRef]
19. Haraguchi, M.; Tretyakov, E.; Gritsan, N.; Romanenko, G.; Gorbunov, D.; Bogomyakov, A.; Maryunina, K.; Suzuki, S.; Kozaki, M.; Shiomi, D.; et al. (Azulene-1,3-diyl)-bis(nitronyl nitroxide) and (Azulene-1,3-diyl)-bis(iminonitroxide) and Their Copper Complexes. *Chem. Asian J.* **2017**, *12*, 2929–2941. [CrossRef]
20. Fidan, I.; Luneau, D.; Ahsen, V.; Hirel, C. Revisiting the Ullman's Radical Chemistry for Phthalocyanine Derivatives. *Chem. A Eur. J.* **2018**, *24*, 5359–5365. [CrossRef]
21. Önal, E.; Fidan, I.; Luneau, D.; Hirel, C. Through the challenging synthesis of tetraphenylporphyrin derivatives bearing nitroxide moieties. *J. Porphyr. Phthalocyanines* **2019**, *23*, 584–588. [CrossRef]
22. Fidan, I.; Önal, E.; Yerli, Y.; Luneau, D.; Ahsen, V.; Hirel, C. Synthetic Access to a Pure Polyradical Architecture: Nucleophilic Insertion of Nitronyl Nitroxide on a Cyclotriphosphazene Scaffold. *ChemPlusChem* **2017**, *82*, 1384–1389. [CrossRef] [PubMed]
23. Morozov, V.; Tretyakov, E. Spin polarization in graphene nanoribbons functionalized with nitroxide. *J. Mol. Model.* **2019**, *25*, 58. [CrossRef] [PubMed]
24. Lecourt, C.; Izumi, Y.; Maryunina, K.; Inoue, K.; Bélanger-Desmarais, N.; Reber, C.; Desroches, C.; Luneau, D. Hypersensitive pressure-dependence of the conversion temperature of hysteretic valence tautomeric manganese–nitronyl nitroxide radical 2D-frameworks. *Chem. Commun.* **2021**, *57*, 2376–2379. [CrossRef]
25. Lecourt, C.; Izumi, Y.; Khrouz, L.; Toche, F.; Chiriac, R.; Bélanger-Desmarais, N.; Reber, C.; Fabelo, O.; Inoue, K.; Desroches, C.; et al. Thermally-induced hysteretic valence tautomeric conversions in the solid state via two-step labile electron transfers in manganese-nitronyl nitroxide 2D-frameworks. *Dalton Trans.* **2020**, *49*, 15646–15662. [CrossRef]
26. Lannes, A.; Suffren, Y.; Tommasino, J.B.; Chiriac, R.; Toche, F.; Khrouz, L.; Molton, F.; Duboc, C.; Kieffer, I.; Hazemann, J.-L.; et al. Room Temperature Magnetic Switchability Assisted by Hysteretic Valence Tautomerism in a Layered Two-Dimensional Manganese-Radical Coordination Framework. *J. Am. Chem. Soc.* **2016**, *138*, 16493–16501. [CrossRef]
27. Zheludev, A.; Barone, V.; Bonnet, M.; Delley, B.; Grand, A.; Ressouche, E.; Rey, P.; Subra, R.; Schweizer, J. Spin density in a nitronyl nitroxide free radical. Polarized neutron diffraction investigation and ab initio calculations. *J. Am. Chem. Soc.* **1994**, *116*, 2019–2027. [CrossRef]
28. Vaz, M.G.; Andruh, M. Molecule-based magnetic materials constructed from paramagnetic organic ligands and two different metal ions. *Coord. Chem. Rev.* **2020**, *427*, 213611. [CrossRef]
29. Luneau, D.; Borta, A.; Chumakov, Y.; Jacquot, J.-F.; Jeanneau, E.; Lescop, C.; Rey, P. Molecular magnets based on two-dimensional Mn(II)–nitronyl nitroxide frameworks in layered structures. *Inorg. Chim. Acta* **2008**, *361*, 3669–3676. [CrossRef]
30. Minguet, M.; Luneau, D.; Lhotel, E.; Villar, V.; Paulsen, C.; Amabilino, D.B.; Veciana, J. An Enantiopure Molecular Ferromagnet. *Angew. Chem. Int. Ed.* **2002**, *41*, 586–589. [CrossRef]
31. Stumpf, H.O.; Pei, Y.; Kahn, O.; Ouahab, L.; Grandjean, D. A Molecular-Based Magnet with a Fully Interlocked Three-Dimensional Structure. *Science* **1993**, *261*, 447–449. [CrossRef] [PubMed]
32. Bernot, K.; Pointillart, F.; Rosa, P.; Etienne, M.; Sessoli, R.; Gatteschi, D. Single molecule magnet behaviour in robust dysprosium–biradical complexes. *Chem. Commun.* **2010**, *46*, 6458–6460. [CrossRef] [PubMed]
33. Coronado, E.; Giménez-Saiz, C.; Recuenco, A.; Tarazón, A.; Romero, F.M.; Camón, A.; Luis, F. Single-Molecule Magnetic Behavior in a Neutral Terbium(III) Complex of a Picolinate-Based Nitronyl Nitroxide Free Radical. *Inorg. Chem.* **2011**, *50*, 7370–7372. [CrossRef]
34. Li, H.; Jing, P.; Lu, J.; Xie, J.; Zhai, L.; Xi, L. Dipyridyl-Decorated Nitronyl Nitroxide–DyIII Single-Molecule Magnet with a Record Energy Barrier of 146 K. *Inorg. Chem.* **2021**, *60*, 7622–7626. [CrossRef] [PubMed]
35. Sun, J.; Wu, Q.; Lu, J.; Jing, P.; Du, Y.; Li, L. Slow relaxation of magnetization in lanthanide–biradical complexes based on a functionalized nitronyl nitroxide biradical. *Dalton Trans.* **2020**, *49*, 17414–17420. [CrossRef] [PubMed]
36. Wang, X.-L.; Li, L.-C.; Liao, D.-Z. Slow Magnetic Relaxation in Lanthanide Complexes with Chelating Nitronyl Nitroxide Radical. *Inorg. Chem.* **2010**, *49*, 4735–4737. [CrossRef]
37. Bernot, K.; Bogani, L.; Caneschi, A.; Gatteschi, D.; Sessoli, R. A Family of Rare-Earth-Based Single Chain Magnets: Playing with Anisotropy. *J. Am. Chem. Soc.* **2006**, *128*, 7947–7956. [CrossRef]
38. Bogani, L.; Sangregorio, C.; Sessoli, R.; Gatteschi, D. Molecular Engineering for Single-Chain-Magnet Behavior in a One-Dimensional Dysprosium-Nitronyl Nitroxide Compound. *Angew. Chem. Int. Ed.* **2005**, *44*, 5817–5821. [CrossRef]
39. Houard, F.; Gendron, F.; Suffren, Y.; Guizouarn, T.; Dorcet, V.; Calvez, G.; Daiguebonne, C.; Guillou, O.; Le Guennic, B.; Mannini, M.; et al. Single-chain magnet behavior in a finite linear hexanuclear molecule. *Chem. Sci.* **2021**, *12*, 10613–10621. [CrossRef]

40. Liu, R.; Li, L.; Wang, X.; Yang, P.; Wang, C.; Liao, D.; Sutter, J.-P. Smooth transition between SMM and SCM-type slow relaxing dynamics for a 1-D assemblage of {Dy(nitronyl nitroxide)2} units. *Chem. Commun.* **2010**, *46*, 2566–2568. [CrossRef]
41. Xie, J.; Li, H.-D.; Yang, M.; Sun, J.; Li, L.-C.; Sutter, J.-P. Improved single-chain-magnet behavior in a biradical-based nitronyl nitroxide-Cu–Dy chain. *Chem. Commun.* **2019**, *55*, 3398–3401. [CrossRef]
42. de Panthou, F.L.; Belorizky, E.; Calemczuk, R.; Luneau, D.; Marcenat, C.; Ressouche, E.; Turek, P.; Rey, P. A New Type of Thermally Induced Spin Transition Associated with an Equatorial.dblarw. Axial Conversion in a Copper(II)-Nitroxide Cluster. *J. Am. Chem. Soc.* **1995**, *117*, 11247–11253. [CrossRef]
43. de Panthou, F.L.; Luneau, D.; Musin, R.; Öhrström, L.; Grand, A.; Turek, P.; Rey, P. Spin-Transition and Ferromagnetic Interactions in Copper(II) Complexes of a 3-Pyridyl-Substituted Imino Nitroxide. Dependence of the Magnetic Properties upon Crystal Packing. *Inorg. Chem.* **1996**, *35*, 3484–3491. [CrossRef]
44. Fedin, M.; Veber, S.; Bagryanskaya, E.; Ovcharenko, V.I. Electron paramagnetic resonance of switchable copper-nitroxide-based molecular magnets: An indispensable tool for intriguing systems. *Coord. Chem. Rev.* **2015**, *289–290*, 341–356. [CrossRef]
45. Luneau, D. Coordination Chemistry of Nitronyl Nitroxide Radicals Has Memory. *Eur. J. Inorg. Chem.* **2020**, *2020*, 597–604. [CrossRef]
46. Tolstikov, S.; Tretyakov, E.; Fokin, S.; Suturina, E.; Romanenko, G.; Bogomyakov, A.; Stass, D.; Maryasov, A.; Fedin, M.; Gritsan, N.; et al. C(sp2)-Coupled Nitronyl Nitroxide and Iminonitroxide Diradicals. *Chem. A Eur. J.* **2014**, *20*, 2793–2803. [CrossRef]
47. Gurskaya, L.; Rybalova, T.; Beregovaya, I.; Zaytseva, E.; Kazantsev, M.; Tretyakov, E. Aromatic nucleophilic substitution: A case study of the interaction of a lithiated nitronyl nitroxide with polyfluorinated quinoline-N-oxides. *J. Fluor. Chem.* **2020**, *237*, 109613. [CrossRef]
48. Addison, A.W.; Rao, T.N.; Reedijk, J.; van Rijn, J.; Verschoor, G.C. Synthesis, structure, and spectroscopic properties of copper(II) compounds containing nitrogen–sulphur donor ligands; the crystal and molecular structure of aqua[1,7-bis(N-methylbenzimidazol-2′-yl)-2,6-dithiaheptane]copper(II) perchlorate. *J. Chem. Soc. Dalton Trans.* **1984**, *7*, 1349–1356. [CrossRef]
49. Hoffmann, S.K.; Hodgson, D.J.; Hatfield, W.E. Crystal structures and magnetic and EPR studies of intradimer and interdimer exchange coupling in [M(en)3]2[Cu2Cl8]Cl2.cntdot.2H2O (M = Co, Rh, Ir) crystals. *Inorg. Chem.* **1985**, *24*, 1194–1201. [CrossRef]
50. Bream, R.A.; Estes, E.D.; Hodgson, D.J. Structural characterization of dichloro[2-(2-methylaminoethyl)pyridine]copper(II). *Inorg. Chem.* **1975**, *14*, 1672–1675. [CrossRef]
51. Estes, E.D.; Estes, W.E.; Hatfield, W.E.; Hodgson, D.J. Molecular structure of bis[dichloro(N,N,N′,N′-tetramethylenediaminecopper(II)], [Cu(tmen)Cl2]2. *Inorg. Chem.* **1975**, *14*, 106–109. [CrossRef]
52. Sinn, E.; Robinson, W.T. X-Ray structure analyses and magnetic correlation of four copper(II) complexes. *J. Chem. Soc. Chem. Commun.* **1972**, *6*, 359–361. [CrossRef]
53. Colomban, C.; Philouze, C.; Molton, F.; Leconte, N.; Thomas, F. Copper(II) complexes of N3O ligands as models for galactose oxidase: Effect of variation of steric bulk of coordinated phenoxyl moiety on the radical stability and spectroscopy. *Inorg. Chim. Acta* **2018**, *481*, 129–142. [CrossRef]
54. Okuniewski, A.; Rosiak, D.; Chojnacki, J.; Becker, B. Coordination polymers and molecular structures among complexes of mercury(II) halides with selected 1-benzoylthioureas. *Polyhedron* **2015**, *90*, 47–57. [CrossRef]
55. Rosiak, D.; Okuniewski, A.; Chojnacki, J. Novel complexes possessing Hg–(Cl, Br, I)···O C halogen bonding and unusual Hg 2 S 2 (Br/I) 4 kernel. The usefulness of τ′4 structural parameter. *Polyhedron* **2018**, *146*, 35–41. [CrossRef]
56. Yang, L.; Powell, D.R.; Houser, R.P. Structural variation in copper(i) complexes with pyridylmethylamide ligands: Structural analysis with a new four-coordinate geometry index, τ4. *Dalton Trans.* **2007**, *9*, 955–964. [CrossRef] [PubMed]
57. Brookhart, M.; Green, M.L. Carbon hydrogen-transition metal bonds. *J. Organomet. Chem.* **1983**, *250*, 395–408. [CrossRef]
58. Crabtree, R.H. Transition Metal Complexation ofσ Bonds. *Angew. Chem. Int. Ed.* **1993**, *32*, 789–805. [CrossRef]
59. Brookhart, M.; Green, M.L.H.; Parkin, G. Agostic interactions in transition metal compounds. *Proc. Natl. Acad. Sci. USA* **2007**, *104*, 6908–6914. [CrossRef]
60. Braga, D.; Grepioni, A.F.; Tedesco, E.; Biradha, K.; Desiraju, G.R. Hydrogen Bonding in Organometallic Crystals. 6. X−H—M Hydrogen Bonds and M—(H−X) Pseudo-Agostic Bonds. *Organometallics* **1997**, *16*, 1846–1856. [CrossRef]
61. Mei, X.; Wang, X.; Wang, J.; Ma, Y.; Li, L.; Liao, D. Dinuclear lanthanide complexes bridged by nitronyl nitroxide radical ligands with 2-phenolate groups: Structure and magnetic properties. *New J. Chem.* **2013**, *37*, 3620–3626. [CrossRef]
62. Tandon, S.S.; Bunge, S.D.; Patel, N.; Wang, E.C.; Thompson, L.K. Self-Assembly of Antiferromagnetically-Coupled Copper(II) Supramolecular Architectures with Diverse Structural Complexities. *Molecules* **2020**, *25*, 5549. [CrossRef] [PubMed]
63. Crawford, V.H.; Richardson, H.W.; Wasson, J.R.; Hodgson, D.J.; Hatfield, W.E. Relation between the singlet-triplet splitting and the copper-oxygen-copper bridge angle in hydroxo-bridged copper dimers. *Inorg. Chem.* **1976**, *15*, 2107–2110. [CrossRef]
64. Thompson, L.K.; Mandal, S.K.; Tandon, S.S.; Bridson, J.N.; Park, M.K. Magnetostructural Correlations in Bis(μ2-phenoxide)-Bridged Macrocyclic Dinuclear Copper(II) Complexes. Influence of Electron-Withdrawing Substituents on Exchange Coupling. *Inorg. Chem.* **1996**, *35*, 3117–3125. [CrossRef]
65. Khan, O. *Molecular Magnetism*; Wiley-VCH: Weinhein, Germany, 1993.
66. Rajendiran, T.M.; Kannappan, R.; Mahalakshmy, R.; Rajeswari, J.; Venkatesan, R.; Rao, P. New unsymmetrical μ-phenoxo bridged binuclear copper(II) complexes. *Transit. Met. Chem.* **2003**, *28*, 447–454. [CrossRef]
67. Caneschi, A.; Gatteschi, D.; Rey, P. The Chemistry and Magnetic Properties of Metal Nitronyl Nitroxide Complexes. *Prog. Inorg. Chem.* **1991**, *39*, 331–429. [CrossRef]

68. Cogne, A.; Laugier, J.; Luneau, D.; Rey, P. Novel Square Planar Copper(II) Complexes with Imino or Nitronyl Nitroxide Radicals Exhibiting Large Ferro- and Antiferromagnetic Interactions. *Inorg. Chem.* **2000**, *39*, 5510–5514. [CrossRef] [PubMed]
69. Chilton, N.F.; Anderson, R.P.; Turner, L.D.; Soncini, A.; Murray, K.S. PHI: A powerful new program for the analysis of anisotropic monomeric and exchange-coupled polynucleard- andf-block complexes. *J. Comput. Chem.* **2013**, *34*, 1164–1175. [CrossRef]
70. Agilent. *CrysAlis PRO*; Agilent Technologies Ltd.: Yarnton, UK, 2014.
71. Sheldrick, G.M. SHELXT—Integrated space-group and crystal-structure determination. *Acta Crystallogr. Sect. A Found. Adv.* **2015**, *71*, 3–8. [CrossRef]
72. Sheldrick, G.M. Crystal structure refinement with SHELXL. *Acta Crystallogr. Sect. C Struct. Chem.* **2015**, *C71*, 3–8. [CrossRef]
73. Dolomanov, O.V.; Bourhis, L.J.; Gildea, R.J.; Howard, J.A.K.; Puschmann, H. OLEX2: A complete structure solution, refinement and analysis program. *J. Appl. Cryst.* **2009**, *42*, 339–341. [CrossRef]
74. Bourhis, L.J.; Dolomanov, O.V.; Gildea, R.J.; Howard, J.A.K.; Puschmann, H. The anatomy of a comprehensive constrained, restrained refinement program for the modern computing environment –Olex2dissected. *Acta Crystallogr. Sect. A Found. Adv.* **2015**, *71*, 59–75. [CrossRef] [PubMed]
75. Pascal, P. Magnetochemical Researches. *Ann. Chim. Phys.* **1910**, *19*, 5–70.
76. Bain, G.A.; Berry, J.F. Diamagnetic Corrections and Pascal's Constants. *J. Chem. Educ.* **2008**, *85*, 532. [CrossRef]

Viewpoint

Highlighting Recent Crystalline Engineering Aspects of Luminescent Coordination Polymers Based on F-Elements and Ditopic Aliphatic Ligands

Richard F. D'Vries [1,*], Germán E. Gomez [2] and Javier Ellena [3]

[1] Facultad de Ciencias Básicas, Universidad Santiago de Cali, Calle 5 # 62-00, Cali 760035, Colombia
[2] Instituto de Investigaciones en Tecnología Química (INTEQUI), Área de Química General e Inorgánica, Facultad de Química, Bioquímica y Farmacia, Chacabuco y Pedernera, Universidad Nacional de San Luis, Almirante Brown, 1455, San Luis 5700, Argentina; germangomez1986@gmail.com
[3] São Carlos Institute of Physics, University of São Paulo, São Carlos CEP 13.566-590, SP, Brazil; javiere@ifsc.usp.br
* Correspondence: richard.dvries00@usc.edu.co

Abstract: Three principal factors may influence the final structure of coordination polymers (CPs): (i) the nature of the ligand, (ii) the type and coordination number of the metal center, and (iii) the reaction conditions. Further, flexible carboxylate aliphatic ligands have been widely employed as building blocks for designing and synthesizing CPs, resulting in a diverse array of materials with exciting architectures, porosities, dimensionalities, and topologies as well as an increasing number of properties and applications. These ligands show different structural features, such as torsion angles, carbon backbone number, and coordination modes, which affect the desired products and so enable the generation of polymorphs or crystalline phases. Additionally, due to their large coordination numbers, using $4f$ and $5f$ metals as coordination centers combined with aliphatic ligands increases the possibility of obtaining different crystal phases. Additionally, by varying the synthetic conditions, we may control the production of a specific solid phase by understanding the thermodynamic and kinetic factors that influence the self-assembly process. This revision highlights the relationship between the structural variety of CPs based on flexible carboxylate aliphatic ligands and f-elements (lanthanide and actinides) and their outstanding luminescent properties such as solid-state emissions, sensing, and photocatalysis. In this sense, we present a structural analysis of the CPs reported with the oxalate ligand, as the one rigid ligand of the family, and other flexible dicarboxylate linkers with $-CH_2-$ spacers. Additionally, the nature of the luminescence properties of the $4f$ or $5f$-CPs is analyzed, and finally, we present a novel set of CPs using a glutarate-derived ligand and samarium, with the formula [2,2'-bipyH][Sm(HFG)$_2$(2,2'-bipy)(H$_2$O)$_2$]•(2,2'-bipy) (**α-Sm**) and [2,2'-bipyH][Sm(HFG)$_2$(2,2'-bipy)(H$_2$O)$_2$] (**β-Sm**).

Keywords: coordination polymers; f-elements; luminescence

1. Introduction

Coordination polymers are composed of a rational combination of metallic centers (connectors) and organic ligands, resulting in extended 1D, 2D, or 3D structures [1,2]. Many classifications have been assigned to these materials, depending on their structure, dimensionality, porosity, catalytic capacity, etc. [1,3,4]. However, in this work, we will focus on the nature of the aliphatic ligand in combination with $4f$ and $5f$ elements to yield a plethora of crystalline coordination polymers (CPs). These materials have been extensivelt studied in recent decades due to their multifunctional properties, which are intimately related to their structural features [5–7]. Properties such as luminescence, catalysis, sensing, gas storage, and drug delivery were thoroughly investigated, becoming mature areas [8–21].

Moreover, the use of metallic connectors from $4f$ and $5f$ metals allows the development of materials with unique optical properties derived from their pure color emission, fine line f-transitions, and variable lifetime values depending on the desirable

applications [12,15,22,23]. Additionally, the high coordination numbers and the oxophilic nature of these metallic centers [24,25] allow us to combine these ions with flexible carboxylate aliphatic ligands and build a "toolbox" to construct families of novel crystalline structures [26]. This work discusses some outstanding examples in structural variety of CPs using ditopic flexible ligands [−OOC-(CH$_2$)n-COO−] and the role of the oxalate ligand as the first member of the dicarboxylate family, focusing on their luminescent properties, as well as applications such as dye photocatalysis and chemical and thermal sensing in *Ln*-CPs or *An*-CPs (*Ln* = lanthanides and *An* = actinides). Additionally, we present two novel crystalline phases of CPs obtained from the combination of hexafluorglutaric acid (H$_2$HFG) and samarium ions.

2. Discussion
2.1. Coordination Polymers Based on Oxalate Linker

In this highlight, we will analyze the entire family of aliphatic dicarboxylate ligands, from oxalate to dodecanedioate, used as ligands in the formation of CPs. The analysis on Cambridge Crystallographic Data Centre (CCDC) during the last two decades shows a small number of entries with a decreasing number of reported structures as the length of the carbon backbone increases (Figure 1). This trend is attributed to a higher degree of freedom and flexibility when chain length increases, which restricts or decreases the chance of crystallization.

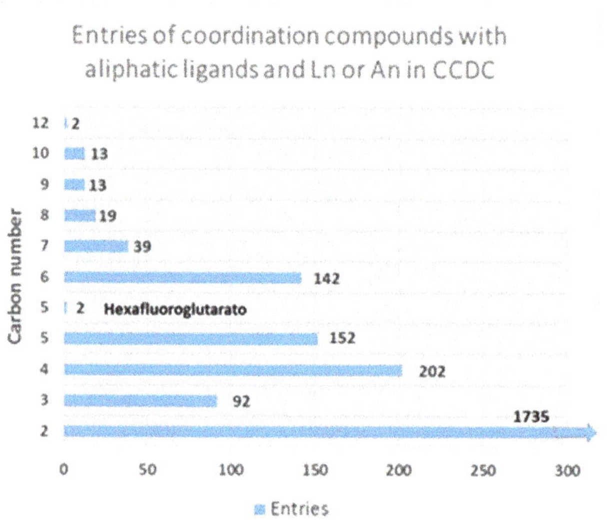

Figure 1. Entries distribution in the CCDC for compounds with aliphatic ditopic ligands and Ln or An metals.

In this sense, it is important to consider the nature of the sp^3 covalent bond along the aliphatic chains, which has the ability to rotate along the C–C bond. This degree of freedom in the simple bond enables the formation of multiple conformers. The amount of conformers for a particular aliphatic linker is directly related to the carbon number and carboxylate coordination modes. The use of ditopic aliphatic ligands in the synthesis of CPs is limited by the stability of the conformer, the size of the ligand and the architecture of the polymer. Additionally, it is important to note the high number of reports with short dicarboxylates (carbon numbers between 2 and 5, Figure 1) forming *Ln*-CPs or *An*-CPs (*Ln* = lanthanides and *An* = actinides) [27–30]. The oxalate anion is not a flexible ligand and contributes with rigidity and stability in the resulting CP. In fact, this molecule has been found in biologic systems [31], in materials such as electronic and luminescent devices [32,33], drugs [34], and minerals [35,36]. Additionally, the use of this ligand in combination with Ln and An metals as a toolbox to obtain CPs dates back to the 1980s, where Kahwa et al. [37] reported a family of isostructural 1D structures with the general formula $K_3[Ln(ox)_3(OH_2)]\cdot 2H_2O$, (Ln = Nd, Sm, Eu, Gd, Tb). Additionally, Alexander et al. [38] reported a 2D terbium oxalate with lanthanide-centered green luminescence at 543 nm by exciting the sample at 369 nm. The observed signals correspond to the most intense forbidden f–f transition of terbium ions exhibiting an optical performance without ligand sensitization, comparable to the commercial green phosphors [38].

The general use of this ligand is extended to mixtures of two or more ligands using oxalate and auxiliary ligands. The combination of two types of linkers is extensively used since it allows the presence of a structural ligand and a functional one allowing the formation of CPs with improved properties such as luminescence or catalysis. Indeed, approximately 69% of the CCDC entries found in this work (Figure 1) present a combination of ligands. One example of this trend is the work by Xu et al. [39], where they used a combination of aromatic ligands as 4-(4-carboxyphenoxy)-isophthalic acid (cphtH$_3$) and 1,10-phenanthroline (phen) as an ancillary to form a highly stable luminescent 3D CPs [39]. These compounds produce green (Tb), white (Sm), blue (Dy) and red (Eu) emissions. Additionally, based on the excellent luminescence of the Eu-MOF produced, the compound was tested for quercetin and Fe^{3+} ion sensing based on quenching processes [39].

Moreover, studies of An-CPs based on uranyl or thorium ions have been explored in the last decade, mostly due to their potential applications as light emitters for sensing and photocatalysis, and the variety of novel achieved architectures [22,40–49]. Furthermore, the rational combination of O-donor with the uranyl cation $[UO_2]^{2+}$ coordination modes has led to the formation of an important number of new organic–inorganic connectivity with different dimensionalities and nuclearities of uranyl-centered building units [50–54]. These assemblies are commonly obtained by employing O-donor-chelating agents such as polycarboxylic acids. In spite of that, the formation of An-CPs is scarce and just 5.9% of all the structures reported in the CCDC search present the oxalate anion (Figure 2). The formation of new structures with these components could open a wide research area of functional compounds derived from the nuclear activity [55,56].

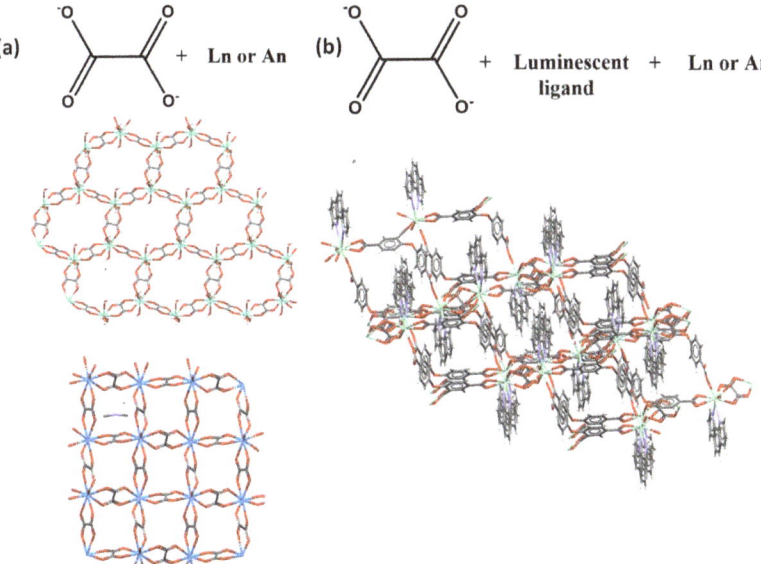

Figure 2. (a) Terbium [38] And uranium [57] 2D oxalate CPs and (b) lanthanide–oxalate–phen-cphtH$_3$ 3D CP [39].

2.2. Flexible Linkers with –CH$_2$– Spacers

The torsion angle of the most used ligands in the synthesis of CP is shown in Figure 3, where the oxalate ligand is a planar ligand, whereas malonate and succinate show a wide range of O–C–C–O torsion angle values. Thus, the degree of freedom around the sp^3 carbon as well as the coordination modes enables the formation of multiple phases with a slight change in the reactions conditions. Cañadillas-Delgado et al. [58], Chrysomallidou et al. [59], and Delgado et al. [29] reported that malonate compounds show different dimensionalities, coordination modes, and carbonyl–carbonyl torsion angles according to the synthetic methodology. These findings describe the synthesis of compounds with a 2D and 3D topology by solvothermal reactions, slow diffusion of the reagents in metasilacate gel and slow evaporation of the solvent. From a crystal engineering point of view, it is possible to modify the structural features using different synthetic methodologies, reaction conditions and use of ancillary ligands [60]. Within the most used synthetic techniques are (i) hydro or solvothermal synthesis, (ii) slow solvent evaporation, (iii) gel diffusion, (iv) mechanochemical, and (v) microwave-assisted synthesis [61–67]. It has been observed that the use of methodologies that involve the application of energy favors the formation of structures that are more compact, with more complex coordination modes and high dimensionalities [68]. In the case of methodologies involving low energy, they generate low-dimensionality structures with simpler coordination modes. On other hand, the use of *guest molecules acting as templates* refers to the presence of organic molecules or solvent molecules giving space for the formation of cavities into the coordination polymer [69]. Guest molecules can be small and isolated species included in the CP structure and non-coordinated to the metal that allow the formation of cavities [67,70–72].

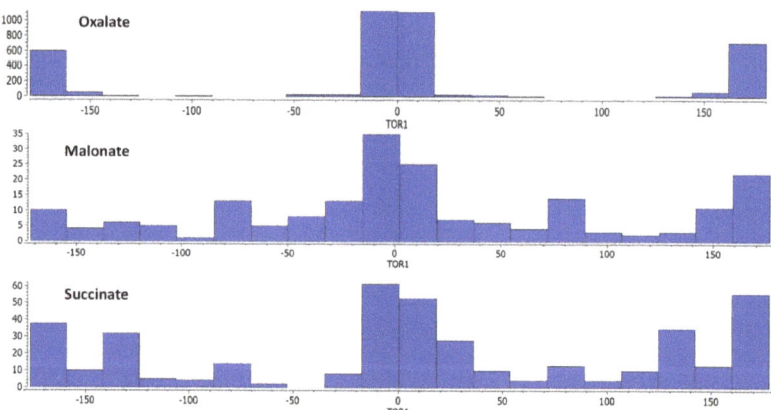

Figure 3. Torsion angle values for the entries in the CCDC for oxalate, malonate and succinate ligands.

In our group, we have carried out studies involving the formation of four different phases of *Ln*-succinate compounds by solvothermal reaction conditions [73], using the *template effect* or *guest molecule acting as a template* of aromatic solvents as toluene, and aromatic organic molecules as 5-sulfosalicylate (5-SSA^{3-}). These guest molecules are directly involved in the formation of CPs with large pores or cavities, unlike using conventional protic or organic solvents. Additionally, interesting reviews dedicated to the formation of CPs based on succinate ligands with 2D and 3D structures involving different topologies and applications, show the use of these type of ligands in the design of multifunctional CPs [74]. The use of aliphatic linkers such as 2-methylsuccinates and 2-phenysuccinates can efficiently separate the lanthanide ions in order to avoid concentration quenching and giving rise lanthanide-centered emissions upon direct excitation into the 4*f* levels. The fine lanthanide emissions were enough to use the materials as solid-state emitters, thermal sensors and chemical sensors for small molecules [42,75]. Additionally, it is important to highlight the variety of coordination modes found reports employing flexible ditopic ligands [55,76,77], as shown in Figure 4.

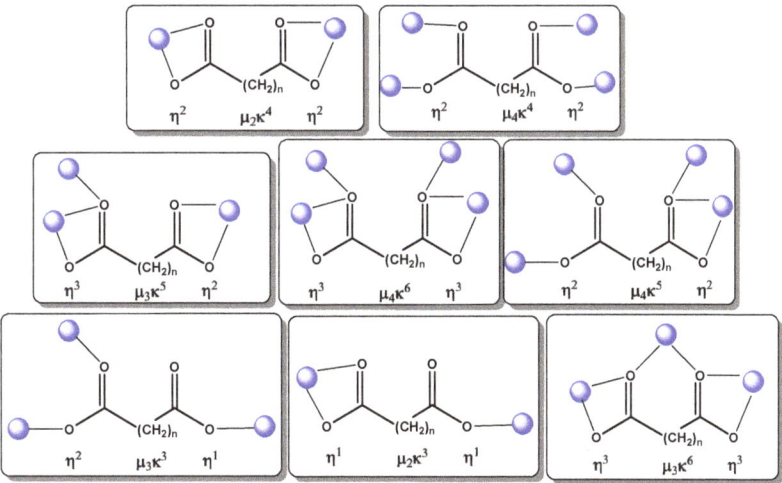

Figure 4. Typical coordination modes observed in ditopic aliphatic ligands.

2.3. The Luminescent Properties of 4f-5f Compounds

According to the vast literature on luminescent materials, their properties can be explored by the following studies [78]: (1) excitation and emission spectra; (2) quantum yield (QYs) determinations, and (3) experimental or observed lifetime (τ) of emission. Additionally, the attenuation of luminescence experienced by certain materials is known as *quenching of luminescence* and can be derived from structural features as well as from external parameters. The contributions of non-radiative pathways mainly include electron transfer quenching, back-energy transfer to the sensitizer as well as quenching by organic vibrations from the linkers (C-H, O-H, N-H) [79]. Among lanthanide ions, Sm^{3+}, Eu^{3+}, Gd^{3+}, Tb^{3+} and Dy^{3+} are preferred ions for optical device implementation and optoelectronics due to their intense, long-lived and fine emissions into the visible region [12,32,80]. Additionally, Er^{3+}, Ho^{3+}, Yb^{3+} and Tm^{3+} ions are suitable ions for up-conversion emissions into visible and near-infrared regions [81,82].

The complete review by Bernini and colleagues highlights the contributions of diverse lanthanide-succinate-derived structures with applications in solid-state lighting, sensing, and catalysis [74]. In this report, the authors mention the impact of aryl and alkyl substitute succinate ligands on the final dimension of the crystalline structure and also on the final property. On the other hand, the improved performance can be assessed by a correct selection of building blocks that allows a correct energy gap between the lanthanide ions and the exited states (singlet or triplet) from the linkers.

In general, materials constructed by aryl-derivate linkers or by incorporating auxiliary aromatic ligands (i.e., 1,10-phenanthroline, 2,2'-bipyridine, etc.) show better emission efficiency than the solely ditopic-based materials [53]. One example is phthalate ligand sensitization to improve Eu/Tb luminescence and metal-to-metal energy transfer in mixed $[Ln(adipate)_{0.5}(phth)(H_2O)_2]$ compounds, being used as thermometers in the 303–423 K range [83].

From 2018, we can mention remarkable contributions employing lanthanide-succinate compounds by Professor Narda's group. They report the synthesis of a bidimensional mixed structure Tb^{3+}@Y-succ-sal (succ = succinate, sal = salycilate) with the particularity of producing reactive oxygen species upon UV excitation in aqueous suspensions, allowing the photodynamic inactivation of *Candida albicans* culture by intersystem crossing mechanisms [84]. In 2022, the same group reported the use of doped $Yb^{3+}/Er^{3+}/Gd^{3+}$@Y-succ-sal systems as sacrificial materials to obtain $Yb^{3+}/Er^{3+}/Gd^{3+}$@Y_2O_3 and finally deposit them onto glass substrates for thin film implementation (Figure 5) [85]. The red emissions derived from up-conversion processes by exciting into the near-infrared region yielded τ values ranging from 144 to 300 µs.

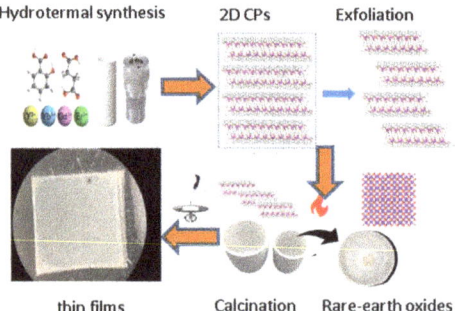

Figure 5. Manufacture of thin films based on bidimensional Y-succ-sal compounds: solvothermal synthesis of layered compounds, exfoliation, calcination to obtain lanthanide-doped Y_2O_3 systems and deposition by spin coating. Adapted from Ref. [85].

Further, uranyl emission is originated from a ligand to metal charge transfer (LMCT) by exciting an electron from non-bonding $5f_\delta$, $5f_\varphi$ uranyl orbitals to uranyl–oxygen bonding orbitals (σ_u, σ_g, π_u, π_g) [86], which is further coupled to "yl" vibrational ($S_{11} \to S_{01}$ and $S_{10} \to S_{0v}$ [ν = 0–4]) states of the U=O axial bond [87]. Its phosphorescence is often characterized by green emission, which manifests as four to six vibronically coupled peaks (up to 12) in the 400–650 nm range.

As shown in lanthanide-flexible compounds, uranyl versions can be explored in interesting applications such as photocatalysis and sensing. This is achievable due to its high water stability, repeatability, and bright emission into the visible region, showing a bright future for uranyl-CPs.

In this case, we can highlight the **UNSL-1** (Universidad Nacional de San Luis) compound, $[(UO_2)_2(phen)(succ)_{0.5}(OH)(\mu_3\text{-}O)(H_2O)] \cdot H_2O$, which corresponds to a 1D coordination polymer formed by tetramer units of uranyl ions, decorated by coordinated 1,10-phenanthroline molecules and connected by succinate ligands into the [−1 0 1] direction [22]. The phen plays the role of a suitable antenna molecule to store UV energy and then transfer it to the uranyl ions, yielding a bright green emission into the visible region (see Figure 6). For these optical features, **UNSL-1** material was employed as a photocatalyst for methylene blue degradation upon sunlight excitation. Additionally, the material was used as a sensor toward metallic ions in aqueous media, exhibiting sensitivity under Fe^{2+} ions.

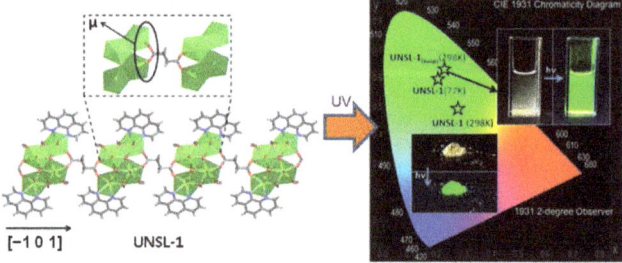

Figure 6. View of the infinite chains along the [−1, 0, 1] direction. CIE x, y chromaticity of **UNSL-1** compound in the suspension and solid states (77 and 298 K), Adapted from Ref. [18].

2.4. Sm-Hexafluoroglutarato CPs

Glutaric acid is a long, flexible carboxylic ligand with –CH$_2$– spacers (polymethylene groups) with a size chain of approximately 7 Å. Similarly, the glutarate presents a wide range of C–C–C–O torsion angles as well as different configurations around the methylene scaffold [88–90]. According to Kumar et al. [30], it is interesting to differentiate between two types of compounds: (i) those without ancillary co-ligands and (ii) those with ancillary co-ligand [30]. Further, there is one report of complexes formed with H$_2$HFG ligand [91], with Eu^{3+} and Tb^{3+} ions giving rise to dimeric assemblies with the formula [Ln$_2$(HFG)$_2$(phen)4(H$_2$O)$_6$]. · HFG · 2H$_2$O (where Ln = Eu or Tb). In the mentioned work, the phen molecule locks the coordination positions in the metal ions, limiting the dimensionality of the final product.

Here, we present a novel set of CPs using a glutarate-derived ligand and samarium, with the formulae [2,2′-bipyH][Sm(HFG)$_2$ (2,2′-bipy) (H$_2$O)$_2$]•(2,2′-bipy) (α-Sm) and [2,2′-bipyH][Sm(HFG)$_2$ (2,2′-bipy) (H$_2$O)$_2$] (β-Sm). The ORTEP diagrams of the asymmetric units of both compounds are shown in Figure 7, and the crystallographic and refinement data are shown in Table 1.

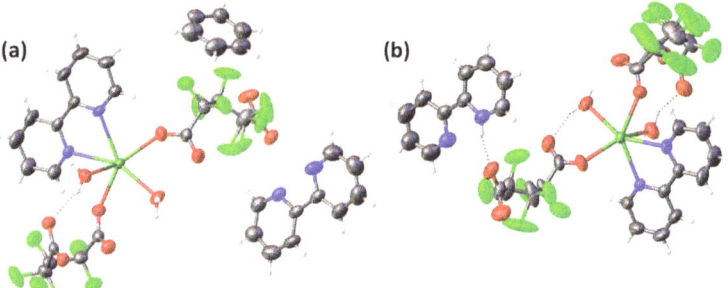

Figure 7. ORTEP-type diagrams with 50% of ellipsoid probability of the compounds (**a**) α-**Sm** and (**b**) β-**Sm**.

Table 1. Crystallographic data and refinement parameters for α-**Sm** and β-**Sm** compounds.

	α-**Sm**	β-**Sm**
Empirical formula	C$_{35}$H$_{24}$F$_{12}$N$_5$O$_{10}$Sm	C$_{30}$H$_{21}$F$_{12}$N$_4$O$_{10}$Sm
Formula weight (g/mol)	1052.94	975.86
Crystal system	triclinic	triclinic
Space group	P$\bar{1}$	P$\bar{1}$
a/Å	10.5754(4)	10.7751(10)
b/Å	12.8624(5)	13.3601(15)
c/Å	15.5452(7)	14.5413(17)
α(°)	70.371(4)	105.535(5)
β(°)	77.786(3)	98.863(5)
γ(°)	76.070(3)	114.104(5)
Volume/Å^3	1913.23(13)	1756.4(3)
Z	2	2
ρ$_{calc}$ mg/mm^3	1.828	1.845
μ/mm^{-1}	1.658	1.797
F(000)	1038	958
2θ range for data collection/°	5.04 to 69.18	6.31 to 51
Reflections collected	49,281	11,855
Independent reflections	15,269	6533
Data/restraints/parameters	15,269/12/570	6533/676/561
Goodness of fit on F^2	1.161	0.969
Final R index [I>2σ(I)]	0.0550	0.0522
Largest diff. Peak/hole/e.Å^{-3}	1.75/−1.66	1.13/−1.00

In both cases, α-**Sm** and β-**Sm** are formed by one crystallographically independent nine-coordinated Sm(III) center surrounded by two nitrogen atoms from a 2,2′-bipy, and seven oxygen atoms belonging to two HFG ligands and two water molecules to form a distorted trigonal prism, square-face tricapped polyhedron. In the first case, it is possible to observe one free 2,2′-bipy and a half protonated 2,2′-bipy (2,2′-bipyH) in the asymmetric unit, while just one protonated (2,2′-bipyH) molecule is presented in β-**Sm**. However, in both cases, the HFG ligand acts as a bridge between metallic centers through the $\mu_2\kappa^3$ and $\mu_2\kappa^2$ coordination modes, giving raise to 1D CPs. In α-**Sm**, chains grow along the [0 0 1] direction, whereas chains in β-**Sm** grow along the [1 1 1] direction (Figure 8). The presence of 2,2′-bipy and 2,2′-bipyH is observed in the inter-chain space, where they play role of counter ions as well as in the stabilization of the crystal packing.

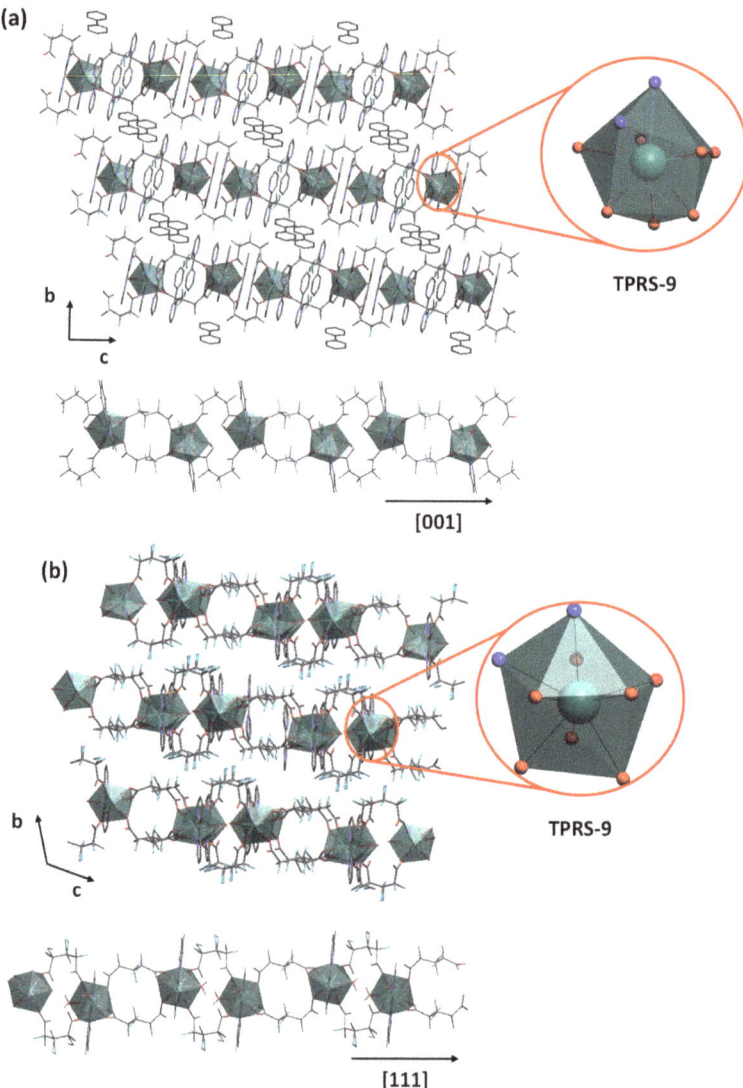

Figure 8. Crystal packing: 1D chains and coordination polyhedron for the compounds (**a**) α-**Sm** and (**b**) β-**Sm**.

As was previously mentioned, the synthesis conditions as well as the flexibility of the ligand and coordination modes determine the presence of different crystalline phases. In this case, the reaction conditions allow the formation of two crystalline phases, where there are clearly observed structural differences around the HFG ligand. Both structures have two HFG ligands coordinated to the metal center, but the carboxylate torsion angle with respect to the C–C scaffold significantly changes, as it is presented in Figure 9.

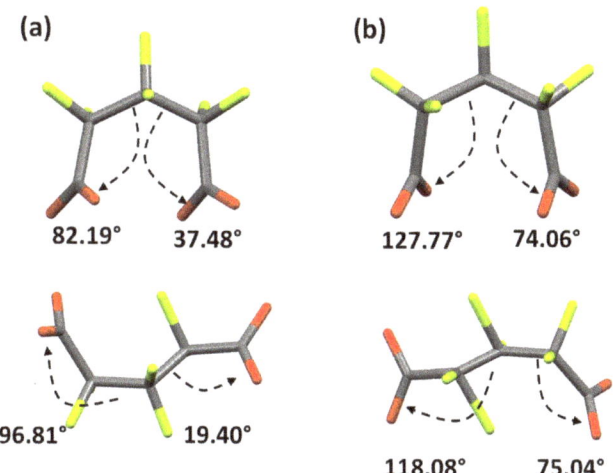

Figure 9. Torsion angles in the HFG ligands observed in the asymmetric unit for the compounds (a) α-Sm and (b) β-Sm.

3. Materials and Methods

A Cambridge Structural Database (CSD) search was performed in CSD 2021.2 (November 2021) using Conquest software (version 2021.20). To compute *Ln* and *An* complexes with ditopic ligands, we used the formula [M(ditopic)] as a query in the general search.

3.1. Synthesis of [2,2′-bipyH][Sm(HFG)₂ (2,2′-bipy) (H₂O)₂]•(2,2-bipy) and [2,2′-bipyH][Sm(HFG)₂ (2,2′-bipy) (H₂O)₂]

Both compounds were synthesized by mixing $Sm(NO_3)_3 \cdot 6H_2O$, H_2HFG and 2,2′-bipyridine in a 0.1:0.15:0.2 millimolar ratio in 4 mL of water. After that, a white suspension was achieved and one drop of concentrated nitric acid (65%) was added in order to obtain a transparent solution. Then, the mixture was left to stand at room temperature for slow solvent evaporation. After three months, the crystalline product was washed with distilled water and dried at room temperature (yield: 45 mg).

3.2. Single-Crystal X-ray Diffraction (SCXRD) for Structure Determination

SCXRD data for **α-Sm** were collected at room temperature (293(2) K) using MoK$_\alpha$ radiation (0.71073 Å) monochromated by graphite on a Rigaku XTALAB-MINI diffractometer. The unit cell determination as well as the final cell parameters were obtained on all reflections using *CrysAlisPro* software [92]. Data collection strategy, integration and scaling were performed using *CrysAlisPro* software [92]. **B-Sm** was collected at room temperature (293(2) K) using MoK$_\alpha$ radiation (0.71073 Å) monochromated by graphite on an Enraf–Nonius Kappa-CCD diffractometer. The initial cell refinements were performed using the software Collect [93] and Scalepack [94], and the final cell parameters were obtained on all reflections. Data reduction was carried out using the software Denzo-SMN and Scalepack [94].

The structures were solved and refined with SHELXT [95], and SHELXL [96], software, respectively, including in Olex2 [97]. In all cases, non-hydrogen atoms were clearly resolved and full-matrix least-squares refinement with anisotropic thermal parameters was performed. In addition, hydrogen atoms were stereochemically positioned and refined using the riding model in all cases [98]. The Mercury [99] program was used for the preparation of artwork. The structures were deposited in the CCDC database under the codes 2159499-2159500. Copies of the data can be obtained, free of charge, via www.ccdc.cam.ac.uk (accessed on 16 May 2022).

4. Conclusions

Although considerable research has been conducted on the design of new CPs, the use of flexible ditopic aliphatic ligands as structural support or even as a sensitizer for luminescence remains a viable strategy to developing a new generation of optically efficient Ln-CP- and An-CP-based materials. Combining aliphatic and aromatic auxiliary ligands or space holders seems to be a promising route for developing novel materials with luminescent properties. Additionally, the synthesis conditions as well as the diverse methodology approaches determine the variations in the promotion of ligand motion, which may result in the formation of different crystal structures and dimensionalities. One example of the use of a flexible ditopic ligand is shown in this work, where two CPs obtained at room temperature employ a hexafluoroglutarate ligand and samarium, $[2,2'$-bipyH][Sm(HFG)$_2$ (2,2'-bipy) (H$_2$O)$_2$]•(2,2'-bipy) (α-Sm) and $[2,2'$-bipyH][Sm(HFG)$_2$ (2,2'-bipy) (H$_2$O)$_2$] (β-Sm), and are reported for the first time herein. Although the chemistry of flexible ligand-based CPs is in constant growth, few studies on their incorporation into composites have been reported. As such, the future of Ln- and An-CPs is bright in terms of mesmerizing crystalline structures and exciting applications over the next decades.

Author Contributions: Conceptualization, R.F.D., G.E.G. and J.E.; methodology, R.F.D. and G.E.G.; software, R.F.D. and G.E.G.; validation, R.F.D., G.E.G. and J.E.; formal analysis, R.F.D. and G.E.G.; investigation, R.F.D. and G.E.G.; resources, R.F.D., G.E.G. and J.E.; data curation, R.F.D.; writing—original draft preparation, R.F.D. and G.E.G.; writing—review and editing, R.F.D., G.E.G. and J.E.; visualization, R.F.D. and G.E.G.; supervision, J.E.; project administration, R.F.D.; funding acquisition, R.F.D., G.E.G. and J.E. All authors have read and agreed to the published version of the manuscript.

Funding: This research received no external funding.

Institutional Review Board Statement: Not applicable.

Informed Consent Statement: Not applicable.

Data Availability Statement: Not applicable.

Acknowledgments: The authors acknowledge the Dirección General de Investigaciones from Universidad Santiago de Cali for financial support (Convocatoria Interna No. 01-2022 and No. 07-2022). This work was supported by the Consejo Nacional de Investigaciones Científicas y Técnicas, CONICET (PICT-2018-03583). G.E.G. is a member of Carrera del Investigador Científico (CIC-CONICET). J.E. is also grateful to Brazilian agencies FAPESP (Process No. 2017/15850-0) and CNPq (Process No. 305190/2017-2).

Conflicts of Interest: The authors declare no conflict of interest.

Sample Availability: Samples of the compounds are available from the authors.

References

1. Horike, S.; Nagarkar, S.S.; Ogawa, T.; Kitagawa, S. A New Dimension for Coordination Polymers and Metal–Organic Frameworks: Towards Functional Glasses and Liquids. *Angew. Chem. Int. Ed.* **2020**, *59*, 6652–6664. [CrossRef]
2. Batten, S.R.; Champness, N.R.; Chen, X.-M.; Garcia-Martinez, J.; Kitagawa, S.; Öhrström, L.; O'Keeffe, M.; Suh, M.P.; Reedijk, J. Coordination polymers, metal–organic frameworks and the need for terminology guidelines. *CrystEngComm* **2012**, *14*, 3001–3004. [CrossRef]
3. Bennett, T.D.; Horike, S. Liquid, glass and amorphous solid states of coordination polymers and metal—Organic frameworks. *Nat. Rev. Mater.* **2018**, *3*, 431–440. [CrossRef]

4. Kitagawa, S.; Matsuda, R. Chemistry of coordination space of porous coordination polymers. *Coord. Chem. Rev.* **2007**, *251*, 2490–2509. [CrossRef]
5. Janiak, C. Engineering coordination polymers towards applications. *Dalton Trans.* **2003**, *14*, 2781–2804. [CrossRef]
6. Liu, J.-Q.; Luo, Z.-D.; Pan, Y.; Kumar Singh, A.; Trivedi, M.; Kumar, A. Recent developments in luminescent coordination polymers: Designing strategies, sensing application and theoretical evidences. *Coord. Chem. Rev.* **2020**, *406*, 213145. [CrossRef]
7. Kuznetsova, A.; Matveevskaya, V.; Pavlov, D.; Yakunenkov, A.; Potapov, A. Coordination Polymers Based on Highly Emissive Ligands: Synthesis and Functional Properties. *Materials* **2020**, *13*, 2699. [CrossRef]
8. Bernini, M.C.; Brusau, E.V.; Narda, G.E.; Echeverria, G.E.; Pozzi, C.G.; Punte, G.; Lehmann, C.W. The Effect of Hydrothermal and Non-Hydrothermal Synthesis on the Formation of Holmium(III) Succinate Hydrate Frameworks. *Eur. J. Inorg. Chem.* **2007**, *2007*, 684–693. [CrossRef]
9. D'Vries, R.F.; Iglesias, M.; Snejko, N.; Alvarez-Garcia, S.; Gutierrez-Puebla, E.; Monge, M.A. Mixed lanthanide succinate-sulfate 3D MOFs: Catalysts in nitroaromatic reduction reactions and emitting materials. *J. Mater. Chem.* **2012**, *22*, 1191–1198. [CrossRef]
10. Manna, S.C.; Zangrando, E.; Bencini, A.; Benelli, C.; Chaudhuri, N.R. Syntheses, Crystal Structures, and Magnetic Properties of [LnIII$_2$(Succinate)$_3$(H$_2$O)$_2$]·0.5H$_2$O [Ln = Pr, Nd, Sm, Eu, Gd, and Dy] Polymeric Networks: Unusual Ferromagnetic Coupling in Gd Derivative. *Inorg. Chem.* **2006**, *45*, 9114–9122. [CrossRef]
11. Legendziewicz, J.; Keller, B.; Turowska-Tyrk, I.; Wojciechowski, W. Synthesis, optical and magnetic properties of homo- and heteronuclear systems and glasses containing them. *New J. Chem.* **1999**, *23*, 1097–1103. [CrossRef]
12. Hasegawa, Y.; Kitagawa, Y. Thermo-sensitive luminescence of lanthanide complexes, clusters, coordination polymers and metal–organic frameworks with organic photosensitizers. *J. Mater. Chem. C* **2019**, *7*, 7494–7511. [CrossRef]
13. Thuéry, P.; Harrowfield, J. Anchoring flexible uranyl dicarboxylate chains through stacking interactions of ancillary ligands on chiral U(vi) centres. *CrystEngComm* **2016**, *18*, 3905–3918. [CrossRef]
14. Mohan, M.; Essalhi, M.; Durette, D.; Rana, L.K.; Ayevide, F.K.; Maris, T.; Duong, A. A Rational Design of Microporous Nitrogen-Rich Lanthanide Metal–Organic Frameworks for CO2/CH4 Separation. *ACS Appl. Mater. Interfaces* **2020**, *12*, 50619–50627. [CrossRef]
15. Gorai, T.; Schmitt, W.; Gunnlaugsson, T. Highlights of the development and application of luminescent lanthanide based coordination polymers, MOFs and functional nanomaterials. *Dalton Trans.* **2021**, *50*, 770–784. [CrossRef]
16. Loukopoulos, E.; Kostakis, G.E. Review: Recent advances of one-dimensional coordination polymers as catalysts. *J. Coord. Chem.* **2018**, *71*, 371–410. [CrossRef]
17. Zhu, W.; Zhao, J.; Chen, Q.; Liu, Z. Nanoscale metal-organic frameworks and coordination polymers as theranostic platforms for cancer treatment. *Coord. Chem. Rev.* **2019**, *398*, 113009. [CrossRef]
18. Lawson, H.D.; Walton, S.P.; Chan, C. Metal–Organic Frameworks for Drug Delivery: A Design Perspective. *ACS Appl. Mater. Interfaces* **2021**, *13*, 7004–7020. [CrossRef]
19. Gomez, G.E.; Roncaroli, F. Photofunctional metal-organic framework thin films for sensing, catalysis and device fabrication. *Inorg. Chim. Acta* **2020**, *513*, 119926. [CrossRef]
20. Zhang, S.; Yang, Q.; Liu, X.; Qu, X.; Wei, Q.; Xie, G.; Chen, S.; Gao, S. High-energy metal–organic frameworks (HE-MOFs): Synthesis, structure and energetic performance. *Coord. Chem. Rev.* **2016**, *307*, 292–312. [CrossRef]
21. Zhao, D.; Yu, S.; Jiang, W.-J.; Cai, Z.-H.; Li, D.-L.; Liu, Y.-L.; Chen, Z.-Z. Recent Progress in Metal-Organic Framework Based Fluorescent Sensors for Hazardous Materials Detection. *Molecules* **2022**, *27*, 2226. [CrossRef]
22. Gomez, G.E.; Onna, D.; D'Vries, R.F.; Barja, B.C.; Ellena, J.; Narda, G.E.; Soler-Illia, G.J.A.A. Chain-like uranyl-coordination polymer as a bright green light emitter for sensing and sunlight driven photocatalysis. *J. Mater. Chem. C* **2020**, *8*, 11102–11109. [CrossRef]
23. D'Vries, R.F.; Álvarez-García, S.; Snejko, N.; Bausá, L.E.; Gutiérrez-Puebla, E.; de Andrés, A.; Monge, M.Á. Multimetal rare earth MOFs for lighting and thermometry: Tailoring color and optimal temperature range through enhanced disulfobenzoic triplet phosphorescence. *J. Mater. Chem. C* **2013**, *1*, 6316–6324. [CrossRef]
24. Gontcharenko, V.E.; Kiskin, M.A.; Dolzhenko, V.D.; Korshunov, V.M.; Taydakov, I.V.; Belousov, Y.A. Mono- and Mixed Metal Complexes of Eu^{3+}, Gd^{3+}, and Tb^{3+} with a Diketone, Bearing Pyrazole Moiety and CHF$_2$-Group: Structure, Color Tuning, and Kinetics of Energy Transfer between Lanthanide Ions. *Molecules* **2021**, *26*, 2655. [CrossRef]
25. Gomez, G.E.; Marin, R.; Carneiro Neto, A.N.; Botas, A.M.P.; Ovens, J.; Kitos, A.A.; Bernini, M.C.; Carlos, L.D.; Soler-Illia, G.J.A.A.; Murugesu, M. Tunable Energy-Transfer Process in Heterometallic MOF Materials Based on 2,6-Naphthalenedicarboxylate: Solid-State Lighting and Near-Infrared Luminescence Thermometry. *Chem. Mater.* **2020**, *32*, 7458–7468. [CrossRef]
26. Bünzli, J.-C.G. Review: Lanthanide coordination chemistry: From old concepts to coordination polymers. *J. Coord. Chem.* **2014**, *67*, 3706–3733. [CrossRef]
27. Ellart, M.; Blanchard, F.; Rivenet, M.; Abraham, F. Structural Variations of 2D and 3D Lanthanide Oxalate Frameworks Hydrothermally Synthesized in the Presence of Hydrazinium Ions. *Inorg. Chem.* **2020**, *59*, 491–504. [CrossRef]
28. Santos, G.C.; de Oliveira, C.A.F.; da Silva, F.F.; Alves, S. Photophysical studies of coordination polymers and composites based on heterometallic lanthanide succinate. *J. Mol. Struct.* **2020**, *1207*, 127829. [CrossRef]
29. Delgado, F.S.; Lorenzo-Luís, P.; Pasán, J.; Cañadillas-Delgado, L.; Fabelo, O.; Hernández-Molina, M.; Lozano-Gorrín, A.D.; Lloret, F.; Julve, M.; Ruiz-Pérez, C. Crystal growth and structural remarks on malonate-based lanthanide coordination polymers. *CrystEngComm* **2016**, *18*, 7831–7842. [CrossRef]

30. Kumar, M.; Qiu, C.-Q.; Zaręba, J.K.; Frontera, A.; Jassal, A.K.; Sahoo, S.C.; Liu, S.-J.; Sheikh, H.N. Magnetic, luminescence, topological and theoretical studies of structurally diverse supramolecular lanthanide coordination polymers with flexible glutaric acid as a linker. *New J. Chem.* **2019**, *43*, 14546–14564. [CrossRef]
31. Mills, E.L.; Pierce, K.A.; Jedrychowski, M.P.; Garrity, R.; Winther, S.; Vidoni, S.; Yoneshiro, T.; Spinelli, J.B.; Lu, G.Z.; Kazak, L.; et al. Accumulation of succinate controls activation of adipose tissue thermogenesis. *Nature* **2018**, *560*, 102–106. [CrossRef] [PubMed]
32. Hasegawa, Y.; Nakanishi, T. Luminescent lanthanide coordination polymers for photonic applications. *RSC Adv.* **2015**, *5*, 338–353. [CrossRef]
33. Yang, H.-W.; Xu, P.; Ding, B.; Wang, X.-G.; Liu, Z.-Y.; Zhao, H.-K.; Zhao, X.-J.; Yang, E.-C. Isostructural Lanthanide Coordination Polymers with High Photoluminescent Quantum Yields by Effective Ligand Combination: Crystal Structures, Photophysical Characterizations, Biologically Relevant Molecular Sensing, and Anti-Counterfeiting Ink Application. *Cryst. Growth Des.* **2020**, *20*, 7615–7625. [CrossRef]
34. Culy, C.R.; Clemett, D.; Wiseman, L.R. Oxaliplatin. *Drugs* **2000**, *60*, 895–924. [CrossRef] [PubMed]
35. Piro, O.E.; Baran, E.J. Crystal chemistry of organic minerals—Salts of organic acids: The synthetic approach. *Crystallogr. Rev.* **2018**, *24*, 149–175. [CrossRef]
36. Dazem, C.L.F.; Amombo Noa, F.M.; Nenwa, J.; Öhrström, L. Natural and synthetic metal oxalates—A topology approach. *CrystEngComm* **2019**, *21*, 6156–6164. [CrossRef]
37. Kahwa, I.A.; Fronczek, F.R.; Selbin, J. The Crystal structure and coordination geometry of potassium-catena-μ-oxalato-bis-oxalato aquo lanthanate(III) dihydrates, $K_3[Ln(Ox)_3(OH_2)] \cdot 2H_2O$, (Ln = Nd, Sm, Eu, Gd, Tb). *Inorg. Chim. Acta* **1984**, *82*, 161–166. [CrossRef]
38. Alexander, D.; Joy, M.; Thomas, K.; Sisira, S.; Biju, P.R.; Unnikrishnan, N.V.; Sudarsanakumar, C.; Ittyachen, M.A.; Joseph, C. Efficient green luminescence of terbium oxalate crystals: A case study with Judd-Ofelt theory and single crystal structure analysis and the effect of dehydration on luminescence. *J. Solid State Chem.* **2018**, *262*, 68–78. [CrossRef]
39. Xu, Q.-W.; Dong, G.; Cui, R.; Li, X. 3D lanthanide-coordination frameworks constructed by a ternary mixed-ligand: Crystal structure, luminescence and luminescence sensing. *CrystEngComm* **2020**, *22*, 740–750. [CrossRef]
40. Xu, C.; Tian, G.; Teat, S.J.; Rao, L. Complexation of U(VI) with Dipicolinic Acid: Thermodynamics and Coordination Modes. *Inorg. Chem.* **2013**, *52*, 2750–2756. [CrossRef]
41. Masci, B.; Thuéry, P. Uranyl complexes with the pyridine-2,6-dicarboxylato ligand: New dinuclear species with μ-η^2,η^2-peroxide, μ_2-hydroxide or μ_2-methoxide bridges. *Polyhedron* **2005**, *24*, 229–237. [CrossRef]
42. Harrowfield, J.M.; Lugan, N.; Shahverdizadeh, G.H.; Soudi, A.A.; Thuéry, P. Solid-State Luminescence and π-Stacking in Crystalline Uranyl Dipicolinates. *Eur. J. Inorg. Chem.* **2006**, *2006*, 389–396. [CrossRef]
43. Masci, B.; Thuéry, P. Pyrazinetetracarboxylic Acid as an Assembler Ligand in Uranyl–Organic Frameworks. *Cryst. Growth Des.* **2008**, *8*, 1689–1696. [CrossRef]
44. Thuéry, P.; Masci, B. Uranyl Ion Complexation by Cucurbiturils in the Presence of Perrhenic, Phosphoric, or Polycarboxylic Acids. Novel Mixed-Ligand Uranyl–Organic Frameworks. *Cryst. Growth Des.* **2010**, *10*, 716–725. [CrossRef]
45. Shu, Y.-B.; Xu, C.; Liu, W.-S. A Uranyl Hybrid Compound Designed from Urea-Bearing Dipropionic Acid. *Eur. J. Inorg. Chem.* **2013**, *2013*, 3592–3595. [CrossRef]
46. Wang, Y.; Yin, X.; Liu, W.; Xie, J.; Chen, J.; Silver, M.A.; Sheng, D.; Chen, L.; Diwu, J.; Liu, N.; et al. Emergence of Uranium as a Distinct Metal Center for Building Intrinsic X-ray Scintillators. *Angew. Chem. Int. Ed.* **2018**, *57*, 7883–7887. [CrossRef]
47. Cantos, P.M.; Frisch, M.; Cahill, C.L. Synthesis, structure and fluorescence properties of a uranyl-2,5-pyridinedicarboxylic acid coordination polymer: The missing member of the UO_2^{2+}-2,n-pyridinedicarboxylic series. *Inorg. Chem. Commun.* **2010**, *13*, 1036–1039. [CrossRef]
48. Frisch, M.; Cahill, C.L. Synthesis, structure and fluorescent studies of novel uranium coordination polymers in the pyridinedicarboxylic acid system. *Dalton Trans.* **2006**, *39*, 4679–4690. [CrossRef]
49. Gomez, G.E.; Ridenour, J.A.; Byrne, N.M.; Shevchenko, A.P.; Cahill, C.L. Novel Heterometallic Uranyl-Transition Metal Materials: Structure, Topology, and Solid State Photoluminescence Properties. *Inorg. Chem.* **2019**, *58*, 7243–7254. [CrossRef]
50. Liu, D.-D.; Wang, Y.-L.; Luo, F.; Liu, Q.-Y. Rare Three-Dimensional Uranyl–Biphenyl-3,3′-disulfonyl-4,4′-dicarboxylate Frameworks: Crystal Structures, Proton Conductivity, and Luminescence. *Inorg. Chem.* **2020**, *59*, 2952–2960. [CrossRef]
51. Yang, W.; Parker, T.G.; Sun, Z.-M. Structural chemistry of uranium phosphonates. *Coord. Chem. Rev.* **2015**, *303*, 86–109. [CrossRef]
52. Andrews, M.B.; Cahill, C.L. Uranyl Bearing Hybrid Materials: Synthesis, Speciation, and Solid-State Structures. *Chem. Rev.* **2013**, *113*, 1121–1136. [CrossRef] [PubMed]
53. Wang, K.-X.; Chen, J.-S. Extended Structures and Physicochemical Properties of Uranyl–Organic Compounds. *Acc. Chem. Res.* **2011**, *44*, 531–540. [CrossRef] [PubMed]
54. Loiseau, T.; Mihalcea, I.; Henry, N.; Volkringer, C. The crystal chemistry of uranium carboxylates. *Coord. Chem. Rev.* **2014**, *266*, 69–109. [CrossRef]
55. Gao, J.; Ye, K.; He, M.; Xiong, W.-W.; Cao, W.; Lee, Z.Y.; Wang, Y.; Wu, T.; Huo, F.; Liu, X.; et al. Tuning metal–carboxylate coordination in crystalline metal–organic frameworks through surfactant media. *J. Solid State Chem.* **2013**, *206*, 27–31. [CrossRef]
56. Lv, K.; Fichter, S.; Gu, M.; März, J.; Schmidt, M. An updated status and trends in actinide metal-organic frameworks (An-MOFs): From synthesis to application. *Coord. Chem. Rev.* **2021**, *446*, 214011. [CrossRef]

57. Duvieubourg-Garela, L.; Vigier, N.; Abraham, F.; Grandjean, S. Adaptable coordination of U(IV) in the 2D-(4,4) uranium oxalate network: From 8 to 10 coordinations in the uranium (IV) oxalate hydrates. *J. Solid State Chem.* **2008**, *181*, 1899–1908. [CrossRef]
58. Cañadillas-Delgado, L.; Pasán, J.; Fabelo, O.; Hernández-Molina, M.; Lloret, F.; Julve, M.; Ruiz-Pérez, C. Two- and Three-Dimensional Networks of Gadolinium(III) with Dicarboxylate Ligands: Synthesis, Crystal Structure, and Magnetic Properties. *Inorg. Chem.* **2006**, *45*, 10585–10594. [CrossRef]
59. Chrysomallidou, K.E.; Perlepes, S.P.; Terzis, A.; Raptopoulou, C.P. Synthesis, crystal structures and spectroscopic studies of praseodymium(III) malonate complexes. *Polyhedron* **2010**, *29*, 3118–3124. [CrossRef]
60. D'Vries, R.F.; Gomez, G.E.; Hodak, J.H.; Soler-Illia, G.J.A.A.; Ellena, J. Tuning the structure, dimensionality and luminescent properties of lanthanide metal–organic frameworks under ancillary ligand influence. *Dalton Trans.* **2016**, *45*, 646–656. [CrossRef]
61. Tong, M.L.; Chen, X.M. Chapter 8—Synthesis of Coordination Compounds and Coordination Polymers. In *Modern Inorganic Synthetic Chemistry*, 2nd ed.; Xu, R., Xu, Y., Eds.; Elsevier: Amsterdam, The Netherlands, 2017; pp. 189–217. [CrossRef]
62. Darwish, S.; Wang, S.-Q.; Croker, D.M.; Walker, G.M.; Zaworotko, M.J. Comparison of Mechanochemistry vs Solution Methods for Synthesis of 4,4'-Bipyridine-Based Coordination Polymers. *ACS Sustain. Chem. Eng.* **2019**, *7*, 19505–19512. [CrossRef]
63. Mirtamizdoust, B. Sonochemical synthesis of nano lead(II) metal-organic coordination polymer; New precursor for the preparation of nano-materials. *Ultrason. Sonochem.* **2017**, *35*, 263–269. [CrossRef]
64. Rizzato, S.; Moret, M.; Merlini, M.; Albinati, A.; Beghi, F. Crystal growth in gelled solution: Applications to coordination polymers. *CrystEngComm* **2016**, *18*, 2455–2462. [CrossRef]
65. Amiaud, T.; Stephant, N.; Dessapt, R.; Serier-Brault, H. Microwave-assisted synthesis of anhydrous lanthanide-based coordination polymers built upon benzene-1,2,4,5-tetracarboxylic acid. *Polyhedron* **2021**, *204*, 115261. [CrossRef]
66. Lin, Z.; Slawin, A.M.Z.; Morris, R.E. Chiral Induction in the Ionothermal Synthesis of a 3-D Coordination Polymer. *J. Am. Chem. Soc.* **2007**, *129*, 4880–4881. [CrossRef] [PubMed]
67. Hu, M.-L.; Masoomi, M.Y.; Morsali, A. Template strategies with MOFs. *Coord. Chem. Rev.* **2019**, *387*, 415–435. [CrossRef]
68. D'Vries, R.F.; de la Peña-O'Shea, V.A.; Benito Hernández, Á.; Snejko, N.; Gutiérrez-Puebla, E.; Monge, M.A. Enhancing Metal–Organic Framework Net Robustness by Successive Linker Coordination Increase: From a Hydrogen-Bonded Two-Dimensional Supramolecular Net to a Covalent One Keeping the Topology. *Cryst. Growth Des.* **2014**, *14*, 5227–5233. [CrossRef]
69. Tanaka, D.; Kitagawa, S. Template Effects in Porous Coordination Polymers. *Chem. Mater.* **2008**, *20*, 922–931. [CrossRef]
70. Xu, N.; Shi, W.; Liao, D.-Z.; Yan, S.-P.; Cheng, P. Template Synthesis of Lanthanide (Pr, Nd, Gd) Coordination Polymers with 2-Hydroxynicotinic Acid Exhibiting Ferro-/Antiferromagnetic Interaction. *Inorg. Chem.* **2008**, *47*, 8748–8756. [CrossRef]
71. Xu, H.; Li, Y. The organic ligands as template: The synthesis, structures and properties of a series of the layered structure rare-earth coordination polymers. *J. Mol. Struct.* **2004**, *690*, 137–143. [CrossRef]
72. Bernini, M.C.; Snejko, N.; Gutierrez-Puebla, E.; Brusau, E.V.; Narda, G.E.; Monge, M.Á. Structure-Directing and Template Roles of Aromatic Molecules in the Self-Assembly Formation Process of 3D Holmium–Succinate MOFs. *Inorg. Chem.* **2011**, *50*, 5958–5968. [CrossRef] [PubMed]
73. D'Vries, R.F.; Camps, I.; Ellena, J. Exploring the System Lanthanide/Succinate in the Formation of Porous Metal–Organic Frameworks: Experimental and Theoretical Study. *Cryst. Growth Des.* **2015**, *15*, 3015–3023. [CrossRef]
74. Bernini, M.C.; Gomez, G.E.; Brusau, E.V.; Narda, G.E. Reviewing Rare Earth Succinate Frameworks from the Reticular Chemistry Point of View: Structures, Nets, Catalytic and Photoluminescence Applications. *Isr. J. Chem.* **2018**, *58*, 1044–1061. [CrossRef]
75. Gomez, G.E.; Brusau, E.V.; Kaczmarek, A.M.; Mellot-Draznieks, C.; Sacanell, J.; Rousse, G.; Van Deun, R.; Sanchez, C.; Narda, G.E.; Soler Illia, G.J.A.A. Flexible Ligand-Based Lanthanide Three-Dimensional Metal–Organic Frameworks with Tunable Solid-State Photoluminescence and OH-Solvent-Sensing Properties. *Eur. J. Inorg. Chem.* **2017**, *2017*, 2321–2331. [CrossRef]
76. Gomez, G.E.; Bernini, M.C.; Brusau, E.V.; Narda, G.E.; Vega, D.; Kaczmarek, A.M.; Van Deun, R.; Nazzarro, M. Layered exfoliable crystalline materials based on Sm-, Eu- and Eu/Gd-2-phenylsuccinate frameworks. Crystal structure, topology and luminescence properties. *Dalton Trans.* **2015**, *44*, 3417–3429. [CrossRef]
77. Janicki, R.; Mondry, A.; Starynowicz, P. Carboxylates of rare earth elements. *Coord. Chem. Rev.* **2017**, *340*, 98–133. [CrossRef]
78. Chen, B.; Yang, Y.; Zapata, F.; Lin, G.; Qian, G.; Lobkovsky, E.B. Luminescent Open Metal Sites within a Metal–Organic Framework for Sensing Small Molecules. *Adv. Mater.* **2007**, *19*, 1693–1696. [CrossRef]
79. Beeby, A.; Clarkson, I.M.; Dickins, R.S.; Faulkner, S.; Parker, D.; Royle, L.; de Sousa, A.S.; Gareth Williams, J.A.; Woods, M. Non-radiative deactivation of the excited states of europium, terbium and ytterbium complexes by proximate energy-matched OH, NH and CH oscillators: An improved luminescence method for establishing solution hydration states. *J. Chem. Soc. Perkin Trans.* **1999**, *2*, 493–504. [CrossRef]
80. Xiong, T.; Zhang, Y.; Amin, N.; Tan, J.-C. A Luminescent Guest@MOF Nanoconfined Composite System for Solid-State Lighting. *Molecules* **2021**, *26*, 7583. [CrossRef]
81. Sun, G.; Xie, Y.; Sun, L.; Zhang, H. Lanthanide upconversion and downshifting luminescence for biomolecules detection. *Nanoscale Horiz.* **2021**, *6*, 766–780. [CrossRef]
82. Bünzli, J.-C.G.; Eliseeva, S.V. Lanthanide NIR luminescence for telecommunications, bioanalyses and solar energy conversion. *J. Rare Earths* **2010**, *28*, 824–842. [CrossRef]
83. Chuasaard, T.; Ngamjarurojana, A.; Surinwong, S.; Konno, T.; Bureekaew, S.; Rujiwatra, A. Lanthanide Coordination Polymers of Mixed Phthalate/Adipate for Ratiometric Temperature Sensing in the Upper-Intermediate Temperature Range. *Inorg. Chem.* **2018**, *57*, 2620–2630. [CrossRef] [PubMed]

84. Godoy, A.A.; Bernini, M.C.; Funes, M.D.; Sortino, M.; Collins, S.E.; Narda, G.E. ROS-generating rare-earth coordination networks for photodynamic inactivation of Candida albicans. *Dalton Trans.* **2021**, *50*, 5853–5864. [CrossRef] [PubMed]
85. Godoy, A.A.; Gomez, G.E.; Miranda, C.D.; Illescas, M.; Barja, B.C.; Vega, D.; Bernini, M.C.; Narda, G.E. Strong Red Up-Conversion Emission in Thin Film Devices Based on Rare-Earth Oxides Obtained from Templating 2D Coordination Networks. *Eur. J. Inorg. Chem.* **2022**, *9*, e202101025. [CrossRef]
86. Denning, R.G. Electronic Structure and Bonding in Actinyl Ions and their Analogs. *J. Phys. Chem. A* **2007**, *111*, 4125–4143. [CrossRef] [PubMed]
87. Thuéry, P.; Harrowfield, J. Uranyl–Organic Frameworks with Polycarboxylates: Unusual Effects of a Coordinating Solvent. *Cryst. Growth Des.* **2014**, *14*, 1314–1323. [CrossRef]
88. Vaidhyanathan, R.; Natarajan, S.; Rao, C.N.R. A chiral mixed carboxylate, [Nd$_4$(H$_2$O)$_2$(OOC(CH$_2$)$_3$COO)$_4$(C$_2$O$_4$)$_2$], exhibiting NLO properties. *J. Solid State Chem.* **2004**, *177*, 1444–1448. [CrossRef]
89. Maouche, R.; Belaid, S.; Benmerad, B.; Bouacida, S.; Daiguebonne, C.; Suffren, Y.; Freslon, S.; Bernot, K.; Guillou, O. Highly Luminescent Europium-Based Heteroleptic Coordination Polymers with Phenantroline and Glutarate Ligands. *Inorg. Chem.* **2021**, *60*, 3707–3718. [CrossRef]
90. Hussain, S.; Khan, I.U.; Harrison, W.T.A.; Tahir, M.N. Crystal structures and characterization of two rare-earth-glutarate coordination networks: One-dimensional [Nd(C$_5$H$_6$O$_4$)(H$_2$O)$_4$]•Cl and three-dimensional [Pr(C$_5$H$_6$O$_4$)(C$_5$H$_7$O$_4$)(H$_2$O)]•H$_2$O. *J. Struct. Chem.* **2015**, *56*, 934–941. [CrossRef]
91. Zhang, Y.; Li, X.; Li, Y. Hydrothermal syntheses, crystal structures and luminescence properties of two lanthanide dinuclear complexes with hexafluoroglutarate. *J. Coord. Chem.* **2009**, *62*, 583–592. [CrossRef]
92. CrysAlisPro. *CrysAlisPro*; Agilent Technologies Ltd.: Yarnton, Oxfordshire, UK, 2014.
93. Hooft, R.W.W. *COLLECT*; Nonius BV: Delft, The Netherlands, 1998.
94. Otwinowski, Z.; Minor, W. Processing of X-ray diffraction data collected in oscillation mode. In *Methods Enzymol*; Academic Press: Cambridge, MA, USA, 1997; Volume 276, pp. 307–326.
95. Sheldrick, G. SHELXT—Integrated space-group and crystal-structure determination. *Acta Crystallogr. Sect. A* **2015**, *71*, 3–8. [CrossRef] [PubMed]
96. Sheldrick, G. Crystal structure refinement with SHELXL. *Acta Crystallogr. Sect. C* **2015**, *71*, 3–8. [CrossRef] [PubMed]
97. Dolomanov, O.V.; Bourhis, L.J.; Gildea, R.J.; Howard, J.A.K.; Puschmann, H. OLEX2: A complete structure solution, refinement and analysis program. *J. Appl. Crystallogr.* **2009**, *42*, 339–341. [CrossRef]
98. Sheldrick, G. A short history of SHELX. *Acta Crystallogr. Sect. A* **2008**, *64*, 112–122. [CrossRef] [PubMed]
99. Macrae, C.F.; Bruno, I.J.; Chisholm, J.A.; Edgington, P.R.; McCabe, P.; Pidcock, E.; Rodriguez-Monge, L.; Taylor, R.; Van De Streek, J.; Wood, P.A. Mercury CSD 2.0—New features for the visualization and investigation of crystal structures. *J. Appl. Crystallogr.* **2008**, *41*, 466–470. [CrossRef]

Article

Fluorescent Zn(II)-Based Metal-Organic Framework: Interaction with Organic Solvents and CO_2 and Methane Capture

Sifani Zavahir [1], Hamdi Ben Yahia [2], Julian Schneider [3], DongSuk Han [1], Igor Krupa [1], Tausif Altamash [2], Mert Atilhan [4], Abdulkarem Amhamed [2,*] and Peter Kasak [1,*]

1. Center for Advanced Materials, Qatar University, Doha P.O. Box 2713, Qatar; fathima.z@qu.edu.qa (S.Z.); dhan@qu.edu.qa (D.H.); igor.krupa@qu.edu.qa (I.K.)
2. Qatar Environment & Energy Research Institute, Hamad Bin Khalifa University, Doha 34110, Qatar; benyahia_hamdi@yahoo.fr (H.B.Y.); taltamash@hbku.edu.qa (T.A.)
3. Department of Materials Science and Engineering, and Center for Functional Photonics (CFP), City University of Hong Kong, 83 Tat Chee Avenue, Hong Kong 999077, China; julianschneider86@gmx.de
4. Department of Chemical and Paper Engineering, Western Michigan University, Kalamazoo, MI 49008, USA; mert.atilhan@wmich.edu
* Correspondence: aamhamed@hbku.edu.qa (A.A.); peter.kasak@qu.edu.qa (P.K.)

Citation: Zavahir, S.; Ben Yahia, H.; Schneider, J.; Han, D.; Krupa, I.; Altamash, T.; Atilhan, M.; Amhamed, A.; Kasak, P. Fluorescent Zn(II)-Based Metal-Organic Framework: Interaction with Organic Solvents and CO_2 and Methane Capture. *Molecules* 2022, 27, 3845. https://doi.org/10.3390/molecules27123845

Academic Editors: Hiroshi Sakiyama and Ana Margarida Gomes da Silva

Received: 10 May 2022
Accepted: 2 June 2022
Published: 15 June 2022

Publisher's Note: MDPI stays neutral with regard to jurisdictional claims in published maps and institutional affiliations.

Copyright: © 2022 by the authors. Licensee MDPI, Basel, Switzerland. This article is an open access article distributed under the terms and conditions of the Creative Commons Attribution (CC BY) license (https://creativecommons.org/licenses/by/4.0/).

Abstract: Adsorption of carbon dioxide (CO_2), as well as many other kinds of small molecules, is of importance for industrial and sensing applications. Metal-organic framework (MOF)-based adsorbents are spotlighted for such applications. An essential for MOF adsorbent application is a simple and easy fabrication process, preferably from a cheap, sustainable, and environmentally friendly ligand. Herein, we fabricated a novel structural, thermally stable MOF with fluorescence properties, namely Zn [5-oxo-2,3-dihydro-5H-[1,3]-thiazolo [3,2-a]pyridine-3,7-dicarboxylic acid (TPDCA)] • dimethylformamide (DMF) •0.25 H_2O (coded as QUF-001 MOF), in solvothermal conditions by using zinc nitrate as a source of metal ion and TPDCA as a ligand easy accessible from citric acid and cysteine. Single crystal X-ray diffraction analysis and microscopic examination revealed the two-dimensional character of the formed MOF. Upon treatment of QUF-001 with organic solvents (such as methanol, isopropanol, chloroform, dimethylformamide, tetrahydrofuran, hexane), interactions were observed and changes in fluorescence maxima as well as in the powder diffraction patterns were noticed, indicating the inclusion and intercalation of the solvents into the interlamellar space of the crystal structure of QUF-001. Furthermore, CO_2 and CH_4 molecule sorption properties for QUF-001 reached up to 1.6 mmol/g and 8.1 mmol/g, respectively, at 298 K and a pressure of 50 bars.

Keywords: MOF; citric acid derivative; TPDCA

1. Introduction

Carbon dioxide sequestration technology is in line with recommendations of the IPCC Special Report, which set a goal for the reduction of global carbon dioxide emissions by about 45% by 2030 and achievement of net-zero by 2050 compared to 2010 levels [1,2]. Additionally, as greenhouse gases have a direct impact on human health and the environment, safe separation of hazardous gases such as methane and carbon dioxide is essential but technically challenging [3–6]. In order to strengthen the renewable gas separation technology, there have been several attempts focused on the use of various biomolecule-based (citric, malic, and lactic acid) materials [7–9]. Similarly, Zn ion-based metal-organic frameworks (MOFs) were gaining interest in gas capturing and separation applications [10–12].

In general, MOFs have drawn keen research interest as a family of crystalline materials; they consist of inorganic metal ions connected with organic ligands such as carboxylate, phosphonate or heteroaromatics and others [13,14]. The coordination ability of the ligands is the main factor that determines the assembly of MOFs into one-, two-, or three-dimensional architectures that subsequently dictate their properties and possible application [15–17]. Due to tunable architecture, high surface area, and porosity [18], MOFs have already found vast utilization in catalysis [19], gas storage [20] and separation [21], (bio)imaging, drug delivery, optoelectronics, and sensing [22,23]. One of the important classes of MOFs is the luminescent type, where fluorescence or phosphorescence results from the absorption of light at a radiative excitation state that leads to photon emission. Inherent porosity and a precise crystal structure are advantages that determine the selectivity of interaction for sensing applications of luminescent MOFs [24], which can be used for sensing explosives, ions, biomolecules, toxic and volatile organic compounds, temperature, pH, etc. [25,26]. Luminescence can originate from the organic linker and metal node [27] as the luminescent part or by inserting luminescent guest molecules [28]. One of the most commonly studied origins of luminescence in MOFs is emission from the organic linker, where the ligand molecule acts as a rigid structural component and simultaneously provides the emissive character of the structure. Thus, such ligands as a building moiety are one of the most crucial components of luminescent MOFs, and there is high demand for cheap, sustainable, and environmentally benign ligands with tailored properties.

Citric acid is a natural, sustainable, cheap, and functional low-molecular-weight carbon source [29]. Recently, there has been extensive research interest in citric acid-based functional materials [30–32], such as citric-based carbon dots; this molecule was also identified as a source of a low-molecular-weight fluorophore formed under certain synthetic conditions [33–36], as well as major luminescent molecular fluorophore [37–39]. This is because citric acid can form stable fluorophore derivatives by reaction with amines [40], α,β-diamines, α-amino acids, and α,β-heteroatom amines [41]. Thus, dicarboxylic acid derivatives as potential ligands for MOFs can be prepared in reaction with some multifunctional natural amino acids. In the reaction of citric acid with natural amino acid L-cysteine, 5-oxo-2,3-dihydro-5H-[1,3]-thiazolo [3,2-a]pyridine-3,7-dicarboxylic acid (TPDCA) [42,43] has been synthesized and further applied as a component of fluorescent biodegradable polymers [44,45] and other soft materials [46–49]. TPDCA also has been identified as a key component for induction of gelation of natural polysaccharide-alginate [50] and related carbon dots [51,52]. In this work, TPDCA was employed as a ligand for the formation of different MOF structures. The crystal structure was confirmed for Zn-based structure as QUF-001, and used for solvent interaction investigation. Moreover, carbon dioxide and methane gases have been successfully tested to quantify the adsorption–desorption range in prepared Zn-based QUF-001 structures at high pressures and temperatures.

2. Results and Discussion

Zn-, Cu- and Fe-based MOFs were synthesized using the TPDCA ligand, and the obtained samples were coded as QUF-001, QUF-002 and QUF-003, respectively (Figure S1). The three MOF samples were prepared using DMF as a solvent as well as a base precursor for carboxylate formation in solvothermal conditions. The three central elements were chosen on the basis of their respective importance in the human body, as they can lead to essential bio-inorganic materials [53–55]. These different central metal nodes were expected to result in slightly different geometry and open channels in the host crystal structure.

X-ray powder diffraction (XRPD) patterns of the prepared samples are depicted in Figure 1a. Only the Zn-MOF (QUF-001) was well-crystallized with sharp diffraction peaks. Therefore, a full pattern matching was performed on this MOF only, using the JANA 2006 program, which confirmed the purity of this compound as depicted in Figure S2. Furthermore, the refined cell parameters were in good agreement with those obtained from single crystal XRD data (Table 1). The Cu-MOF (QUF-002), although formed, was not as well-crystallized as the Zn-based one, as its XRPD pattern (Figure 1a) showed significant peak broadening. The crystal structure of Cu-MOF could not be solved from XRPD, and all attempts to grow single crystals failed. No diffraction peaks were observed for Fe-based QUF-003, which indicated that the sample was amorphous and the Fe-MOF was not formed.

This information is in line with the visual observation where the solid powder of QUF-003 appeared to be a precipitate rather than a polycrystalline substance.

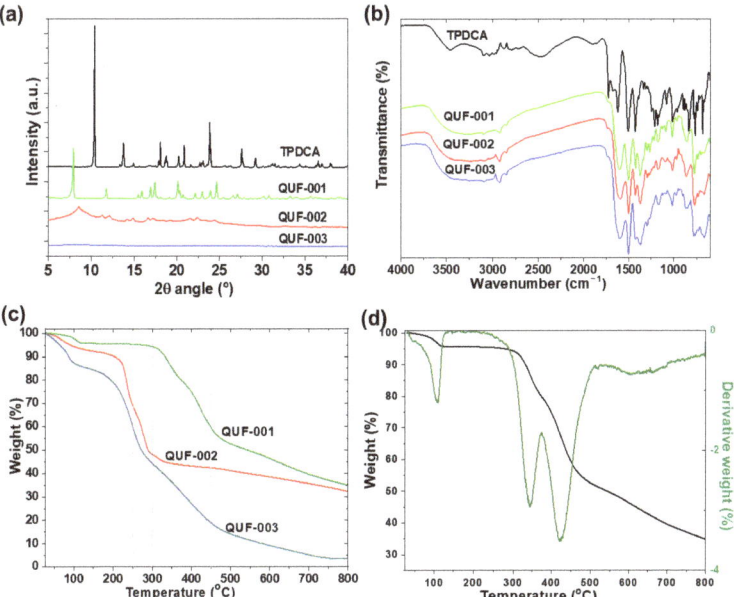

Figure 1. (**a**) XRPD patterns and (**b**) FTIR spectra of TPDCA ligand, QUF-001, QUF-002 and QUF-003. (**c**) TGA curves of QUF-001, QUF-002 and QUF-003. (**d**) TGA and DTA curves of QUF-001.

Table 1. Crystal data and structure refinement of QUF-001.

Crystal Data	
Chemical formula	$C_{12}H_{12}N_2O_6SZn \times 0.25H_2O$
M_r	382.2
Crystal system, space group	Triclinic, P-1
Temperature (K)	293
a, b, c (Å)	6.4459 (3), 10.4427 (5), 11.2947 (5)
α, β, γ (°)	88.693 (2), 83.751 (3), 77.741 (2)
V (Å^3)	738.52 (6)
Z	2
Radiation type	Mo $K\alpha$
μ (mm^{-1})	1.84
Crystal size (mm)	$0.06 \times 0.03 \times 0.01$
Refinement	
$R[F^2 > 2\sigma(F^2)], wR(F^2), S$	0.042, 0.106, 1.09
No. of reflections	3269
No. of parameters	203
No. of restraints	0
H-atom treatment	H-atom parameters constrained
$\Delta\rho_{max}, \Delta\rho_{min}$ (e Å^{-3})	0.93, −0.62

FTIR spectra of the three MOF samples had similar peaks; they were partly different from that of the TPDCA ligand pattern as depicted in Figure 1b. TPDCA showed several FTIR bands not present in the MOF structures, centered at 3450, 2500, and 1724 cm^{-1}, which could be assigned to O-H stretching, intermolecular bonding from O-H stretching, and C=O stretching of carboxylic acid groups, respectively. The absence of such bands in the MOF structures was due to transformation of the carboxylic acid group of TPDCA to carboxylate during slow formation of the base from DMF decomposition, as well as due to formation of multidentate ligand making coordination bonds with the metal central node. FTIR absorption signals at 3042, 1630 1517, 1430, 1070 and 680 cm^{-1} in TPDCA were attributed to aromatic C-H stretching, amidic C=O stretching, C-N stretching, C-O stretching, C-N bending and S-C stretching modes [56], respectively. The amidic C=O stretching vibration overlapped with the carboxylate stretching vibration at 1607, 1603 and 1604 cm^{-1} for Zn-, Cu- and Fe-based structures [56], respectively. Moreover, a broad peak related to hydrogen bonding was centered at 3300 cm^{-1}. On the other hand, the ligand TPDCA differed in absorption peaks for hydrogen bonding at 3460 cm^{-1} and an additional absorption peak belonging to the carboxylic acid functionality at 1732 cm^{-1}. This observation assumed that the coordination of the TPDCA ligand to the central metal ion in the three MOF structures was due to similar carboxylate functionalities.

The thermogravimetric analysis (TGA) data for the three samples are given in Figure 1c. The three samples exhibited significantly different thermal behaviors, which was expected since their XRPD patterns were significantly different. QUF-003 was the least heat-stable sample, with ~15%, ~55% and ~90% weight losses at 150 °C, 300 °C, and 800 °C, respectively. This indicated a large water or solvent inclusion that was related to loss at 150 °C and further organic residue degradation at relatively low temperature until 300 °C. Additionally, stable residue of only about 10% after heating to 800 °C indicated low abundance of metallic-inorganic components in sample. Sample QUF-002 contained 6% moisture and volatile compounds, and further decomposition started at 200 °C, reaching 50% of initial mass. Residual mass of 67% after heating to 800 °C indicated Cu as the metallic component in MOF, whereas QUF-001 was the most stable sample, with only ~4%, ~45% and ~65% of weight lost at 300 °C, 500 °C, and 800 °C, respectively. The first-order derivative curve of the TGA of QUF-001 is given in Figure 1d, and shows three weight loss points. The first one at 110 °C with an initial weight loss of 4% can be attributed to the removal of moisture/trapped H$_2$O and DMF molecules from the channels and pores of this MOF. Two distinctly separated peaks at 330 °C and 420 °C accounting for ~45% weight loss remained during the thermal decomposition of the sample, and it could be assumed that led to the collapse of the crystal structure.

Thus, the analysis based on the techniques discussed above suggests that out of the three samples prepared in this study, only QUF-001 based on Zn exhibited high crystallinity and good thermal stability. Hence, QUF-001 was further employed to evaluate the solvent inclusion properties and gas storage capacity of the MOF based on the TPDCA ligand.

SEM images of QUF-001 demonstrate a well-ordered lamellar-like structure as visualized in Figure 2a–d. The material possesses a two-dimensional character and lamellar structure with low magnitude cracks between them, indicating lower interaction in interlamellar spacing. On the other hand, the polycrystalline form and X-ray single-crystal refinement measurements performed on QUF-001 revealed that the compound $C_{12}H_{12}N_2O_6SZn \bullet 0.25H_2O$ crystallizes in the triclinic crystal system with the space group P-1 (Table 1 and Figure 2e). Most of the atomic positions were found by the direct method using SIR2004 [57]. With isotropic atomic displacement parameters (ADPs), the residual factors converged to the value $R(F) = 0.0987$ and $wR(F^2) = 0.2061$ for 89 refined parameters and 1762 observed reflections. At this stage of the refinement, the chemical formula $C_{12}N_2O_6SZn$ could not be equilibrated yet. After adding H atoms and applying restrictions on their positions and (ADPs), the chemical formula became $C_{12}H_{12}N_2O_6SZn$, and the residual factors decreased only slightly to $R(F) = 0.0956$ and $wR(F^2) = 0.1971$. By refining the anisotropic ADPs of all atoms except the H atoms, the residual factors converged to the value $R(F) = 0.0473$, $wR(F^2) = 0.1103$ and S= 1.59 for 199 refined parameters. The Fourier difference showed a very weak electron density residue along the MOF tunnels. Therefore, a water molecule was included in the crystal structure. The refinement of the occupancy of oxygen from the water molecule showed a significant decrease from 1 to 0.25. Consequently the occupancies of the water molecule were restricted to 0.25, leading to the chemical formula $C_{12}H_{12}N_2O_6SZn \bullet 0.25H_2O$, for which the final residual factors were for $R(F)$, $wR(F^2)$ and S values 0.042, 0.106, 1.09, respectively, as provided in Tables 1 and S1. Further crystallographic data for the atomic positions and anisotropic ADPs are given in Tables S2 and S3, respectively. Further details on the structural refinement may be obtained from the Cambridge Crystallographic Data Centre (CCDC), by quoting the Registry No. CCDC 2120295 [58].

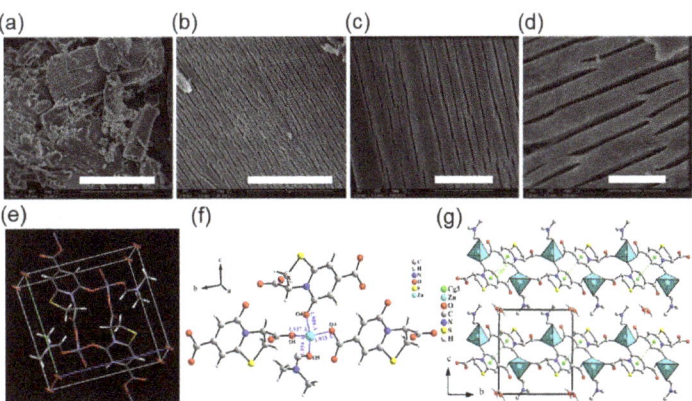

Figure 2. SEM images of QUF-001 at different magnifications, namely (**a**) 25,000×, (**b**) 50,000×, (**c**) 100,000× and (**d**) 200,000×, with scale bars indicating 5 microns, 3 microns, 1 micron and 500 nm, respectively. (**e**) Scheme showing the Zn(II) coordination in the QUF-001 unit cell and (**f**) the coordination sphere of the zinc cation. (**g**) View of the layered structure of QUF-001 along the *a*-axis. The green dashed lines correspond to π–π interactions between the six-membered rings forming TPDCA linkers. $Cg3$ is the centroid of the six-membered ring.

Since a structural disorder of the water molecules was observed along the MOF tunnels, it could indicate that the *P*-1 symmetry was higher than the true symmetry. Therefore, a second refinement was conducted using the space group *P*1. The atomic positions from the first refinement were used as a starting model. By reducing the symmetry from *P*-1 to *P*1, we doubled the number of atoms. The refinement led to residual factors very similar to those from the first refinement; however, most of the atoms displayed a non-positive, definite ADP matrix. Consequently, the MOF structure was considered to be centrosymmetric (space group *P*-1), where each carboxylate group formed from carboxylic acid by decomposition of DMF binds to one Zn(II) in a monodentate fashion. Zn(II) ions have a tetrahedral coordination geometry with two oxygen atoms from a particular carboxylate from TPDCA molecules and two coordination complexes from amidic oxygen atoms from TPDCA and dimethylformamide molecules, as shown in Figure 2b. Each zinc atom interconnects three TPDCA molecules at different positions and one DMF molecule (Figure 2f), forming a 2D framework in the (001) plane (Figure 2g). The Zn-O distances for O1, O3, O4 and O5 are 1.937(3) 1.933(3), 1.987(3) and 1.955(3) Å, respectively, and the angles between carbon, oxygen and zinc atoms for C1-O1-Zn1, C7-O3-Zn1, C9-O4-Zn1 and C10-O5-Zn1 are 118.8(3), 118.3(3), 132.0(2) and 121.5(3)°, respectively. A packing diagram of QUF-001, viewed down the *a*-axis, is given in Figure 2g. Coordination in crystallographic unit consists of two 14-atom rings, and a two-dimensional structure with Zn(II) centered coordination is due to a peripheral DMF molecule oriented to coordination lamellar structure, as shown in Figure 2d. It is worth noting that no hydrogen bonds connecting the different layers were observed. Even the offset π–π interactions between TPDCA molecules [$Cg3$-$Cg3i$ = 3.758(2) Å, interplanar distance = 3.3618 (16) Å, slippage = 1.679 Å, α = 0°, $Cg3$ was the centroid of the six-membered ring, symmetry code (i): 1-X,-Y,1-Z] existed only within a single layer and not between layers, which confirms that the structure is bi-dimensional in the (001) plane (see the green dashed lines in Figure 2d).

After confirming the crystallographic structure of Zn-MOF, QUF-001 was activated to remove possible intercalated solvent molecules. During the MOF fabrication, the solvent trapping into pores was rather obvious. Thus, activation is important to have complete accessibility of the pores and to obtain the guest free pores. Thermal activation is a simple and effective method that works well with the vast majority of MOF materials. For QUF-001, heating at 90 °C for 16 h was found to be the perfect temperature versus time combination. The structural integrity of the MOF lattice during such activation was confirmed by XRPD as given in Figure 3a, with complete retention of the crystallinity. At higher temperatures, the framework of QUF-001 tended to collapse. TGA analysis as provided in Figure 3b shows clear benefits of the activation process: about 4% of initial weight loss was observed in the as-synthesized Zn-MOF below 120 °C, while the material tested for thermal stability after activation showed very high thermal resistance with less than 1% (~1%) weight loss up to 275 °C. It is noteworthy that the crystal structure of QUF-001 contained also a molecule of DFM with a boiling point of 156 °C, and this was about 19% of the total weight. However, TGA analysis showed stability up to 300 °C with weight loss of 4% corresponding to water released at around 100 °C. This assumed that the coordinated DMF molecule was not released and was strongly coordinated in the Zn coordination sphere and released from samples only after 300 °C, with the peak from derivate TA at 330°C as shown in Figure 1d. A further peak from DTA at 420 °C can be attributed to thermal degradation of the TPDCA segment in QUF-001.

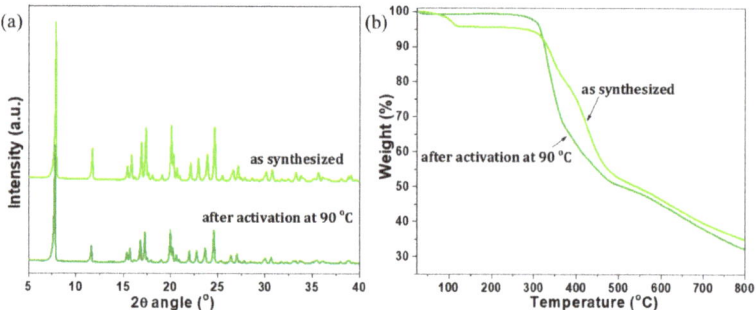

Figure 3. (a) XRPD patterns and (b) TGA curves of QUF-001, as synthesized and after activation at 90 °C for 16 h.

Inclusion of different organic solvents into the activated sample of QUF-001 was examined. In the testing process, activated QUF-001 particles with 10–12 nm average particle size distribution (calculated based on Scherrer equation) were kept mechanically agitated for a certain time in the respective solvent; the progress of the inclusion was followed by performing XRPD measurements after Day 1 and Day 3, with the data summarized in Figure 4.

After Day 1, QUF-001 mixed with the solvents acetonitrile, tetrahydrofuran (THF), benzene and dichloromethane (DCM) showed peak splitting around $2\theta = 7.92°$, as can be seen in Figure 4, curves a, b, c, and d. For hexadecane, no peak splitting was observed at low angle, indicating the presence of a single phase (Figure 4, curve e). Furthermore, since the (001) peak was strongly shifted to lower angle with $2\theta = 7.78°$, the obtained phase was probably a pure intercalated one. With methanol and isopropanol, a broadening of the (001) peak indicated a loss in crystallinity (Figure 4, curves f and g).

After Day 3, QUF-001 mixed with the solvents acetonitrile, THF, benzene and hexadecane still showed peak splitting around $2\theta = 7.92°$, as can be seen in Figure 4, curves a, b, c and e. This peak splitting was most probably due to the coexistence of the initial and the intercalated phase. Furthermore, the positions of the new peak (below $2\theta = 7.92°$) were at a lower angle compared to the (001) peak of QUF-001. This indicated an enlargement of the inter-reticular distance d_{001} and the c cell parameter, which was most probably due to the inclusion of the solvent in the QUF-001 structure. It should be mentioned that this process is reversible and intercalated solvent can penetrate and introduce the other phase formation. Similarly, such solvent stimulus response studies on MOF were performed previously on Zn-, Cd- and other metal-based MOFs [59–61]. With DCM and methanol, a broadening of the peak (001) was observed due to the loss in crystallinity (Figure 4, curves d and f). With isopropanol, a significant change in the pattern was observed compared to the other samples, and the (001) peak disappeared. This could be due either to the decomposition of the QUF-001 phase and formation of the ligand TPDCA (Figure S3), or to a structural change.

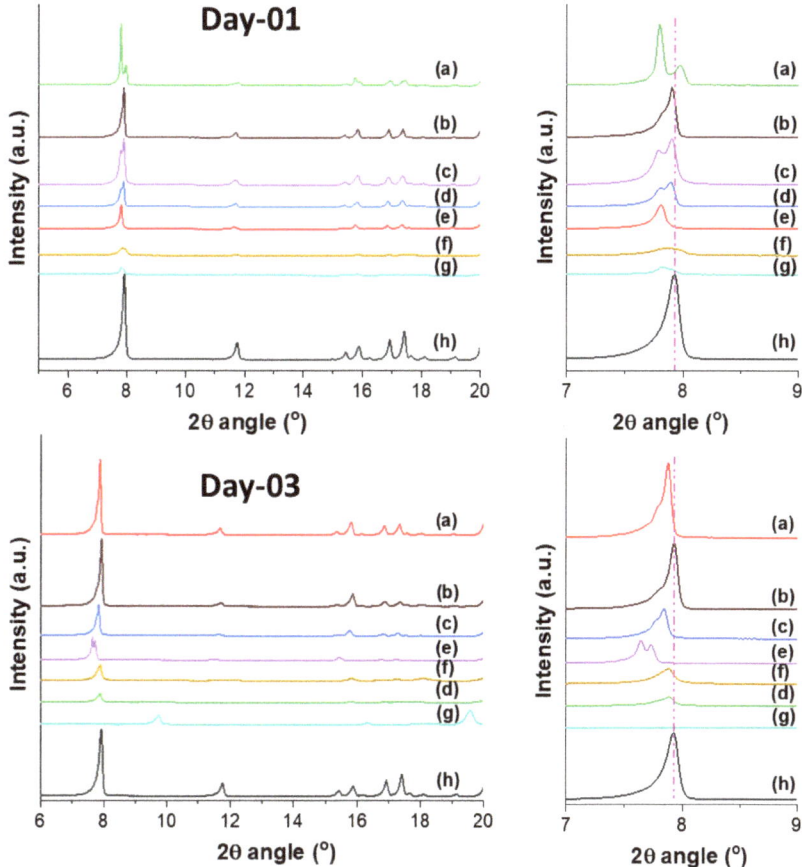

Figure 4. XRPD patterns taken from the QUF-001 mixed with different solvents after Day 1 (upper frames) and Day 3 (bottom frames). The solvents are: (**a**) acetonitrile, (**b**) THF, (**c**) benzene, (**d**) DCM, (**e**) hexadecane, (**f**) methanol, (**g**) isopropanol; patterns (**h**) belong to pure (no solvent) activated QUF-001. Frames on the right-hand side show enlarged view of the (001) low 2θ angle peak(s).

As the next step, luminescence properties of QUF-001 were examined after incorporation of the solvents. TPDCA ligand itself exhibited a photoluminescence (PL) emission peak at 450 nm while as-prepared QUF-001 has a peak at 475 nm upon excitation at 350 nm (Figure S4). It is apparent from Figure 5 that after mixing QUF-001 with different solvents, alteration in the position of peak maxima appeared.

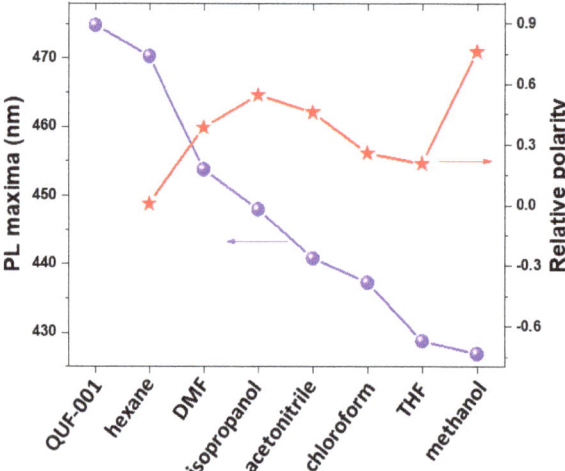

Figure 5. Trend in the PL maxima peak positions for activated QUF-001 and QUF-001 mixed with different solvents (violet cycles). All samples excited at 350 nm and compared with the solvent polarity (red stars).

The effect of the used solvent on PL spectra was observed for QUF-001 and showed PL emission maxima of methanol located at 426 nm, followed further in order by THF (429.7 nm), chloroform (436 nm), acetonitrile (441 nm), isopropanol (446 nm), DMF (454 nm) and hexane (470 nm) (Figure 5).

Hexane was the most non-polar solvent employed in the present study and had PL maxima at 470 nm. Alternatively, methanol was the most polar and had PL maxima located at 426 nm; PL maxima of QUF-001 with other solvents were in between, but not in a directional order. Hence, it is reasonable to assume PL emission maxima depend on the collective effects of solvent polarity, hydrogen bonding, size of the solvent molecule and the interaction of hetero atoms in the solvent molecule with the hetero atoms in the QUF-001 MOF structure. This can be ascribed to a solvatochromic effect; however, it should be pointed out that XRD indicated changes that could have resulted from new phase formation. Thus, the changes in luminescence properties do not need to be the result of solvatochromism of QUF-001 but may also be due to other phases with different optical properties. Nevertheless, this approach can be potentially employed for sensing materials, since it gives distinct spectral lines for the individual solvent media they are in that are not overlapping.

We then studied the methane storage and CO_2 capture capacity of QUF-001. Pre-programmed gas sorption–desorption measurements made from 0–50 bar pressure for adsorption and reversing back to zero for desorption resulted in a cumulative 20 data points for a complete cycle. According to N_2 gas adsorption–desorption measurements shown in Figure S5, QUF-001 exhibited slight hysteresis in low and high vapor pressure regions, with a Langmuir surface area of 2.85 m^2/g. We noticed that the sample had a BET surface area of 1.9078 m^2/g, which was relatively low. It also had a BJH desorption pore diameter of ~40 nm and t-plot pore volume of 0.000610 cm^3/g. The measured surface area was low for a porous material with open networks of MOF type. As can be seen from the SEM images of the MOF (Figure 2a), solvent inclusion and gas storage behavior may arise from the lamellar channels in the structure.

We then investigated the sorption capacity of CO_2 and CH_4 gases onto QUF-001 as depicted in Figure 6. High-pressure experiments were performed at 298 and 318 K isotherms. A complete adsorption–desorption cycle passed through stepwise pressure increases and decreases with each adsorbate was carried out from vacuum to 50 bars and back to vacuum at the end of the measurements to observe the hysteresis behavior. At each isotherm, there were a total of 12 adsorption and 8 desorption data points collected for QUF-001, which are presented in Figures S6–S9. At first glance, all of those curves demonstrated a smooth increasing trend with increasing pressure. However, adsorption and desorption data points showed almost the same values, which showed that there was no hysteresis and no significant changes occurred in the samples during the overall pressure loop. We noticed that in order to obtain reliable data, peripheral conditions such as humidity, ambient pressure, and temperature had to be considered [7]. To prevent potential irreversible structure collapse and reduction of the surface area as well as pore volume due to moisture within the measurement chamber, samples were degassed. Furthermore, a Drierite column was used to pre-dry the gases. A similar sorption–desorption overlapped trend of variation was observed while studying a $(NH_4)_2Mg(H_2P_2O_7)_2 \bullet 2H_2O$ single-crystal sample with CO_2 and CH_4 at isotherms 298 and 318 K [62]. Figures S9 and S10 confirm the expected thermodynamic trend of variation, namely that the sorptivity of both used gases decreased upon increasing the temperature and increased upon increasing the pressure. However, all of the isotherms with CO_2 and CH_4 with QUF-001 were completely reversible and the absence of hysteresis confirmed the advantages of reusability and cost efficiency of this MOF under primary vacuum [3,7,63]. Sorption results showed that CO_2 (1.6413 mmol/g) sorption was significantly lower than that for CH_4 (8.0907 mmol/g) at temperature 298 K and pressure of 50 bars, which meant that QUF-001 had higher affinity to capture CH_4 as compared to CO_2. Clear evidence of QUF-001 CH_4 sorption selectivity over CO_2 gas at each temperature and pressure showed that the use of this MOF may be beneficial for the chemical and petroleum industries in terms of CH_4 separation from CH_4/CO_2 mixtures [7]. From Figure 6, at temperature 298 K, between 45 to 50 bar pressure, the sorption curve seems flattened. Typically, MOFs, covalent organic frameworks and covalent organic polymers follow type IV adsorption isotherms, showing finite multi-layer adsorption corresponding to complete filling of the capillaries and pores [64]. The adsorption isotherm profiles (Figure 6) rather fell within the type III behavior, indicating weak substrates and the formation of multilayers. Here, there was no flattish region in the curve assuming lack of a monolayer. In one of our studies, the CO_2 sorption capacities of $Rb_2Co(H_2P_2O_7)_2 \bullet 2H_2O$ were higher than those of the currently investigated sample, but CH_4 sorption efficiency was 3.5-fold better in QUF-001. Although there was no clear superiority among either CO_2, CH_4 or other gases for MOF structures, the trend in ranking the sorption performance of such gases showed that the CO_2 capture performance of MOFs was higher than that for methane [65–67]. We also showed a similar trend through gas sorption demonstrations on MOF-5 previously [68,69]. Moreover, on comparing the CO_2 sorption data of this work with hydroxy metal carbonates $M(CO_3)_x(OH)_y$ (M = Zn, Zn-Mg, Mg, Mg-Cu, Cu, Ni, and Pb) [70], $Rb_2Co(H_2P_2O_7)_2 \bullet 2H_2O$ showed higher values at 35 bar and 318 K, although the hydroxy metal carbonates were measured at 316 K.

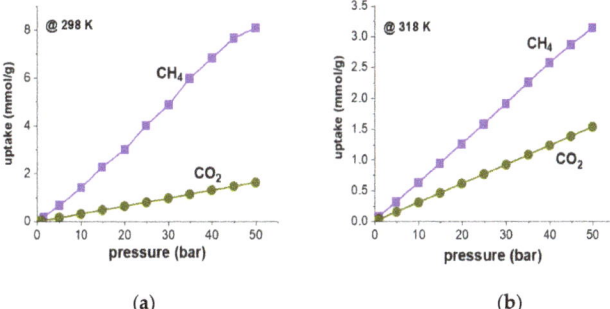

(a)   (b)

Figure 6. Plots of the CO_2 (violet squares) and CH_4 gas (green cycles) adsorption behavior of QUF-001and the uptake of these two gases as a function of the pressure, at two different temperatures of 298 K (**a**) and 318 K (**b**).

3. Materials and Methods

3.1. Chemicals

Citric acid, L-cysteine, N,N-dimethylformamide (DMF), $Zn(NO_3)_2\bullet 6H_2O$, $Cu(NO_3)_2\bullet 3H_2O$, $Fe(NO_3)_3\bullet 9H_2O$, 37% hydrochloric acid, absolute ethanol, acetone, acetonitrile, benzene, methanol, tetrahydrofuran (THF), dichloromethane (DCM), hexadecane (HD), propanol, chloroform and isopropanol were purchased at the highest purity level available from Sigma Aldrich and used as received without further purification. Deionized water was obtained from Millipore system.

3.2. Preparation of MOFs

(Step 1) Synthesis of the ligand, TPDCA.

TPDCA ligand was prepared according to the procedure published previously [39] (Figure S1). A 28.7 g amount of anhydrous citric acid was mixed with 18.1 g of L-cysteine in a flat-bottom flask and autoclaved at 150 °C for 4 h. The reaction product was separated and recrystallized in acetone. The structure of the TPDCA was confirmed by ^{1}H NMR (Figure S10) and ^{13}C NMR (Figure S11) prior to employing it in the subsequent MOF preparation.

^{1}H NMR (δ,d_6 DMSO, 600 MHz) ppm: 13.58 (broad singlet, 2H, -COO-H), 6.56 (doublet, J = 1.5 Hz, 1H, Ar-H), 6.51 (doublet, J = 1.5 Hz, 1H, Ar-H), 5.43 (doublet- doublet, J = 1.5 & 8.5 Hz, 1H, HOOC-C-H), 3.87 (doublet- doublet, J = 8.5 & 11.6 Hz, 1H, H-C-H), 3.57 (doublet-doublet, J = 1.5 & 11.6 Hz, 1H, H-C-H).

^{13}C NMR ^{1}H NMR (δ,d_6 DMSO, 150 MHz) ppm: 169.2, 165.6, 160.6, 149.9, 142.7, 115.0, 97.94, 62.6, 31.6.

(Step 2) Synthesis of the MOFs: In a typical synthesis process, 50.0 mg of the TPDCA ligand and 61.72 mg of $Zn(NO_3)_2.6H_2O$ for QUF-001, 50.01 mg of $Cu(NO_3)_2$ for QUF-002 and 83.63 mg of $Fe(NO_3)_3$ for QUF-003 were placed into a dry flask. Then, 10 mL of DMF was added and acidified with a drop of concentrated HCl acid. The solution was incubated in an oven at 80 °C for 96 h under Ar atmosphere. The solids formed were separated by centrifugation and dried under vacuum at 60 °C for 16 h. Reaction yield was 72, 52 and 45% for QUF-001, QUF-002 and QUF-003, respectively.

3.3. Characterization

X-ray powder diffraction (XRPD) patterns of the samples were recorded on a PANalytical empyrean machine equipped with Cu Kα radiation as the X-ray source [71]. Measurements were made between 5 and 40° 2θ angles. Fourier transform infra-red (FTIR) measurements were performed on a PerkinElmer Frontier device with ZnSe ATR unit in the wavenumber region from 4000–500 cm^{-1} in the transmittance mode at a scan rate of 64 scans per cycle [72]. Thermal stability of the samples was assessed using thermogravimetric analysis (TGA) employing 10 mg sample on a TGA 4000 device by PerkinElmer under N_2 atmosphere [73]. Scanning electron microscopy (SEM) imaging was performed on a ZEISS SIGMA 500 VP FE SEM device under different magnifications [74].

3.4. Single Crystal X-ray Diffraction Measurements

Single crystals of QUF-001 suitable for X-ray diffraction were selected on the basis of the size and sharpness of their diffraction spots. Data collection was carried out on a D8 venture diffractometer using MoKα radiation. Data processing and all refinements were performed with the Jana2006 program package [75]. A multi-scan-type absorption correction was applied using SADABS [76], and the crystal shape was determined using a microscope.

3.5. Solvent Inclusion

Two milligrams of activated QUF-001 were placed in a glass vial filled with 2 mL of the respective solvent. Solvents used were acetonitrile, benzene, methanol, THF, DCM, hexadecane, and isopropanol. The contents in the vials were aged for 1 day and 3 days.

For XRD measurements, solid powder was separated from the solvent by decanting, and the solid was spread on a Petri dish to facilitate air drying for 2 h; dried powders were collected and characterized.

3.6. Gas Sorption Measurements

Ten milligrams of QUF-001 were dried overnight at 90 °C for activation of the MOF material and stored in a desiccator connected to vacuum until the crystals were used. Gas adsorption–desorption tests were performed for pure CO_2 and CH_4 gases and the mixture of CO_2 and CH_4 on a Rubotherm magnetic suspension sorption apparatus (MSA), which operated on the basis of Archimedes' buoyancy principle. Pressure in the sample bucket was increased from vacuum to a predefined high-pressure value (0–50 bar) in the adsorption process, and was reversed for the desorption. At the very beginning of the test, the sample bucket was held at vacuum for 10 h to ensure complete surface degassing. Gas adsorption–desorption measurements were performed at two different temperatures, 298 K and 318 K. Buoyancy correction was carried out for the sorption measurements as well. The details of the correction process were explained in detail previously [6].

Pressure transducers (Paroscientific, Redmond, WA, USA) worked from vacuum up to 350 bar with an uncertainty of 0.01% of the full scale ($u(P)$ 0.035 bar), whereas the temperature sensor (Minco PRT, Fridley, MN, USA) had a measurement accuracy of 0.5 K ($u(T) = 0.05$ K).

4. Conclusions

In summary, we successfully synthesized fluorescent and thermally stable Zn(II)-based MOF as a probe for solvent and gas adsorption. Fluorescent organic ligand TPDCA obtained from easily accessible, sustainable precursors citric acid and cysteine was applied for the first time for MOF structure fabrication. Single crystal analysis of the QUF-001 sample confirmed the two-dimensional lamellar structure of this MOF, where TPDCA ligand coordinated to Zn(II) central atoms, along with the incorporation of DMF. Solvent molecules inclusion led to tunability of the diffraction pattern as well as shifts in emission maxima, which indicated the solvent inclusion within the interlamellar space in the two-dimensional QUF-001 structure. Gas sorption properties of QUF-001 for CO_2 and CH_4 were examined and determined at 1.6 mmol/g and 8.1 mmol/g, respectively, at temperature 298 K and the pressure of 50 bars. Thus, our study offers a platform for the application of cheap and accessible ligand TPDCA for fluorescent MOFs with tailored properties.

Supplementary Materials: The following supporting information can be downloaded at: https://www.mdpi.com/article/10.3390/molecules27123845/s1, Figure S1. Synthesis of TPDCA and related MOF. Figure S2. Full pattern matching of the XRPD pattern of the QUF-001 compound. Figure S3. XRPD patterns of (a) TPDCA, (b) QUF-001 with the solvent isopropanol and (c) pure QUF-001. Figure S4. Emission spectra of TPDCA (red line) and QUF-001 (black line), excited at 375 nm. Figure S5. N_2 gas adsorption–desorption measurements of QUF-001. Figure S6: CO_2 absorption–desorption plot at 25 °C for QUF-001. Figure S7: CO_2 absorption–desorption plot at 45 °C for QUF-001.Figure S8: CH_4 absorption–desorption plot at 25 °C for QUF-001. Figure S9: CH_4 absorption–desorption plot at 45 °C for QUF-001.Figure S10. ^{1}H spectrum of TPDCA. Figure S11. ^{13}C NMR spectra of TPDCA. Table S1. Structure refinement for QUF-001. Table S2. Fractional atomic coordinates and isotropic atomic displacement parameters (Å2) for C12H12N2O6SZn × 0.25H2O. Table S3. Anisotropic displacement parameters (Å2) for QUF-001. The anisotropic displacement factor exponent takes the form: $-2\pi^2[(ha^*)2U_{11}+...+2hka^*b^*U_{12}]$.

Author Contributions: P.K., conceptualization; P.K. and A.A., methodology; S Z., J.S., H.B.Y. and M.A., formal analysis; P.K., D.H., I.K. and A.A., resources. P.K. and A.A. evaluated data. T.A. performed the gas adsorption experiments, data analysis and interpretation. P.K., H.B.Y., S.Z. and T.A. wrote the first version of the manuscript. All authors revised manuscript, discussed the results and contributed to the final version of the paper. All authors have read and agreed to the published version of the manuscript.

Funding: This work was made possible by NPRP 12-Cluster grant # [NPRP12C-0821-190017] and NPRP grant # NPRP13S-0202-200228 from the Qatar National Research Fund (a member of Qatar Foundation). The findings achieved herein are solely the responsibility of the authors. This research was made possible by a grant from the Qatar National Research Fund under its National Priorities Research Program (award number NPRP12S-0311-190299) and by financial support from the ConocoPhillips Global Water Sustainability Center (GWSC) and Qatar Petrochemical Company (QAPCO). The paper's content is solely the responsibility of the authors and does not necessarily represent the official views of the Qatar National Research Fund or ConocoPhillips and QAPCO.

Institutional Review Board Statement: Not applicable.

Informed Consent Statement: Not applicable.

Data Availability Statement: Not applicable.

Acknowledgments: SEM and NMR were accomplished in the Central Laboratories unit, Qatar University.

Conflicts of Interest: The authors declare no conflict of interest.

References

1. Livingston, J.E.; Rummukainen, M. Taking science by surprise: The knowledge politics of the IPCC Special Report on 1.5 degrees. *Environ. Sci. Policy* **2020**, *112*, 10–16. [CrossRef]
2. Djalante, R. Key assessments from the IPCC special report on global warming of 1.5 C and the implications for the Sendai framework for disaster risk reduction. *Prog. Disaster Sci.* **2019**, *1*, 100001. [CrossRef]
3. Amhamed, A.; Atilhan, M.; Berdiyorov, G. Permeabilities of CO_2, H_2S and CH_4 through choline-based ionic liquids: Atomistic-scale simulations. *Molecules* **2019**, *24*, 2014. [CrossRef] [PubMed]
4. Al-Tamreh, S.A.; Ibrahim, M.H.; El-Naas, M.H.; Vaes, J.; Pant, D.; Benamor, A.; Amhamed, A. Electroreduction of carbon dioxide into formate: A comprehensive review. *ChemElectroChem* **2021**, *8*, 3207–3220. [CrossRef]
5. Tariq, M.; Soromenho, M.R.; Rebelo, L.P.N.; Esperança, J.M. Insights into CO_2 hydrates formation and dissociation at isochoric conditions using a rocking cell apparatus. *Chem. Eng. Sci.* **2022**, *249*, 117319. [CrossRef]
6. Rozyyev, V.; Thirion, D.; Ullah, R.; Lee, J.; Jung, M.; Oh, H.; Atilhan, M.; Yavuz, C.T. High-capacity methane storage in flexible alkane-linked porous aromatic network polymers. *Nat. Energy* **2019**, *4*, 604–611. [CrossRef]
7. Altamash, T.; Amhamed, A.I.; Aparicio, S.; Atilhan, M. Combined Experimental and Theoretical Study on High Pressure Methane Solubility in Natural Deep Eutectic Solvents. *Ind. Eng. Chem. Res.* **2019**, *58*, 8097–8111. [CrossRef]
8. Altamash, T.; Nasser, M.S.; Elhamarnah, Y.; Magzoub, M.; Ullah, R.; Qiblawey, H.; Aparicio, S.; Atilhan, M. Gas solubility and rheological behavior study of betaine and alanine based natural deep eutectic solvents (NADES). *J. Mol. Liq.* **2018**, *256*, 286–295. [CrossRef]
9. Altamash, T.; Nasser, M.S.; Elhamarnah, Y.; Magzoub, M.; Ullah, R.; Anaya, B.; Aparicio, S.; Atilhan, M. Gas solubility and rheological behavior of natural deep eutectic solvents (NADES) via combined experimental and molecular simulation techniques. *Chem. Select* **2017**, *2*, 7278–7295. [CrossRef]
10. Senkovska, I.; Kaskel, S. High pressure methane adsorption in the metal-organic frameworks $Cu_3(btc)_2$, $Zn_2(bdc)_2dabco$, and $Cr_3F(H_2O)_2O(bdc)_3$. *Microporous Mesoporous Mater.* **2008**, *112*, 108–115. [CrossRef]
11. Becker, T.M.; Heinen, J.; Dubbeldam, D.; Lin, L.-C.; Vlugt, T.J. Polarizable force fields for CO_2 and CH_4 adsorption in M-MOF-74. *J. Phys. Chem. C* **2017**, *121*, 4659–4673. [CrossRef] [PubMed]
12. Ursueguía, D.; Díaz, E.; Ordóñez, S. Metal-Organic Frameworks (MOFs) as methane adsorbents: From storage to diluted coal mining streams concentration. *Sci. Total Environ.* **2021**, *790*, 148211. [CrossRef] [PubMed]
13. Paz, F.A.A.; Klinowski, J.; Vilela, S.M.; Tome, J.P.; Cavaleiro, J.A.; Rocha, J. Ligand design for functional metal–organic frameworks. *Chem. Soc. Rev.* **2012**, *41*, 1088–1110.
14. Lin, Z.-J.; Lü, J.; Hong, M.; Cao, R. Metal–organic frameworks based on flexible ligands (FL-MOFs): Structures and applications. *Chem. Soc. Rev.* **2014**, *43*, 5867–5895. [CrossRef]
15. Cui, Y.; Li, B.; He, H.; Zhou, W.; Chen, B.; Qian, G. Metal–organic frameworks as platforms for functional materials. *Acc. Chem. Res.* **2016**, *49*, 483–493. [CrossRef] [PubMed]
16. Razavi, S.A.A.; Morsali, A. Linker functionalized metal-organic frameworks. *Coord. Chem. Rev.* **2019**, *399*, 213023. [CrossRef]
17. Furukawa, H.; Cordova, K.E.; O'Keeffe, M.; Yaghi, O.M. The chemistry and applications of metal-organic frameworks. *Science* **2013**, *341*, 1230444. [CrossRef]
18. Ji, Z.; Wang, H.; Canossa, S.; Wuttke, S.; Yaghi, O.M. Pore chemistry of metal–organic frameworks. *Adv. Funct. Mater.* **2020**, *30*, 2000238. [CrossRef]
19. Goetjen, T.A.; Liu, J.; Wu, Y.; Sui, J.; Zhang, X.; Hupp, J.T.; Farha, O.K. Metal–organic framework (MOF) materials as polymerization catalysts: A review and recent advances. *Chem. Commun.* **2020**, *56*, 10409–10418. [CrossRef]
20. Eddaoudi, M.; Kim, J.; Rosi, N.; Vodak, D.; Wachter, J.; O'Keeffe, M.; Yaghi, O.M. Systematic design of pore size and functionality in isoreticular MOFs and their application in methane storage. *Science* **2002**, *295*, 469–472. [CrossRef]

21. Konstas, K.; Osl, T.; Yang, Y.; Batten, M.; Burke, N.; Hill, A.J.; Hill, M.R. Methane storage in metal organic frameworks. *J. Mater. Chem.* **2012**, *22*, 16698–16708. [CrossRef]
22. Wales, D.J.; Grand, J.; Ting, V.P.; Burke, R.D.; Edler, K.J.; Bowen, C.R.; Mintova, S.; Burrows, A.D. Gas sensing using porous materials for automotive applications. *Chem. Soc. Rev.* **2015**, *44*, 4290–4321. [CrossRef]
23. Dhakshinamoorthy, A.; Li, Z.; Garcia, H. Catalysis and photocatalysis by metal organic frameworks. *Chem. Soc. Rev.* **2018**, *47*, 8134–8172. [CrossRef]
24. Dou, Z.; Yu, J.; Cui, Y.; Yang, Y.; Wang, Z.; Yang, D.; Qian, G. Luminescent metal–organic framework films as highly sensitive and fast-response oxygen sensors. *J. Am. Chem. Soc.* **2014**, *136*, 5527–5530. [CrossRef]
25. Altamash, T.; Ahmed, W.; Rasool, S.; Biswas, K.H. Intracellular Ionic Strength Sensing Using NanoLuc. *Int. J. Mol. Sci.* **2021**, *22*, 677. [CrossRef]
26. Skorjanc, T.; Shetty, D.; Valant, M. Covalent organic polymers and frameworks for fluorescence-based sensors. *ACS Sens.* **2021**, *6*, 1461–1481. [CrossRef]
27. Allendorf, M.D.; Bauer, C.A.; Bhakta, R.; Houk, R. Luminescent metal–organic frameworks. *Chem. Soc. Rev.* **2009**, *38*, 1330–1352. [CrossRef]
28. Xiong, T.; Zhang, Y.; Amin, N.; Tan, J.-C. A Luminescent Guest@ MOF Nanoconfined Composite System for Solid-State Lighting. *Molecules* **2021**, *26*, 7583. [CrossRef]
29. Salihu, R.; Abd Razak, S.I.; Zawawi, N.A.; Kadir, M.R.A.; Ismail, N.I.; Jusoh, N.; Mohamad, M.R.; Nayan, N.H.M. Citric acid: A green cross-linker of biomaterials for biomedical applications. *Eur. Polym. J.* **2021**, *146*, 110271. [CrossRef]
30. Shan, D.; Hsieh, J.T.; Bai, X.; Yang, J. Citrate-Based Fluorescent Biomaterials. *Adv. Healthcare Mater.* **2018**, *7*, 1800532. [CrossRef] [PubMed]
31. Xiao, L.; Sun, H. Novel properties and applications of carbon nanodots. *Nanoscale Horiz.* **2018**, *3*, 565–597. [CrossRef]
32. Chung, Y.J.; Kim, J.; Park, C.B. Photonic carbon dots as an emerging nanoagent for biomedical and healthcare applications. *ACS Nano* **2020**, *14*, 6470–6497. [CrossRef] [PubMed]
33. Schneider, J.; Reckmeier, C.J.; Xiong, Y.; von Seckendorff, M.; Susha, A.S.; Kasák, P.; Rogach, A.L. Molecular fluorescence in citric acid-based carbon dots. *J. Phys. Chem. C* **2017**, *121*, 2014–2022. [CrossRef]
34. Xiong, Y.; Zhang, X.; Richter, A.F.; Li, Y.; Döring, A.; Kasák, P.; Popelka, A.; Schneider, J.; Kershaw, S.V.; Yoo, S.J. Chemically synthesized carbon nanorods with dual polarized emission. *ACS Nano* **2019**, *13*, 12024–12031. [CrossRef]
35. Liang, T.; Liu, E.; Li, M.; Ushakova, E.V.; Kershaw, S.V.; Rogach, A.L.; Tang, Z.; Qu, S. Morphology control of luminescent carbon nanomaterials: From dots to rolls and belts. *ACS Nano* **2020**, *15*, 1579–1586. [CrossRef]
36. Qu, D.; Sun, Z. The formation mechanism and fluorophores of carbon dots synthesized via a bottom-up route. *Mater. Chem. Front.* **2020**, *4*, 400–420. [CrossRef]
37. Wang, Y.; Zhuang, Q.; Ni, Y. Facile microwave-assisted solid-phase synthesis of highly fluorescent nitrogen–sulfur-codoped carbon quantum dots for cellular imaging applications. *Chem. Eur. J.* **2015**, *21*, 13004–13011. [CrossRef]
38. Reckmeier, C.; Schneider, J.; Susha, A.; Rogach, A. Luminescent colloidal carbon dots: Optical properties and effects of doping. *Opt. Express* **2016**, *24*, A312–A340. [CrossRef]
39. Xiong, Y.; Zhu, M.; Zhao, Z.; Schneider, J.; Huang, H.; Kershaw, S.V.; Zhi, C.; Rogach, A.L. A Building Brick Principle to Create Transparent Composite Films with Multicolor Emission and Self-Healing Function. *Small* **2018**, *14*, 1800315. [CrossRef]
40. Reckmeier, C.J.; Schneider, J.; Xiong, Y.; Häusler, J.; Kasák, P.; Schnick, W.; Rogach, A.L. Aggregated molecular fluorophores in the ammonothermal synthesis of carbon dots. *Chem. Mater.* **2017**, *29*, 10352–10361. [CrossRef]
41. Xiong, Y.; Schneider, J.; Ushakova, E.V.; Rogach, A.L. Influence of molecular fluorophores on the research field of chemically synthesized carbon dots. *Nano Today* **2018**, *23*, 124–139. [CrossRef]
42. Kasprzyk, W.; Bednarz, S.; Żmudzki, P.; Galica, M.; Bogdał, D. Novel efficient fluorophores synthesized from citric acid. *RSC Adv.* **2015**, *5*, 34795–34799. [CrossRef]
43. Shi, L.; Yang, J.H.; Zeng, H.B.; Chen, Y.M.; Yang, S.C.; Wu, C.; Zeng, H.; Yoshihito, O.; Zhang, Q. Carbon dots with high fluorescence quantum yield: The fluorescence originates from organic fluorophores. *Nanoscale* **2016**, *8*, 14374–14378. [CrossRef] [PubMed]
44. Yang, J.; Zhang, Y.; Gautam, S.; Liu, L.; Dey, J.; Chen, W.; Mason, R.P.; Serrano, C.A.; Schug, K.A.; Tang, L. Development of aliphatic biodegradable photoluminescent polymers. *Proc. Natl. Acad. Sci. USA* **2009**, *106*, 10086–10091. [CrossRef]
45. Kasprzyk, W.; Bednarz, S.; Bogdał, D. Luminescence phenomena of biodegradable photoluminescent poly (diol citrates). *Chem. Commun.* **2013**, *49*, 6445–6447. [CrossRef] [PubMed]
46. Wang, H.X.; Yang, Z.; Liu, Z.G.; Wan, J.Y.; Xiao, J.; Zhang, H.L. Facile Preparation of Bright-Fluorescent Soft Materials from Small Organic Molecules. *Chem. Eur. J.* **2016**, *22*, 8096–8104. [CrossRef]
47. Chen, H.; Yan, X.; Feng, Q.; Zhao, P.; Xu, X.; Ng, D.H.; Bian, L. Citric acid/cysteine-modified cellulose-based materials: Green preparation and their applications in anticounterfeiting, chemical sensing, and UV shielding. *ACS Sustain. Chem. Eng.* **2017**, *5*, 11387–11394. [CrossRef]
48. Kim, J.P.; Xie, Z.; Creer, M.; Liu, Z.; Yang, J. Citrate-based fluorescent materials for low-cost chloride sensing in the diagnosis of cystic fibrosis. *Chem. Sci.* **2017**, *8*, 550–558. [CrossRef]
49. Zhang, C.; Kim, J.P.; Creer, M.; Yang, J.; Liu, Z. A smartphone-based chloridometer for point-of-care diagnostics of cystic fibrosis. *Biosens. Bioelectron.* **2017**, *97*, 164–168. [CrossRef] [PubMed]

50. Kasak, P.; Danko, M.; Zavahir, S.; Mrlik, M.; Xiong, Y.; Yousaf, A.B.; Lai, W.-F.; Krupa, I.; Tkac, J.; Rogach, A.L. Identification of molecular fluorophore as a component of carbon dots able to induce gelation in a fluorescent multivalent-metal-ion-free alginate hydrogel. *Sci. Rep.* **2019**, *9*, 15080. [CrossRef] [PubMed]
51. Langer, M.; Paloncyova, M.; Medved', M.; Otyepka, M. Molecular fluorophores self-organize into C-dot seeds and incorporate into C-dot structures. *J. Phys. Chem. Lett.* **2020**, *11*, 8252–8258. [CrossRef]
52. Langer, M.; Hrivnák, T.s.; Medved', M.; Otyepka, M. Contribution of the molecular fluorophore IPCA to excitation-independent photoluminescence of carbon dots. *J. Phys. Chem. C* **2021**, *125*, 12140–12148. [CrossRef]
53. Mutailipu, M.; Li, F.; Jin, C.; Yang, Z.; Poeppelmeier, K.R.; Pan, S. Strong Nonlinearity Induced by Coaxial Alignment of Polar Chain and Dense [BO_3] Units in $CaZn_2(BO_3)_2$. *Angew. Chem.* **2022**, *134*, e202202096. [CrossRef]
54. Festa, R.A.; Thiele, D.J. Copper: An essential metal in biology. *Curr. Biol.* **2011**, *21*, R877–R883. [CrossRef]
55. Flynn Jr, C.M. Hydrolysis of inorganic iron (III) salts. *Chem. Rev.* **1984**, *84*, 31–41. [CrossRef]
56. Stuart, B.H. *Infrared Spectroscopy: Fundamentals and Applications*; John Wiley & Sons: Chichester, UK, 2004.
57. Burla, M.; Camalli, M.; Cascarano, G.; Giacovazzo, C.; Polidori, G.; Spagna, R.T.; Viterbo, D. SIR88–a direct-methods program for the automatic solution of crystal structures. *J. Appl. Crystallogr.* **1989**, *22*, 389–393. [CrossRef]
58. Groom, C.R.; Bruno, I.J.; Lightfoot, M.P.; Ward, S.C. The Cambridge structural database. *Acta Crystallogr. Sect. B Struct. Sci. Cryst. Eng. Mater.* **2016**, *72*, 171–179. [CrossRef] [PubMed]
59. Fan, W.W.; Cheng, Y.; Zheng, L.Y.; Cao, Q.E. Reversible phase transition of porous coordination polymers. *Chem. A Eur. J.* **2020**, *26*, 2766–2779. [CrossRef] [PubMed]
60. Liu, X.L.; Fan, W.W.; Lu, Z.X.; Qin, Y.; Yang, S.X.; Li, Y.; Liu, Y.X.; Zheng, L.Y.; Cao, Q.E. Solvent-Driven Reversible Phase Transition of a Pillared Metal–Organic Framework. *Chem. A Eur. J.* **2019**, *25*, 5787–5792. [CrossRef] [PubMed]
61. Fernandez-Bartolome, E.; Martinez-Martinez, A.; Resines-Urien, E.; Piñeiro-Lopez, L.; Costa, J.S. Reversible single-crystal-to-single-crystal transformations in coordination compounds induced by external stimuli. *Coord. Chem. Rev.* **2022**, *452*, 214281. [CrossRef]
62. Essehli, R.; Sabri, S.; El-Mellouhi, F.; Aïssa, B.; Ben Yahia, H.; Altamash, T.; Khraisheh, M.; Amhamed, A.; El Bali, B. Single crystal structure, vibrational spectroscopy, gas sorption and antimicrobial properties of a new inorganic acidic diphosphates material $(NH_4)_2Mg(H_2P_2O_7)_2 \cdot 2H_2O$. *Sci. Rep.* **2020**, *10*, 8909. [CrossRef] [PubMed]
63. Abotaleb, A.; El-Naas, M.H.; Amhamed, A. Enhancing gas loading and reducing energy consumption in acid gas removal systems: A simulation study based on real NGL plant data. *J. Nat. Gas Sci.* **2018**, *55*, 565–574. [CrossRef]
64. Deng, L.; Dong, X.; An, D.-L.; Weng, W.-Z.; Zhou, Z.-H. Gas Adsorption of Mixed-Valence Trinuclear Oxothiomolybdenum Glycolates. *Inorg. Chem.* **2020**, *59*, 4874–4881. [CrossRef] [PubMed]
65. Qazvini, O.T.; Babarao, R.; Telfer, S.G. Selective capture of carbon dioxide from hydrocarbons using a metal-organic framework. *Nat. Commun.* **2021**, *12*, 197. [CrossRef]
66. Ribeiro, R.P.; Esteves, I.A.; Mota, J.P. Adsorption of Carbon Dioxide, Methane, and Nitrogen on Zn (dcpa) Metal-Organic Framework. *Energies* **2021**, *14*, 5598. [CrossRef]
67. Jiang, J.; Furukawa, H.; Zhang, Y.-B.; Yaghi, O.M. High methane storage working capacity in metal–organic frameworks with acrylate links. *J. Am. Chem. Soc.* **2016**, *138*, 10244–10251. [CrossRef] [PubMed]
68. Jung, J.Y.; Karadas, F.; Zulfiqar, S.; Deniz, E.; Aparicio, S.; Atilhan, M.; Yavuz, C.T.; Han, S.M. Limitations and high pressure behavior of MOF-5 for CO_2 capture. *Phys. Chem. Chem. Phys.* **2013**, *15*, 14319–14327. [CrossRef]
69. Deniz, E.; Karadas, F.; Patel, H.A.; Aparicio, S.; Yavuz, C.T.; Atilhan, M. A combined computational and experimental study of high pressure and supercritical CO_2 adsorption on Basolite MOFs. *Microporous Mesoporous Mater.* **2013**, *175*, 34–42. [CrossRef]
70. Karadas, F.; Yavuz, C.T.; Zulfiqar, S.; Aparicio, S.; Stucky, G.D.; Atilhan, M. CO_2 adsorption studies on hydroxy metal carbonates M (CO_3) x (OH) y (M= Zn, Zn–Mg, Mg, Mg–Cu, Cu, Ni, and Pb) at high pressures up to 175 bar. *Langmuir* **2011**, *27*, 10642–10647. [CrossRef]
71. Caddeo, F.; Loche, D.; Casula, M.F.; Corrias, A.J.S.r. Evidence of a cubic iron sub-lattice in t-$CuFe_2O_4$ demonstrated by X-ray Absorption Fine Structure. *Sci. Rep.* **2018**, *8*, 797. [CrossRef]
72. Wu, S.; Li, Z.; Li, M.-Q.; Diao, Y.; Lin, F.; Liu, T.; Zhang, J.; Tieu, P.; Gao, W.; Qi, F. 2D metal–organic framework for stable perovskite solar cells with minimized lead leakage. *Nat. Nanotechnol.* **2020**, *15*, 934–940. [CrossRef]
73. Altamash, T.; Khraisheh, M.; Qureshi, M.F. Investigating the effects of mixing ionic liquids on their density, decomposition temperature, and gas absorption. *Chem. Eng. Res. Des.* **2019**, *148*, 251–259. [CrossRef]
74. Nguyen, H.B.; Thai, T.Q.; Saitoh, S.; Wu, B.; Saitoh, Y.; Shimo, S.; Fujitani, H.; Otobe, H.; Ohno, N. Conductive resins improve charging and resolution of acquired images in electron microscopic volume imaging. *Sci. Rep.* **2016**, *6*, 23721. [CrossRef] [PubMed]
75. Petříček, V.; Dušek, M.; Palatinus, L. Crystallographic computing system JANA2006: General features. *Z. Für Krist. Cryst. Mater.* **2014**, *229*, 345–352. [CrossRef]
76. Sheldrick, G.M. *SADABS version 2014/5*; University of Göttingen: Göttingen, Germany, 2014.

Article

Mixed-Valent Trinuclear Co^{III}-Co^{II}-Co^{III} Complex with 1,3-Bis(5-chlorosalicylideneamino)-2-propanol

Masahiro Mikuriya [1,*], Yuko Naka [1], Ayumi Inaoka [1], Mika Okayama [1], Daisuke Yoshioka [1], Hiroshi Sakiyama [2], Makoto Handa [3] and Motohiro Tsuboi [1]

[1] School of Biological and Environmental Sciences, Kwansei Gakuin University, Uegahara 1, Gakuen, Sanda 669-1330, Japan; poninma@gmail.com (Y.N.); ayumi.inaoka.570@gmail.com (A.I.); sym12286@gmail.com (M.O.); yoshi0431@gmail.com (D.Y.); tsuboimot@kwansei.ac.jp (M.T.)
[2] Department of Science, Faculty of Science, Yamagata University, 1-4-12 Kojirakawa, Yamagata 990-8560, Japan; saki@sci.kj.yamagata-u.ac.jp
[3] Department of Chemistry, Graduate School of Natural Science and Technology, Shimane University, 1060 Nishikawatsu, Matsue 690-8504, Japan; handam@riko.shimane-u.ac.jp
* Correspondence: junpei@kwansei.ac.jp

Abstract: A mixed-valent trinuclear complex with 1,3-bis(5-chlorosalicylideneamino)-2-propanol (H$_3$clsalpr) was synthesized, and the crystal structure was determined by the single-crystal X-ray diffraction method at 90 K. The molecule is a trinuclear Co^{III}-Co^{II}-Co^{III} complex with octahedral geometries, having a tetradentate chelate of the Schiff-base ligand, bridging acetate, monodentate acetate coordination to each terminal Co^{3+} ion and four bridging phenoxido-oxygen of two Schiff-base ligands, and two bridging acetate-oxygen atoms for the central Co^{2+} ion. The electronic spectral feature is consistent with the mixed valent Co^{III}-Co^{II}-Co^{III}. Variable-temperature magnetic susceptibility data could be analyzed by consideration of the axial distortion of the central Co^{2+} ion with the parameters $\Delta = -254$ cm^{-1}, $\lambda = -58$ cm^{-1}, $\kappa = 0.93$, $tip = 0.00436$ cm^3 mol^{-1}, $\theta = -0.469$ K, $g_z = 6.90$, and $g_x = 2.64$, in accordance with a large anisotropy. The cyclic voltammogram showed an irreversible reduction wave at approximately -1.2 V·vs. Fc/Fc$^+$, assignable to the reduction of the terminal Co^{3+} ions.

Keywords: trinuclear complex; cobalt complex; mixed-valent complex; Schiff-base ligand

1. Introduction

Schiff-base ligands have been synthesized extensively because such organic compounds may be easily accessible for constructing various kinds of multidentate ligands and useful for reacting with main-group and transition metal ions, including lanthanides and actinides, to form a number of metal complexes, which are useful as model compounds in basic chemistry as well as in a wide range of applications [1–10]. Pentadentate Schiff-base ligands, 1,3-bis(salicylideneamino)-2-propanol (H$_3$salpr), and its substituted derivatives have been developed as dinucleating ligands for constructing adjacent coordination sites, as shown in the case of 1,3-bis(5-chlorosalicylideneamino)-2-propanol in Figure 1a [11–48]. X-ray crystal structure analysis was performed for some free Schiff-base ligands [11–16]. In the crystals, the Schiff bases take a "bent" [11–15,34] or "folded" [16] structure with the salicylideneaminomethyl moieties being close to planar. When the Schiff-base ligands are coordinated to two metal atoms with two phenolic-oxygen, two imino-nitrogen, and one bridging alkoxido-oxygen donor atom with a pair of tridentate O, N, O-chelates, Mn^{III}$_2$, Fe^{III}$_2$, Co^{III}$_2$, Ni^{II}$_2$, and Cu^{II}$_2$ complexes were reported as this type of dinuclear species [17–29]. On the other hand, these Schiff-base ligands form mononuclear metal species with a tetradentate O, N, N, O-chelate as shown in the case for 1,3-bis(5-chlorosalicylideneamino)-2-propanol in Figure 1b, where the central alcohol group remains protonated and does not participate in coordination to the metal center. Such mononuclear Mn^{III}, Ni^{II}, Cu^{II}, and Pd^{II} complexes were found in the literature [30–36], and a hydrolyzed product of mononuclear Co^{III} species was also derived from a reaction of a Schiff-base ligand with cobalt salt [37]. In the other case, the mononuclear species are further connected to form trinuclear Cu$^{II}_3$ [36], Zn$^{II}_3$ [38], Cd$^{II}_3$ [39], and Co^{II}Co$^{III}_2$ [40]; tetranuclear Mn$^{II}_2$Mn$^{III}_2$ [42,43], Mn$^{III}_4$ [44], Co$^{II}_4$ [39,45], Co$^{II}_2$Co$^{III}_2$ [46], Ni$^{II}_4$ [39,46], and Zn$^{II}_4$ [39,41]; hexanuclear Co$^{II}_4$Co$^{III}_2$ [47] and Cu$^{II}_6$ [47], octanuclear Mn$^{II}_2$Mn$^{III}_6$ [48], and polynuclear Mn^{III} complexes [25,35]. Among these oligonuclear and polynuclear metal complexes, we have focused on the trinuclear metal systems important as the first step to polynucleation, especially the Co^{II}Co$^{III}_2$ species because of the mixed-valent state. Another interesting point of cobalt complexes stems from the fact that cobalt(II) complexes have attracted much attention as good candidates for single-molecule magnets [49,50]. Although the trinuclear Co^{II}Co$^{III}_2$ complex was prepared for 1,3-bis(salicylideneamino)-2-propanol, the reported magnetic susceptibility data were not analyzed [40], and there are no reports on such complexes with their substituted derivatives. This type of linear cobalt species was found in some related Co^{II}Co$^{III}_2$ and Co$^{II}_3$ complexes and divided into five groups as shown in Figure 2: (a) Co^{III}(octahedral)-Co^{II}(tetrahedral)- Co^{III}(octahedral) in [Co^{II}{Co^{III}(µ-L^1)X$_2$}$_2$] (H$_2$L^1 = 1,3-bis(5-methyl-3-formylpyrazolylmethinimino)propane-2-ol, X = Cl, Br) [51]; (b) Co^{III}(octahedral)-Co^{II}(octahedral)-Co^{III}(octahe dral) in [Co^{II}{Co^{III}(µ-L^2)(µ-SO$_3$)(C$_3$H$_7$OH)}$_2$] (H$_2$L^2 = propane-1,3-dihylbis(α-methylsalicylideneiminate) [52], [Co^{II}{Co^{III}(µ-L^3)(µ-CH$_3$COO)(NCS)}$_2$] (H$_2$L^3 = 1,6-bis(2- hydroxy phenyl)-2,5-diazahexa-1,5-diene) [53], [Co^{II}{Co^{III}(µ-L^4)(µ-CH$_3$COO)(NCS)}$_2$] (H$_2$L^4 = 1,7-bis(2-hydroxyphenyl)-2,6-diazahepta-1,6-diene) [54], [Co^{II}{Co^{III}(µ-L^5)(µ-CH$_3$COO)(CH$_3$COO)}$_2$] (H$_2$L^5 = 1,6-bis(2-hydroxyphenyl)-2,5-diazahexa-1,5-diene or 2,7-bis(2-hydroxyphenyl)-2,6-diazaocta-2,6-diene) [55], and [Co^{II}{Co^{III}(µ-L^6)(µ-CH$_3$COO)(CH$_3$COO)}$_2$] (H$_2$L^6 = N,N'-bis(salicylidene)-meso-1,2-diphenylethylenediamine) [56]; (c) Co^{II}(octahedral)-Co^{II}(tetrahedral)-Co^{II}(octahedral) in [Co^{II}{Co^{II}(µ-L^7)$_2$}$_2$]X$_2$ (HL^7 = 2-[(3-aminopropyl)amino]ethanethiol; X = SCN, ClO$_4$, NO$_3$, Cl, Br, I) and [Co^{II}{Co^{II}(µ-L^8)$_2$}$_2$]X$_2$ (HL^8 = 1-[(3-aminopropyl)amino]-2-methylpropane-2-thiol; X = NO$_3$, ClO$_4$, Cl, Br, I) [57,58]; (d) Co^{II}(square-pyramidal)-Co^{II}(octahedral)-Co^{II}(square-pyramidal) in [Co^{II}{Co^{II}(µ-L^9)(µ-CH$_3$COO)}$_2$] (H$_2$L^9 = 5,5'-dimethoxy-2,2'-[(ethylene)dioxybis(nitrilomethylidyne)]diphenol [59]; and (e) Co^{II}(octahedral)-Co^{II}(octahedral)-Co^{II}(octahedral) in [Co^{II}{Co^{II}(µ-L^{10})(µ-CH$_3$COO)(CH$_3$COCH$_3$)}$_2$] (L^{10} = 4,4'-dichloro-2,2'-[(propane-1,3-dyldioxy)bis(nitrilomethylidyne)]diphenol) [60]. The Co^{III} oxidation state may come from the oxidation of Co^{II} by atmospheric oxygen during the reaction of Co^{II} salt and organic ligand [40,51,53–55]. The bridging µ-acetato ligand is favorable to form a trinuclear Co^{III}-Co^{II}-Co^{III} complex. From our synthesis experience, chloroderivatives of Schiff-base ligands are promising for obtaining single crystals for X-ray crystallographic work [61]. To date, only two examples, (Et$_4$N)[Mn^{II}Mn^{III}(clsalpr)$_2$] [24] and [Mn$^{III}_2$(clsalpr)$_2$(CH$_3$OH)] [25], are known as metal complexes with the chloro-derivative of H$_3$salpr, 1,3-bis(5-chlorosalicylideneamino)-2-propanol (H$_3$clsalpr), which were structurally revealed by X-ray crystallography. In these complexes, the H$_3$clsalpr ligand is fully deprotonated and works as a pentadentate ligand to two manganese centers to form dinuclear manganese complexes. In this study, we synthesized a new mixed-valent cobalt complex with a linear trinuclear Co^{II}Co$^{III}_2$ core by the reaction of H$_3$clsalpr (Figure 1) with cobalt(II) acetate tetrahydrate. The isolated complex was characterized by elemental analyses, IR and UV–vis spectroscopies, variable-temperature magnetic susceptibility measurements, and single-crystal X-ray structure analysis, elucidating the molecular structure of [Co$_3$(Hclsalpr)$_2$(CH$_3$COO)$_4$].

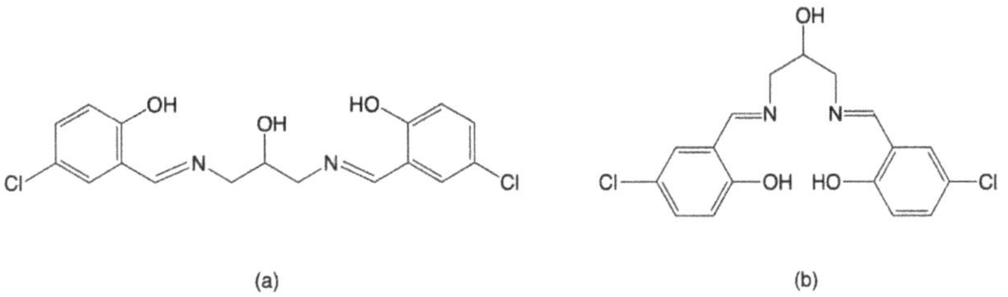

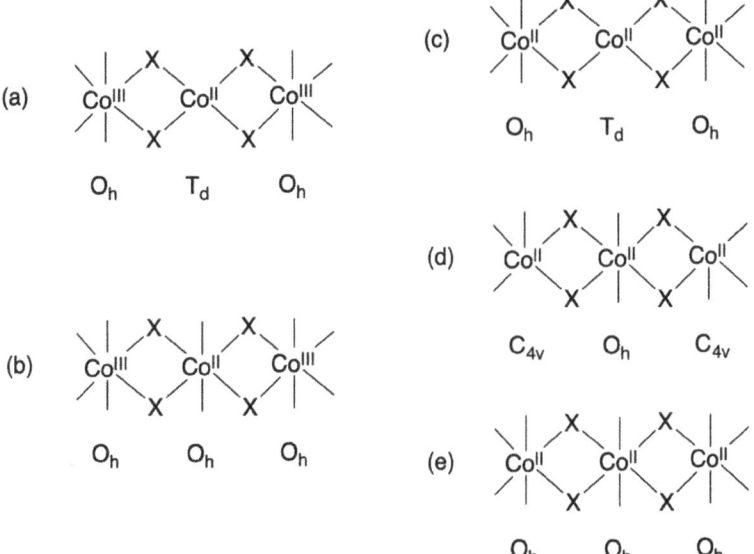

Figure 1. Schiff-base ligand 1,3-bis(5-chlorosalicylideneamino)-2-propanol as (**a**) pentadentate dinucleating ligand and (**b**) tetradentate mononucleating ligand.

Figure 2. (**a**–**e**) Trinuclear cobalt complexes with a linear array of Co^{III}-Co^{II}-Co^{III} or Co^{II}-Co^{II}-Co^{II}.

2. Results and Discussion

2.1. Synthesis of the Trinuclear Cobalt Complex

The present complex was prepared by the reaction of 1,3-bis(5-chlorosalicylideneamino)-2-propanol (H_3clsalpr) and cobalt(II) acetate tetrahydrate in acetonitrile at ambient temperature (Figure 3). As the cobalt salt, we selected cobalt(II) acetate tetrahydrate, aiming at the bridging property of acetate ions to form a trinuclear species. For a favorable condition for trinuclear formation, we reacted H_2clsalpr with $Co(CH_3COO)_2 \cdot 4H_2O$ in a 1:3 molar ratio under aerobic conditions, although we could isolate the same complex with a lower yield when the reaction was performed in a 1:1 or 1:2 molar ratio. The elemental analysis data of the obtained complex are in agreement with the trinuclear formulation of $[Co_3(Hclsalpr)_2(CH_3COO)_4]$. The oxidation of the two Co^{2+} ions to Co^{3+} ions may be accomplished by atmospheric oxygen acting as an oxidant, as usually observed for the synthesis of the related trinuclear $Co^{II}Co^{III}{}_2$ complexes [40,51,53–56]. The synthetic method is similar to that of $[Co_3(Hsalpr)_2(CH_3COO)_4]$ [40]. However, the reported method was slightly complicated with the further addition of an aqueous solution of $NaN(CN)_2$ to the reaction solution.

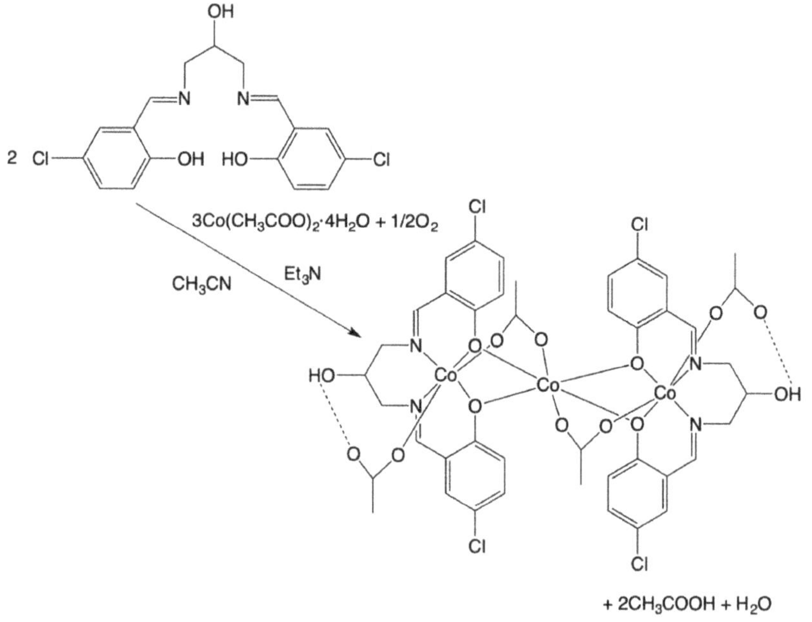

Figure 3. Synthetic scheme of the trinuclear cobalt complex [Co$_3$(Hclsalpr)$_2$(CH$_3$COO)$_4$].

2.2. Infrared Spectra of the Trinuclear Cobalt Complex

In the infrared spectrum of the complex, the C=N stretching band was observed at 1634 cm^{-1} due to the presence of the Schiff-base ligand. The lower energy shift compared with that of the free Schiff-base ligand (H$_3$clsalpr: νC=N at 1646 cm^{-1}) suggests the coordination of the imino-nitrogen atom of the Schiff-base ligand in the cobalt complex. The complex shows two sets of antisymmetric stretching ν_{as}(COO) and symmetric stretching ν_s(COO) bands at 1590 and 1389 cm^{-1}, respectively, with a Δ value of 201 cm^{-1}, and at 1562 and 1415 cm^{-1}, respectively, with a Δ value of 147 cm^{-1}. The former and the latter may be ascribed to the typical IR spectral features of the monodentate and bridging acetate ligands, respectively [62,63]. These spectral features are similar to those of [Co$_3$(Hsalpr)$_2$(CH$_3$COO)$_4$] [40].

2.3. Electronic Spectra of the Trinuclear Cobalt Complex

The solid-state diffuse reflectance spectra exhibit a broad band with a lower-energy side shoulder at 354 nm, which may be ascribed to the CT transition band of the phenolate to metal as shown in Figure 4 [64]. The bands at 568 and 640 nm may be ascribed to d-d transitions ($^1A_{1g} \rightarrow {}^1B_{2g}, {}^1A_{1g} \rightarrow {}^1A_{2g}, {}^1A_{1g} \rightarrow {}^1E_g$) of an octahedral Co^{III} with a low-spin state [64,65]. Furthermore, the spectra show a broad band at approximately 1260 nm, which can be ascribed to the d-d transition ($^4T_{1g} \rightarrow {}^4T_{2g}$) due to an octahedral Co^{II} with a high-spin state [65]. The complex dissolves in THF. The solution spectra are similar to those of the solid-state spectra, showing a d-d absorption band (ε = 538 dm^3 cm^{-1} mol^{-1}) at 564 nm with a shoulder (ε = 320 dm^3 cm^{-1} mol^{-1}) at 632 nm, although absorption in the near-IR region could not be detected. Similar absorption spectra were reported for [Co$_3$(Hsalpr)$_2$(CH$_3$COO)$_4$] [40].

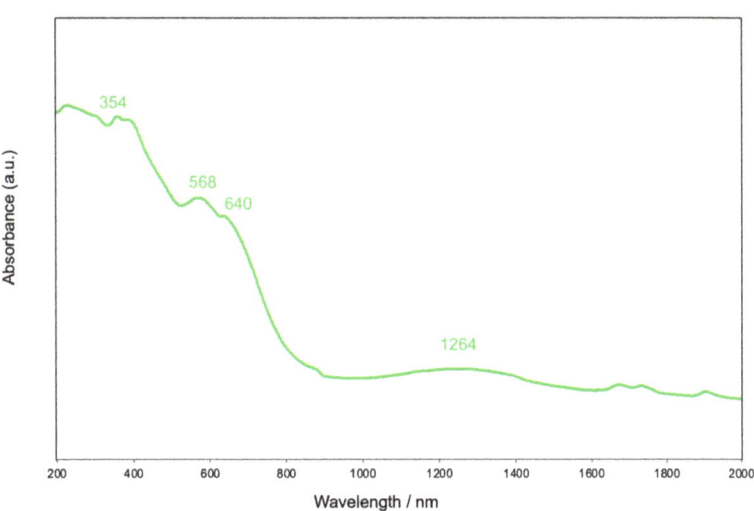

Figure 4. Diffuse reflectance spectra of [Co$_3$(Hclsalpr)$_2$(CH$_3$COO)$_4$] (green line).

2.4. Crystal Structure of the Trinuclear Cobalt Complex

Single crystals of the complex suitable for X-ray crystal structure analysis were grown by the slow evaporation of the THF solution of the complex. Crystallographic data are collected in Table 1. Selected bond distances and angles are given in Table 2. The complex crystallized in the monoclinic system. A perspective drawing of the structure is depicted in Figure 5. The molecule is a centrosymmetric trinuclear cobalt complex, where the Co1 atom is located at the crystallographical inversion center. The two Schiff-base ligands work as anionic tetradentate ligands Hclsalpr^{2-} to the terminal two cobalt atoms, Co2 and Co2^i, where the superscript i denotes the equivalent position $(1 - x, 1 - y, 1 - z)$, and the alcoholate hydrogen atom is not deprotonated, but two phenolate H atoms of each Schiff-base ligand are deprotonated. The Co1 atom is coordinated by two sets of two phenoxido-O atoms of Hclsalpr^{2-} ligands (O1, O3, O1^i, O3^i;) and μ-acetato-O atoms (O4 and O4^i) to form an octahedral geometry with Co-O distances of 2.0493(16)–2.1318(15) Å. It should be noted that the axial bond lengths (2.1318(15) Å) are longer than the equatorial bond lengths (2.0493(16) and 2.0851(16) Å), showing an axial distortion around the Co1 atom. The Co2 atom is coordinated by two phenoxido-O atoms (O1 and O3) and two imino-N atoms (N1 and N2) of the tetradentate Schiff-base ligand in trans geometry [66] to occupy the equatorial site. The axial site is occupied by the O atoms of the μ-bridging acetate (O5) and monodentate acetate (O6). The Co-O and Co-N bond distances are in the range of 1.8943(15)–1.9263(16) Å, significantly shorter than those of the Co1 atom. The difference between the bond distances around the Co1 and Co2 (Co2^i) atoms suggests that the Co1 atom is in a high-spin state of Co^{2+} ion and that the Co2 and Co2^i atoms are in a low-spin Co^{3+} ion state [67]. The bond valence sum calculation supports the mixed-valent Co^{III}-Co^{II}-Co^{III} state [68,69]. This is in agreement with the spectral feature in the diffuse reflectance spectra of the present complex. The alcoholate H atom of O2 is hydrogen bonded to the monodentate acetate-O atom O7 [O2-H ... O7 2.659(2) Å]. In the crystal, there are four THF molecules in the asymmetric unit, and these molecules are oriented around the trinuclear molecule (Figure 6). The trinuclear structure is similar to that of the reported trinuclear cobalt complex [Co$_3$(Hsalpr)(CH$_3$COO)$_4$], which lacks a center of symmetry, where the distortion around the Co^{2+} ion is more distorted compared with the present complex [40]. In these complexes, the two bridging acetate groups and four μ-phenoxido-O atoms of the Schiff-base ligands play an important role in connecting the two tetradentate Co(Hclsalpr)$_2$ moieties. This motif was also found in the trinuclear zinc(II) complex [Zn$_3$(Hsalpr)$_2$(CH$_3$COO)$_2$] [38] and heterometallic trinuclear complexes [(CH$_3$OH)$_2$H$^+$][NaMn$_2$(Hsalpr)$_2$(CH$_3$COO)$_2$] [25] and [Zn{Cu(salpd-μ-O,O')(μ-CH$_3$COO)}$_2$] [H$_2$salpd = propane-1,3-diylbis(salicylideneimine)] [10] as well as trinuclear cobalt complexes, [Co^{II}{Co^{III}(μ-L^5)(μ-CH$_3$COO)(CH$_3$COO)}$_2$] [55] and [Co^{II}{Co^{III}(μ-L^6)(μ-CH$_3$COO)(CH$_3$COO)}$_2$] [56].

Table 1. Crystallographic data and structure refinement.

Complex	[Co$_3$(Hclsalpr)$_2$(CH$_3$COO)$_4$]·8THF
Chemical formula	C$_{74}$H$_{104}$Cl$_4$Co$_3$N$_4$O$_{22}$
FW	1720.20
Temperature, T (K)	90
Crystal system	monoclinic
Space group	C2/c
a (Å)	20.244 (2)
b (Å)	14.0864 (16)
c (Å)	27.646 (3)
β (°)	98.9610 (10)
V (Å^3)	7787.5 (14)
Z	4
D_{calcd} (g cm^{-3})	1.467
Crystal size (mm)	0.07 × 0.50 × 0.62
μ (mm^{-1})	0.845
θ range for data collection (°)	1.49–28.66
Reflections collected/unique	23,114/9103
[R1($I > 2\sigma(I)$); wR2(all data)] [a]	$R_1 = 0.0419$ $\omega R_2 = 0.0953$
GOF	1.003

[a] $R1 = \sum ||F_o| - |F_c||/\sum |F_o|$; $\omega R2 = [\sum \omega(F_o^2 - F_c^2)^2/\sum(F_o^2)^2]^{1/2}$.

Table 2. Selected bond distances (Å) and angles (°).

[Co$_3$(Hclsalpr)$_2$(CH$_3$COO)$_4$]·8THF	
Co1···Co2 3.0383(4)	Co2···Co2^i 6.0765(8)
Co1-O1 2.1318(15)	Co1-O3 2.0851(16)
Co1-O4 2.0493(16)	Co2-O1 1.9263(16)
Co2-O3 1.9220(15)	Co2-O5 1.9144(16)
Co2-O6 1.8943(15)	Co2-N1 1.9188(19)
Co2-N2 1.9173(19)	N1-C7 1.284(3)
N1-C8 1.476(3)	N2-C11 1.283(3)
N2-C10 1.473(3)	
O1-Co1-O1$^{i\ (a)}$ 180.0	O1-Co1-O3 76.23(6)
O1-Co1-O3^i 103.77(6)	O1-Co1-O4 84.59(6)
O1-Co1-O4^i 95.41(6)	O3-Co1-O4 85.33(6)
O3-Co1-O4^i 94.67(6)	O1-Co2-O3 85.13(7)
O1-Co2-O5 91.70(7)	O1-Co2-O6 86.01(7)
O1-Co2-N1 89.97(7)	O1-Co2-N2 175.73(7)
O3-Co2-O5 91.70(7)	O3-Co2-O6 86.35(7)
O3-Co2-N1 175.08(7)	O3-Co2-N2 90.60(7)
O5-Co2-O6 176.96(7)	O5-Co2-N1 87.98(7)
O5-Co2-N2 88.60(7)	O6-Co2-N1 93.76(7)
O6-Co2-N2 93.74(7)	N1-Co2-N2 94.30(7)

[a] i: the equivalent position (1 − x, 1 − y, 1 − z).

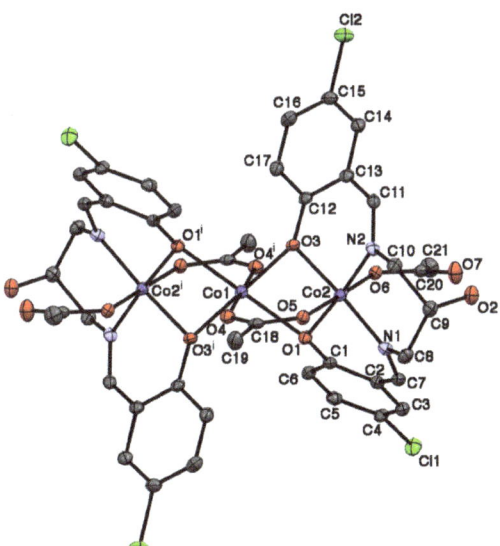

Figure 5. The ORTEP view of the molecular structure of [Co$_3$(Hclsalpr)$_2$(CH$_3$COO)$_4$] with thermal ellipsoids (50% probability level). The hydrogen atoms have been omitted for clarity.

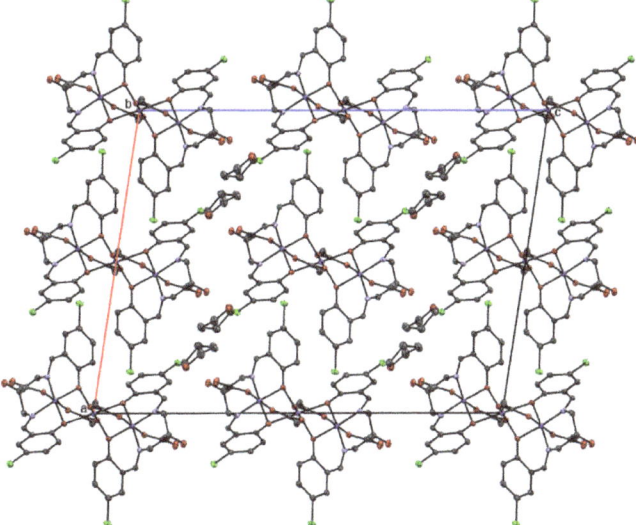

Figure 6. Packing diagram of [Co$_3$(Hclsalpr)$_2$(CH$_3$COO)$_4$]·8THF. The hydrogen atoms have been omitted for clarity.

2.5. Magnetic Properties of the Trinuclear Cobalt Complex

The present complex is expected to be paramagnetic because of the presence of the high-spin Co^{2+} ion at the central position of the trinuclear cobalt molecule, although the terminal two Co^{3+} ions are in a diamagnetic low-spin state. The magnetic susceptibility data for the complex are depicted in Figure 7 as the temperature variation of the $\chi_M T$ product. The effective magnetic moment at 300 K is 5.73 μ_B per trinuclear molecule, which corresponds to the theoretical value of 5.20 μ_B for a magnetically isolated $S = 3/2$ spin with the contribution of orbital angular momentum ($L = 3$). The magnetic moment gradually decreases with decreasing temperature, reaching a value of 3.82 μ_B at 4.5 K. This magnetic behavior is similar to that of the related linear Co^{III}-Co^{II}-Co^{III} complex with 1,3-bis(salicylideneamino)-2-propanol [40]. The decrease in the magnetic moments may be ascribed to the axial distortion around the Co^{2+} ion, which was observed in the crystal structure. The axial splitting parameter Δ was defined as the splitting of the local $^4T_{1g}$ state of the octahedral Co^{2+} ion in the absence of spin–orbit coupling and introduced to the magnetic data analysis [70–72]. The magnetic data were simulated with the axial splitting parameter Δ, the spin-orbit coupling parameter λ, the orbital reduction factor κ for the Co^{2+} ion ($H = \Delta(L_z^2 - 2/3) - (3/2)\kappa\lambda L \cdot S + \beta[-(3/2)\kappa L_u + g_e S_u] \cdot H_u$ ($u = x, z$)), the temperature-independent paramagnetism tip for the Co centers, and the Weiss constant θ for intermolecular magnetic interactions by using the MagSaki(A)W1.0.11 program [72]. Magnetic susceptibility equations are shown below (Equations (1)–(6)), where $E_n^{(0)}$, $E_{u,n}^{(1)}$, and $E_{u,n}^{(2)}$ ($n = \pm 1 - \pm 6$, $u = x, z$) represent the zero-field energies, first-order Zeeman coefficients, and second-order Zeeman coefficients of the local $^4T_1$ ground state for the octahedral Co^{2+} ion. From this, the anisotropic g-factors, g_z and g_x, could be simulated using these parameters [72]. The simulation gave the following parameter values: $\Delta = -254$ cm^{-1}, $\lambda = -58$ cm^{-1}, $\kappa = 0.93$, $tip = 0.00436$ cm^3 mol^{-1}, and $\theta = -0.469$ K. A large value of the tip may be ascribed to the presence of three cobalt atoms in the molecule. The g values were simulated as $g_z = 6.90$ and $g_x = 2.64$. This result suggests that the magnetic behavior of the present complex can be interpreted by the axial distortion of the central Co^{2+} ion and thus proposed to be considerably anisotropic. If we apply the present magnetic analysis to the reported magnetic data of $[Co_3(Hsalpr)_2(CH_3COO)_4]$ [40], we obtain the following parameter values: $\Delta = -950$ cm^{-1}, $\lambda = -131$ cm^{-1}, $\kappa = 0.93$, $tip = 0.00082$ cm^3 mol^{-1}, $\theta = -0.67$ K, $g_z = 7.71$, and $g_x = 1.94$, as shown in Figure 8. The magnetic analysis suggests that a considerable anisotropic character may also be found in $[Co_3(Hsalpr)_2(CH_3COO)_4]$ and that the larger negative Δ and λ values may reflect a greater degree of axial distortion around the Co^{2+} ion in the crystal structure [40].

$$\chi_M = \frac{\chi_z + 2\chi_x}{3} \quad (1)$$

$$\chi_z = N\frac{F_1}{F_2} + tip \quad (2)$$

$$\chi_x = N\frac{F_3}{F_2} + tip \quad (3)$$

$$F_1 = \sum_{n=\pm 1}\left(\frac{E_{z,n}^{(1)2}}{k(T-\theta)} - 2E_{z,n}^{(2)}\right)\exp\left[\frac{-E_n^{(0)}}{kT}\right] + \sum_{n\neq\pm 1}\left(\frac{E_{z,n}^{(1)2}}{kT} - 2E_{z,n}^{(2)}\right)\exp\left[\frac{-E_n^{(0)}}{kT}\right] \quad (4)$$

$$F_2 = \sum_n \exp\left[\frac{-E_n^{(0)}}{kT}\right] \quad (5)$$

$$F_3 = \sum_{n=\pm 1}\left(\frac{E_{x,n}^{(1)2}}{k(T-\theta)} - 2E_{x,n}^{(2)}\right)\exp\left[\frac{-E_n^{(0)}}{kT}\right] + \sum_{n\neq\pm 1}\left(\frac{E_{x,n}^{(1)2}}{kT} - 2E_{x,n}^{(2)}\right)\exp\left[\frac{-E_n^{(0)}}{kT}\right] \quad (6)$$

2.6. Cyclic Voltammogram of the Trinuclear Cobalt Complex

The redox behavior of the complex was studied by cyclic voltammetry. The cyclic voltammogram (Figure 9) showed an irreversible reduction wave at approximately -1.2 V vs. Fc/Fc$^+$, which may be assigned to the reduction of the terminal Co^{3+} ions in the reduction of the Co^{III}-Co^{II}-Co^{III} species. The corresponding oxidation wave can be observed at approximately -0.3 V vs. Fc/Fc$^+$. No oxidation wave was observed until $+1.0$ V vs. Fc/Fc$^+$ on the oxidation side. This result suggests that the trinuclear complex may not be maintained in the redox reaction, meaning that the stable form of the trinuclear species should be the Co^{III}-Co^{II}-Co^{III} mixed-valent state. A similar irreversible reduction wave was observed in $[Co^{II}\{Co^{III}(\mu\text{-}L^1)X_2\}_2]$ [51] and $[Co^{III}_2(\text{nitrosalpr})_2(CH_3OH)]$ (H_3nitrosalpr = 1,3-bis(5-nitrosalicylideneamino)-2-propanol) [29].

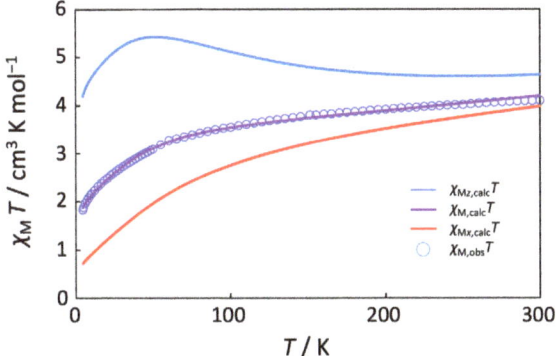

Figure 7. Variable temperature of $\chi_M T$ for $[Co_3(Hclsalpr)_2(CH_3COO)_4]$ (blue circle). The solid lines were calculated and drawn with the parameter values described in the text.

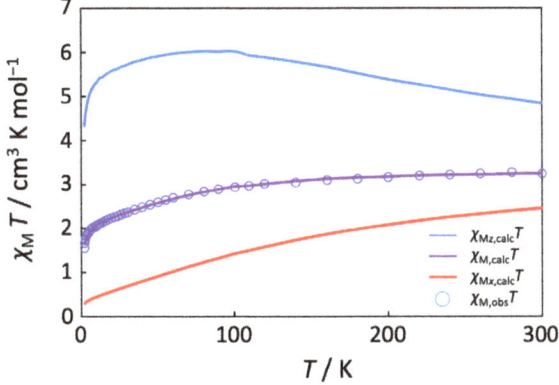

Figure 8. Variable temperature of $\chi_M T$ for $[Co_3(Hsalpr)_2(CH_3COO)_4]$ (blue circle) from the data in [40]. The solid lines were calculated and drawn with the parameter values described in the text.

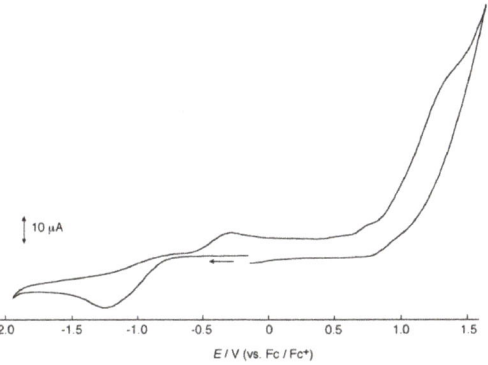

Figure 9. Cyclic voltammograms of $[Co_3(Hclsalpr)_2(CH_3COO)_4]$ in THF ([complex] = 1×10^{-3} M; [TBAP] = 0.2 M; scan rate = 100 mV s^{-1}).

3. Materials and Methods

All reagents and metal salts were obtained from commercial sources and used without further purification.

The Schiff-base ligand H_3clsalpr was prepared by the methods described in the literature [13,15,16]. An amount of 1,3-Diamino-2-propanol (2.177 g, 0.024 mol) and 5-chlorosalicylaldehyde (7.566 g, 0.048 mol) were dissolved in methanol (45 cm^3). The solution was refluxed for 3 h and then left at room temperature overnight. The resulting yellow crystals were filtered off and recrystallized from methanol. Yield, 4.149 g (46%). IR (KBr, cm^{-1}): ν(OH) 3130, ν(Ar-H) 3045, ν(C-H) 2891, ν(C=N) 1646.

Synthesis of [Co$_3$(Hclsalpr)$_2$(CH$_3$COO)$_4$]: To an acetonitrile solution (4 cm^3) of H$_3$clsalpr (36.7 mg, 0.1 mmol), Co(CH$_3$COO)$_2$·4H$_2$O (74.7 mg, 0.3 mmol) and five drops of triethylamine were added. The solution was allowed to stand in a refrigerator, producing dark-brown crystals of **1** in 47% yield (26.6 mg) after several days. Anal. Found: C, 41.49; H, 3.90; N, 4.56%. Calcd for C$_{42}$H$_{48}$Cl$_4$Co$_3$N$_4$O$_{18}$ ([Co$_3$(Hclsalpr)$_2$(CH$_3$COO)$_4$]·4H$_2$O): C, 41.50; H, 3.98; N, 4.61%. IR (KBr, cm^{-1}): ν(OH) 3426, 3216, ν(Ar-H) 3020, ν(C-H) 2927, ν(C=N) 1634; ν_{as}(COO) 1590, 1562, ν_s(COO) 1415, 1389. Diffuse reflectance spectra: λ_{max} 354, 568, 640, 1264 nm.

Analytical data of C, H, and N were obtained on a Thermo Finnigan FLASH EA1112 series CHNO-S analyzer (Thermo Finnigan, Milan, Italy). IR spectra were obtained by KBr discs of samples on a JASCO MFT-2000 FT-IR spectrometer (JASCO, Tokyo, Japan). Powder reflectance spectra were obtained on a Shimadzu Model UV-3100 UV-vis-NIR spectrophotometer (Shimadzu, Kyoto, Japan). Magnetic susceptibility measurements were obtained on a Quantum Design SQUID susceptometer (MPMS-XL7, Quantum Design North America, San Diego, CA, USA) with a magnetic field of 0.5 T over a temperature range of 4.5–300 K. The magnetic susceptibility χ_M is the molar magnetic susceptibility per mole of [Co$_3$(Hclsalpr)$_2$(CH$_3$COO)$_4$] unit and was corrected for the diamagnetic contribution calculated from Pascal's constants [73]. Cyclic voltammograms were measured in THF solutions containing tetra-n-butylammonium perchlorate (TBAP) on a BAS 100BW Electrochemical Workstation (Bioanalytical Systems, West Lafayette, IN, USA) with a glassy carbon electrode, a platinum wire counter electrode, and an Ag/Ag$^+$ reference electrode. Ferrocene (Fc) was used as an internal standard. All the potentials are quoted relative to Fc$^+$/Fc.

X-ray crystallographic data were collected on a Bruker Smart APEX CCD diffractometer (Bruker, Billerica, MA, USA) using graphite monochromated Mo-Kα radiation. The structures were solved by intrinsic phasing methods and refined by full-matrix least-squares methods. The hydrogen atoms were included at their geometrical positions calculated geometrically. All of the calculations were carried out using the SHELXTL software package [74]. Crystallographic data have been deposited with Cambridge Crystallographic Data Centre: Deposit number CCDC-2175785. Copies of the data can be obtained free of charge via http://www.ccdc.cam.ac.uk/conts/retrieving.html (accessed on 30 May 2022) (or from the Cambridge Crystallographic Data Centre, 12, Union Road, Cambridge, CB2 1EZ, UK; Fax: +44 1223 336033; e-mail: deposit@ccdc.cam.ac.uk).

4. Conclusions

In this study, new trinuclear cobalt complex was synthesized by the reaction of 1,3-bis(5-chlorosalicylideneamino)-2-propanol (H$_3$clsalpr) with cobalt(II) acetate tetrahydrate. The X-ray structure analysis revealed that a linear trinuclear Co^{III}-Co^{II}-Co^{III} complex was formed with two partially deprotonated Schiff-base ligands Hclsalpr^{2-}, two bridging acetate ligands, and two monodentate acetate ligands. The electronic absorption spectra and cyclic voltammetry data suggest that the mixed-valent oxidation state is stable. The temperature dependence of the magnetic susceptibilities is in accordance with the magnetic property of the central Co^{2+} ion becoming considerably anisotropic due to the axial distortion of the coordination geometry. This anisotropic property could also be found in the related trinuclear complex [Co$_3$(Hsalpr)$_2$(CH$_3$COO)$_4$]. The anisotropic magnetic behavior of the mixed-valent Co^{III}-Co^{II}-Co^{III} complexes is interesting as a potential application for single-molecule magnets. Further study to pursue such magnetic relaxation properties is planned in our laboratories.

Author Contributions: Conceptualization, M.M.; methodology, M.M.; investigation, Y.N., A.I., M.O., D.Y., H.S. and M.H.; data curation, M.M.; writing—original draft preparation, M.M.; writing—review and editing, M.M. and M.T. All authors have read and agreed to the published version of the manuscript.

Funding: This research received no external funding.

Institutional Review Board Statement: Not applicable.

Informed Consent Statement: Not applicable.

Data Availability Statement: Not applicable.

Conflicts of Interest: The authors declare no conflict of interest.

Sample Availability: Not applicable.

References

1. Sinn, E.; Harris, C.M. Schiff Base Metal Complexes as Ligands. *Coord. Chem. Rev.* **1969**, *4*, 391–422. [CrossRef]
2. Calligaris, M.; Nardin, G.; Randaccio, L. Structural aspects of metal complexes with some tetradentate Schiff bases. *Coord. Chem. Rev.* **1972**, *7*, 385–403. [CrossRef]
3. Mederos, A.; Dominguez, S.; Hernandez-Molina, R.; Sanchiz, J.; Brito, F. Coordinating ability of ligands derived from phenylenediamines. *Coord. Chem. Rev.* **1999**, *193–195*, 857–911. [CrossRef]
4. Sakamoto, M.; Manseki, K.; Okawa, H. d-f Heteronuclear complexes: Synthesis, structures and physicochemical aspects. *Coord. Chem. Rev.* **2001**, *219–221*, 379–414. [CrossRef]
5. Vigato, P.A.; Tamburini, S. The challenge of cyclic and acyclic schiff bases and related derivatives. *Coord. Chem. Rev.* **2004**, *248*, 1717–2128. [CrossRef]
6. Radecka-Paryzek, W.; Patroniak, V.; Lisowski, J. Metal complexes of polyaza and polyoxaaza Schiff base macrocycles. *Coord. Chem. Rev.* **2005**, *249*, 2156–2175. [CrossRef]
7. Hernandez-Molina, R.; Mederos, A. Acyclic and Macrocyclic Schiff Base Ligands. In *Comprehensive Coordination Chemistry II*; McCleverty, J.A., Meyer, T.J., Eds.; Elsevier: Amsterdam, The Netherlands, 2004; Volume 1, pp. 411–446.
8. Lewinski, J.; Prochowicz, D. Assemblies Based on Schiff Base Chemistry. In *Comprehensive Supramolecular Chemistry II*; Atwood, J.L., Gokel, G.W., Barbour, L.J., Eds.; Elsevier: Amsterdam, The Netherlands, 2017; Volume 6, pp. 279–304.
9. Liu, X.; Hamon, J.-R. Recent developments in penta-, hexa- and heptadentate Schiff base ligands and their metal complexes. *Coord. Chem. Rev.* **2019**, *389*, 94–118. [CrossRef]
10. Fukuhara, C.; Tsuneyoshi, K.; Matsumoto, N.; Kida, S.; Mikuriya, M.; Mori, M. Synthesis, and Characterization of Trinuclear Schiff-base Complexes containing Sulphur Dioxide or Hydrogensulphite Ions as Bridging Groups. Crystal Structure of [Zn{(μ-CH$_3$CO$_2$)(salpd-μ-O,O')Cu}$_2$] [salpd = propane-1,3-diylbis(salicylideneiminate)]. *J. Chem. Soc. Dalton Trans.* **1990**, *19*, 3473–3479. [CrossRef]
11. Liang, H.; Wang, X.-J.; Cong, Y.-L.; Li, S.-T.; Zeng, J.-Q. Synthesis and Crystal Structure of Schiff Base N,N'-Bis(2-Hydroxybenzylidene)-2-Hydroxy-1,3-propanediamine. *Jiegou Huaxue Chin. J. Struct. Chem.* **1997**, *16*, 141–143.
12. Elmali, A. Conformation and structure of 1,3-bis(2-hydroxy-5-bromosalicylideneamine)propan-2-ol. *J. Chem. Crystallogr.* **2000**, *30*, 473–477. [CrossRef]
13. Mikuriya, M.; Koyama, Y.; Mitsuhashi, R. Synthesis and Crystal Structure of 1,3-Bis(5-chloro-3-methoxysalicylideneamino)-2-propanol Trihydrate. *X-ray Struct. Anal. Online* **2019**, *35*, 33–34. [CrossRef]
14. Su, W.; Tian, D.; Yu, Z.; Yang, L.; Niu, Y. Crystal structure of 2-ethoxy-6-((E)-((3-(((E)-3-ethoxy-2-hydroxybenzylidene)amino)-2-hydroxypropyl)imino)methyl)phenolate, C$_{21}$H$_{26}$N$_2$O$_5$. *Zeitschrift für Kristallographie New Cryst. Struct.* **2020**, *235*, 53–54.
15. Mikuriya, M.; Tsuchimoto, N.; Koyama, Y.; Mitsuhashi, R.; Tsuboi, M. Crystal Structure of 1,3-Bis(3,5-dibromosalicylideneamino)-2-propanol. *X-ray Struct. Anal. Online* **2022**, *38*, 3–5. [CrossRef]
16. Mikuriya, M.; Naka, Y.; Yoshioka, D. Synthesis and Crystal Structure of 1,3-Bis(5-nitrosalicylideneamino)-2-propanol. *X-ray Struct. Anal. Online* **2016**, *32*, 11. [CrossRef]
17. Mazurek, W.; Kennedy, B.J.; Murray, K.S.; O'Connor, M.J.; Rodgers, J.R.; Snow, M.R.; Wedd, A.G.; Zwack, P.R. Magnetic interactions in Metal Complexes of Binucleating Ligands. 2. Synthesis and Properties of Binuclear Copper(II) Compounds Containing Exogenous Ligands That Bridge through Two Arms. Crystal and Molecular Structure of a Binuclear μ-Pyrazolato-N,N'-Bridged Dicopper(II) Complex of 1,3-Bis(salicylideneamino)propan-2-ol. *Inorg. Chem.* **1985**, *24*, 3258–3264.
18. Nishida, Y.; Kida, S. Crystal Structures and Magnetism of Binuclear Copper(II) Complexes with Alkoxide Bridges. Importance of Orbital Comlementarity in Spin Coupling through Two Different Bridging Groups. *J. Chem. Soc. Dalton Trans.* **1986**, *15*, 2633–2640. [CrossRef]

19. Butcher, R.J.; Diven, G.; Erickson, G.; Mockler, G.M.; Sinn, E. Copper Complexes of Binucleating N,N'-Hydroxyalkyldiaminebis (salicylidine) Ligands containing a Cu-O-Cu Bridge and an Exogenous Bridge. *Inorg. Chim. Acta* **1986**, *111*, L55–L56. [CrossRef]
20. Mikuriya, M.; Sasaki, T.; Anjiki, A.; Ikenoue, S.; Tokii, T. Binuclear Nickel(II) Complexes of Schiff Bases Derived from Salicylaldehydes and 1,n-Diamino-n'-hydroxyalkanes (n,n' = 3,2; 4,2; and 5,3) Having an Endogenous Alkoxo Bridge and a Pyrazolato Exogenous Bridge. *Bull. Chem. Soc. Jpn.* **1992**, *65*, 334–339. [CrossRef]
21. Bertoncello, K.; Fallon, G.D.; Murray, K.S.; Tiekink, E.R.T. Manganese(III) Complexes of a Binucleating Schiff-Base Ligand Based on the 1,3-Diaminopropan-2-ol Backbone. *Inorg. Chem.* **1991**, *30*, 3562–3568. [CrossRef]
22. Mikuriya, M.; Yamato, Y.; Tokii, T. 1,3-Bis(salicylideneamino)-2-propanol as the Ligand for Manganese(III) Ions. *Bull. Chem. Soc. Jpn.* **1992**, *65*, 1466–1468. [CrossRef]
23. Gelasco, A.; Pecoraro, V.L. [Mn(III)(2-OHsalpn)]$_2$ Is an Efficient Functional Model for the Manganese Catalases. *J. Am. Chem. Soc.* **1993**, *115*, 7928–7929. [CrossRef]
24. Gelasco, A.; Kirk, M.L.; Kampf, J.W.; Pecoraro, V.L. The [Mn$_2$(2-OHsalpn)$_2$]$^{2-,-,0,+}$ System: Synthesis, Structure, Spectroscopy, and Magnetism of the First Structurally Characterized Dinuclear Manganese Series Containing Four Distinct Oxidation States. *Inorg. Chem.* **1997**, *36*, 1829–1837. [CrossRef] [PubMed]
25. Bonadies, J.A.; Kirk, M.L.; Lah, M.S.; Kessissoglou, D.P.; Hatfield, W.E.; Pecoraro, V.L. Structurally Diverse Manganese(III) Schiff Base Complexes: Chains, Dimers, and Cages. *Inorg. Chem.* **1989**, *28*, 2037–2044. [CrossRef]
26. Kou, Y.-Y.; Zhao, Q.; Wang, X.-R.; Li, M.-L.; Ren, X.-H. Synthesis of three new binuclear Mn complexes: Characterization and DNA binding and cleavage properties. *J. Coord. Chem.* **2019**, *72*, 2393–2408. [CrossRef]
27. Mikuriya, M.; Koyama, Y.; Yoshioka, D.; Mitsuhashi, R. Dinuclear Manganese(III) Complex with a Schiff-base Having a Di-μ-acetato-μ-alkoxido-bridged Core. *X-ray Struct. Anal. Online* **2020**, *36*, 7–9. [CrossRef]
28. Lan, Y.; Novitchi, G.; Clérac, R.; Tang, J.-K.; Madhu, N.T.; Hewitt, I.J.; Anson, C.E.; Brooker, S.; Powell, A. Di-, tetra- and hexanuclear iron(III), manganese(II/III) and copper(II) complexes of Schiff-base ligands derived from 6-substituted-2-formylphenols. *Dalton Trans.* **2009**, *38*, 1721–1727. [CrossRef]
29. Mikuriya, M.; Naka, Y.; Yoshioka, D.; Handa, M. Synthesis and Crystal Structures of Dinuclear Cobalt(III) Complexes with 1,3-Bis(5-nitrosalicylideneamino)-2-propanol and 1,3-Bis(3-nitrosalicylideneamino)-2-propanol. *X-ray Struct. Anal. Online* **2016**, *32*, 55–58. [CrossRef]
30. Grundhoefer, J.P.; Hardy, E.E.; West, M.M.; Curtiss, A.B.; Gorden, A.E.V. Mononuclear Cu(II) and Ni(II) complexes of bis(naphthalen-2-ol) Schiff base ligands. *Inorg. Chim. Acta* **2019**, *484*, 125–132. [CrossRef]
31. Donmez, E.; Kara, H.; Karakas, A.; Unver, H.; Elmali, A. Synthesis, molecular structure, spectroscopic studies and second-order nonlinear optical behavior of N,N'-(2-hydroxy-propane-1,3-diyl)-bis(5-nitrosalicylaldiminato-N,O)-copper(II). *Spectrochim. Acta* **2007**, *A66*, 1141–1146. [CrossRef]
32. Kaczmarek, M.T.; Skrobanska, M.; Zabiszak, M.; Walesa-Chorab, M.; Kubicki, M.; Jastrzab, R. Coordination properties of N,N'-bis(5-methylsalicylidene)-2-hydroxy-1,3-propanediamine with d- and f-electron ions: Crystal structure, stability in solution, spectroscopic and spectroelectrochemical studies. *RSC Adv.* **2018**, *8*, 30994–31007. [CrossRef]
33. Azam, M.; Hussain, Z.; Warad, I.; Al-Resayes, S.; Khan, M.S.; Shakir, M.; Trzesowska-Kruszynska, A.; Kruszynski, R. Novel Pd(II)-salen complexes showing high in vitro anti-proliferative effects against human hepatoma cancer by modulating specific regulatory genes. *Dalton Trans.* **2012**, *41*, 10854–10864. [CrossRef] [PubMed]
34. Das, K.; Massera, C.; Garribba, E.; Frontera, A. Synthesis, structural and DFT interpretation of a Schiff base assisted Mn(III) derivative. *J. Mol. Struct.* **2020**, *1199*, 126985. [CrossRef]
35. Martinez, D.; Motevalli, M.; Watkinson, M. Is there really a diagnostically useful relationship between the carbon-oxygen stretching frequencies in metal carboxylate complexes and their coordination mode? *Dalton Trans.* **2010**, *39*, 446–455. [CrossRef] [PubMed]
36. Dieng, M.; Thiam, I.; Gaye, M.; Sall, A.S.; Barry, A.H. Synthesis, Crystal Structures and Spectroscopic Properties of a Trinuclear [Cu$_3$(HL)$_2$(NO$_3$)$_2$](H$_2$O)(CH$_2$CH$_2$OH) Complex and a [Mn(HL)(CH$_3$COOH)]$_n$ Polymer With H$_3$L=N,N'-(2-hydroxypropane-1,3-dihyl)-bis-(salicylaldimine). *Acta Chim. Slov.* **2006**, *53*, 417–423.
37. Mikuriya, M.; Tsuchimoto, N.; Koyama, Y.; Mitsuhashi, R.; Tsuboi, M. Crystal Structure of a Hydrolyzed Product of the Cobalt(III) Complex with 1-(3,5-Dichlorosalicylideneamino)-3-amino-2-propanol. *X-ray Struct. Anal. Online* **2022**, *38*, 9–11. [CrossRef]
38. Dey, D.; Kaur, G.; Patra, M.; Choudhury, A.R.; Kole, N.; Biswas, B. A perfectly linear trinuclear zinc-Schiff base complex: Synthesis, luminescence property and photocatalytic activity of zinc oxide nanoparticle. *Inorg. Chim. Acta* **2014**, *421*, 335–341. [CrossRef]
39. Jiang, L.; Zhang, D.-Y.; Suo, J.-J.; Gu, W.; Tian, J.-L.; Liu, X.; Yan, S.-P. Synthesis, magnetism and spectral studies of six defective dicubane tetranuclear {M$_4$O$_6$} (M = Ni^{II}, Co^{II}, Zn^{II}) and three trinuclear Cd^{II} complexes with polydentate Schiff base ligands. *Dalton Trans.* **2016**, *45*, 10233–10248. [CrossRef]
40. Baca-Solis, E.; Bernés, S.; Vazquez-Lima, H.; Boulon, M.-E.; Winpenny, R.E.P.; Reyes-Ortega, Y. Synthesis, Electronic, Magnetic and Structural Characterization of New Trinuclear Mixed-Valence Co^{III}-Co^{II}-Co^{III} Complex. *ChemistrySelect* **2016**, *1*, 6866–6871. [CrossRef]
41. Mikuriya, M.; Ikenoue, S.; Nukada, R.; Lim, J.-W. Synthesis and Structural Characterization of Tetranuclear Zinc(II) Complexes with a Linear Array. *Bull. Chem. Soc. Jpn.* **2001**, *74*, 101–102. [CrossRef]

42. Mikuriya, M.; Kudo, S.; Matsumoto, C.; Kurahashi, S.; Tomohara, S.; Koyama, Y.; Yoshioka, D.; Mitsuhashi, R. Mixed-valent tetranuclear manganese complexes with pentadentate Schiff-base ligand having a Y-shaped core. *Chem. Pap.* **2018**, *72*, 853–862. [CrossRef]
43. Mikuriya, M.; Kurahashi, S.; Tomohara, S.; Koyama, Y.; Yoshioka, D.; Mitsuhashi, R.; Sakiyama, H. Synthesis, Crystal Structures, and Magnetic Properties of Mixed-Valent Tetranuclear Complexes with Y-Shaped $Mn^{II}_2Mn^{III}_2$ Core. *Magnetochemistry* **2019**, *5*, 8. [CrossRef]
44. Bagai, R.; Abboud, K.A.; Christou, G. Ligand-induced distortion of a tetranuclear manganese butterfly complex. *Dalton Trans.* **2006**, *35*, 3306–3312. [CrossRef] [PubMed]
45. Kou, Y.-Y.; Li, M.-L.; Ren, X.-H. Synthesis, structure, and DNA binding/cleavage of two novel binuclear Co(II) complexes. *Spectrochim. Acta* **2018**, *A205*, 435–441. [CrossRef] [PubMed]
46. Banerjee, S.; Nandy, M.; Sen, S.; Mandal, S.; Rosair, G.M.; Slawin, A.M.Z.; Garcia, C.J.G.; Clemente-Juan, J.M.; Zangrando, E.; Guidolin, N.; et al. Isolation of four new Co^{II}/Co^{III} and Ni^{II} complexes with a pentadentate Schiff base ligand: Synthesis, structural descriptions and magnetic studies. *Dalton Trans.* **2011**, *40*, 1652–1661. [CrossRef] [PubMed]
47. Yang, F.-L.; Shao, F.; Zhu, G.-Z.; Shi, Y.-H.; Gao, F.; Li, X.-L. Structures and Magnetostructural Correlation Analyses for Two Novel Hexanuclear Complexes Based on a Pentadentate Schiff Base Ligand. *ChemistrySelect* **2017**, *2*, 110–117. [CrossRef]
48. Mikuriya, M.; Koyama, Y.; Kamioka, C.; Mitsuhashi, R.; Tsuboi, M. Mixed-valent Manganese Complex with a Schiff-base Having a Di-μ_4-oxido-di-μ_3-oxido-di-μ_3-carboxylato-hexa-μ-carboxylato-bridged $Mn^{II}_2Mn^{III}_6$ Core. *X-ray Struct. Anal. Online* **2022**, *38*, 33–35. [CrossRef]
49. Aromi, G.; Brechin, E.K. Synthesis of 3d Metallic Single-Molecule Magnets. In *Structure and Bonding*; Winpenny, R., Ed.; Springer: Berlin, Germany, 2006; Volume 122.
50. Mitsuhashi, R.; Pedersen, K.S.; Ueda, T.; Suzuki, T.; Bendix, J.; Mikuriya, M. Field-induced single-molecule magnet behavior in ideal trigonal antiprismatic cobalt(II) complexes: Precise geometrical control by a hydrogen-bonded rigid metalloligand. *Chem. Commun.* **2018**, *54*, 8869–8872. [CrossRef]
51. Pal, S.; Barik, A.K.; Gupta, S.; Roy, S.; Mandal, T.N.; Harza, A.; Fallah, M.S.E.; Butcher, R.J.; Peng, S.M.; Lee, G.-H.; et al. Anion dependent formation of linear trinuclear mixed valance cobalt(III/II/III) complexes and mononuclear cobalt(III) complexes of a pyrazole derived ligand—Synthesis, characterization and X-ray structures. *Polyhedron* **2008**, *27*, 357–365. [CrossRef]
52. Fukuhara, C.; Asato, E.; Shimoji, T.; Katsura, K.; Mori, M.; Matsumoto, K.; Ooi, S. Mixed Oxidation State Trinuclear Cobalt Complexes with Bridging Sulphito and Schiff-base Ligands. Part 1. Preparation of the Complexes $[Co^{II}(\mu\text{-}SO_3)_2(\mu\text{-}L)_2Co^{III}_2(ROH)_2]$ (L = Schiff base anion, R = alkyl) and Structure Determination of $[Co^{II}(\mu\text{-}SO_3)_2(\mu\text{-}a,a'\text{-}Me_2\text{-}salpd)_2Co^{III}_2(Pr^nOH)_2]\cdot 2Pr^nOH$. *J. Chem. Soc. Dalton Trans.* **1987**, *16*, 1305–1311. [CrossRef]
53. Chattopadhyay, S.; Bocelli, G.; Musatti, A.; Ghosh, A. First oxidative synthetic route of a novel linear mixed valence Co(III)—Co(II)—Co(III) complex with bridging acetate and salen. *Inorg. Chem. Commun.* **2006**, *9*, 1053–1057. [CrossRef]
54. Banerjee, S.; Chen, J.-T.; Lu, C.-Z. A new trinuclear mixed-valence Co(II)-Co(III) complex stabilized by a bis(salicylidene) based ligand. *Polyhedron* **2007**, *26*, 686–694. [CrossRef]
55. Chattopadhyay, S.; Drew, M.G.B.; Ghosh, A. Methylene Spacer-Regulated Structural Variation in Cobalt(II/III) Complexes with Bridging Acetate and Salen- or Salpn-Type Schiff-Base Ligands. *Eur. J. Inorg. Chem.* **2008**, *2008*, 1693–1701. [CrossRef]
56. Welby, J.; Rusere, L.N.; Tanski, J.M.; Tyler, L.A. Synthesis and crystal structures of mono-, di- and trinuclear cobalt complexes of a salen type ligand. *Inorg. Chim. Acta* **2009**, *362*, 1405–1411. [CrossRef]
57. Kotera, T.; Fujita, A.; Mikuriya, M.; Tsutsumi, H.; Handa, M. Cobalt complexes with 2-[(3-aminopropyl)amino]ethanethiol. *Inorg. Chem. Commun.* **2003**, *6*, 322–324. [CrossRef]
58. Mikuriya, M.; Fujita, A.; Kotera, T.; Yoshioka, D.; Sakiyama, H.; Handa, M.; Tsuboi, M. Synthesis, Crystal Structures, Electronic Spectra, and Magnetic Properties of Bis(μ-Thiolato)-Bridged Trinuclear Co^{II} Complexes with Tridentate-*N,N,S*-Thiolates. *Adv. Sustain. Sci. Eng. Technol.* **2021**, *3*, 0210102-01–0210102-13.
59. Dong, X.-Y.; Sun, Y.-X.; Wang, L.; Li, L. Synthesis and structure of a penta- and hexa-coordinated tri-nuclear cobalt(II) complex. *J. Chem. Res.* **2012**, *36*, 387–390. [CrossRef]
60. Li, X.-Y.; Kang, Q.-P.; Liu, L.-Z.; Ma, J.-C.; Dong, W.-K. Trinuclear Co(II) and Mononuclear Ni(II) Salamo-Type Bisoxime Coordination Compounds. *Crystals* **2018**, *8*, 43. [CrossRef]
61. Mikuriya, M.; Yamazaki, Y. Dinuclear Manganese(III) Complex with Cyclam-based Dodecadentate Ligand Bearing Schiff-base Pendants (Cyclam = 1,4,8,11-Tetraazacyclotetradecane). *Chem. Lett.* **1995**, *24*, 373–374. [CrossRef]
62. Nakamoto, K. *Infrared and Raman Spectra of Inorganic and Coordination Compounds, Part B: Application in Coordination, Organometallic, and Bioinorganic Chemistry*, 6th ed.; John Wiley & Sons: Hoboken, NJ, USA, 2009.
63. Wada, S.; Saka, K.; Yoshioka, D.; Mikuriya, M. Synthesis, Crystal Structures, and Magnetic Properties of Dinuclear and Hexanuclear Copper(II) Complexes with Cyclam-Based Macrocyclic Ligands Having Four Schiff-Base Pendant Arms. *Bull. Chem. Soc. Jpn.* **2010**, *83*, 364–374. [CrossRef]
64. Mikuriya, M.; Masuda, N.; Kakuta, Y.; Minato, S.; Inui, T.; Yoshioka, D. Mononuclear cobalt(III) complexes with *N*-salicylidene-2-hydroxy-5-bromobenzylamine and *N*-salicylidene-2-hydroxy-5-chlorobenzylamine. *Chem. Pap.* **2016**, *70*, 126–130. [CrossRef]
65. Murakami, Y.; Sakata, K. *Kireto Kagaku*; Ueno, K., Ed.; Nankodo: Tokyo, Japan, 1976; Volume 1. (In Japanese)
66. Lawrance, G.A. *Introduction to Coordination Chemistry*; John Wiley & Sons: Chichester, UK, 2010; pp. 113–115.

67. Grobelny, R.; Melnik, M.; Mrozinski, J. *Cobalt Coordination Compounds: Classification and Analysis of Crystallographic and Structural Data*; Dolnoslaskie Wydawnictwo Edukacyjne: Wroclaw, Poland, 1996.
68. Brown, I.D.; Wu, K.K. Empirical parameters for calculating cation-oxygen bond valences. *Acta Crystallogr. Sect. B* **1976**, *32*, 1957–1959. [CrossRef]
69. Brown, I.D.; Altermatt, D. Bond-valence parameters obtained from a systematic analysis of the inorganic crystal structure database. *Acta Crystallogr. Sect. B* **1985**, *41*, 244–247. [CrossRef]
70. Sakiyama, H.; Ito, R.; Kumagai, H.; Inoue, K.; Sakamoto, M.; Nishida, Y.; Yamasaki, M. Dinuclear Cobalt(II) Complexes of an Acyclic Phenol-Based Dinucleating Ligand with Four Methoxyethyl Chelating Arms—First Magnetic Analysis in an Axially Distorted Octahedral Field. *Eur. J. Inorg. Chem.* **2001**, *2001*, 2027–2032. [CrossRef]
71. Sakiyama, H. Development of MagSaki software for magnetic analysis of dinuclear high-spin cobalt(II) complexes in an axially distorted octahedral field. *J. Chem. Softw.* **2001**, *7*, 171–178. [CrossRef]
72. Sakiyama, H. Development of MagSaki(A) software for the magnetic analysis of dinuclear high-spin cobalt(II) complexes considering anisotropy in exchange interaction. *J. Comput. Chem. Jpn.* **2007**, *6*, 123–134. [CrossRef]
73. Kahn, O. *Molecular Magnetism*; VCH: Cambridge, UK, 1993.
74. Sheldrick, G.M. A short history of SHELX. *Acta Crystallogr. Sect. A* **2008**, *64*, 112–122. [CrossRef]

Article

Selective Formation of Unsymmetric Multidentate Azine-Based Ligands in Nickel(II) Complexes

Kennedy Mawunya Hayibor [1], Yukinari Sunatsuki [2] and Takayoshi Suzuki [1,3,*]

[1] Graduate School of Natural Science and Technology, Okayama University, Okayama 700-8530, Japan
[2] Advanced Science Research Center, Okayama University, Okayama 700-8530, Japan
[3] Research Institute for Interdisciplinary Science, Okayama University, Okayama 700-8530, Japan
* Correspondence: suzuki@okayama-u.ac.jp

Abstract: A mixture of 2-pyridine carboxaldehyde, 4-formylimidazole (or 2-methyl-4-formylimidazole), and NiCl$_2$·6H$_2$O in a molar ratio of 2:2:1 was reacted with two equivalents of hydrazine monohydrate in methanol, followed by the addition of aqueous NH$_4$PF$_6$ solution, afforded a Ni^{II} complex with two unsymmetric azine-based ligands, [Ni(HL^H)$_2$](PF$_6$)$_2$ (**1**) or [Ni(HL^{Me})$_2$](PF$_6$)$_2$ (**2**), in a high yield, where HL^H denotes 2-pyridylmethylidenehydrazono-(4-imidazolyl)methane and HL^{Me} is its 2-methyl-4-imidazolyl derivative. The spectroscopic measurements and elemental analysis confirmed the phase purity of the bulk products, and the single-crystal X-ray analysis revealed the molecular and crystal structures of the Ni^{II} complexes bearing an unsymmetric HL^H or HL^{Me} azines in a tridentate κ^3 N, N', N'' coordination mode. The HL^H complex with a methanol solvent, **1**·MeOH, crystallizes in the orthorhombic non-centrosymmetric space group $P2_12_12_1$ with Z = 4, affording conglomerate crystals, while the HL^{Me} complex, **2**·H$_2$O·Et$_2$O, crystallizes in the monoclinic and centrosymmetric space group $P2_1/n$ with Z = 4. In the crystal of **2**·H$_2$O·Et$_2$O, there is intermolecular hydrogen-bonding interaction between the imidazole N–H and the neighboring uncoordinated azine-N atom, forming a one-dimensional polymeric structure, but there is no obvious magnetic interaction among the intra- and interchain paramagnetic Ni^{II} ions.

Keywords: (pyridyl)(imidazolyl)azines; aldazines; kryptoracemate; crystal structure

Citation: Hayibor, K.M.; Sunatsuki, Y.; Suzuki, T. Selective Formation of Unsymmetric Multidentate Azine-Based Ligands in Nickel(II) Complexes. *Molecules* **2022**, *27*, 6788. https://doi.org/10.3390/molecules27206788

Academic Editor: Barbara Modec

Received: 30 September 2022
Accepted: 7 October 2022
Published: 11 October 2022

Publisher's Note: MDPI stays neutral with regard to jurisdictional claims in published maps and institutional affiliations.

Copyright: © 2022 by the authors. Licensee MDPI, Basel, Switzerland. This article is an open access article distributed under the terms and conditions of the Creative Commons Attribution (CC BY) license (https://creativecommons.org/licenses/by/4.0/).

1. Introduction

Azines are a class of organic molecules with diimine functionality, R^{1}R^{2}C=N–N=CR^{3}R^4, which are often regarded as analogs of 1,3-butadiene (R^{1}R^{2}C=CH–CH=CR^{3}R^4) due to the resemblance and similarity in their functional groups [1,2], and represent a well-known class of compounds with interesting chemical properties [3] and applications in several and diverse fields [2]. In the area of synthetic organic chemistry, they serve as excellent synthons for obtaining heterocyclic compounds such as purines, pyrazoles, and pyrimidines by undergoing [2,3] criss-cross cycloaddition in the presence of dienophiles [4–6]. They are good pharmaceutical and biological agents, for example, antimalarial, anticonvulsant, and antioxidant properties [7–10]. In the agricultural field, azines exhibit herbicidal properties [11,12]. Azine-based polymers are successfully used for chemosensors [13–15] and they also possess promising cathodic abilities for organic batteries [16–18] and could be also employed for 2D field effect transistors in their crystallized phase as silicon [19]. Some azine-based polymers possess promising cathodic abilities for organic batteries and are vital in optoelectronic devices. Furthermore, compounds containing the diimine (C=N–N=C) linkers have been investigated as highly luminescent frameworks [20–22] as well as good photosensitizers in solar cells [23].

The coordinating ability of the diimine group in the azines is interesting due to the flexibility of the N–N bond. A series of transition-metal complexes of azine-based ligands, which possess interesting structural motifs and other properties, have been reported [2,24].

Pyridyl ketazine is one of the unique azine-based ligands that have been explored by several researchers for its coordinating abilities and modes, thus paving way for the investigation of several azine-based ligands for their unique coordination chemistry [25–27]. Our previous studies reported a series of mono- and dinuclear iron(II) complexes with imidazole-4-carbaldehyde azine and other imidazole groups as azine ligands showcasing different modes of coordination towards the iron(II) in the various complexes [28–30].

Recently, we have investigated the chemistry of pyridyl and imidazolyl azine-based ligands, where an unprecedented selective synthesis of the unsymmetric (2-pyridyl)(2-methyl-4-imidazolyl)azine is obtained in an excellent yield with an interesting bonding mode [30]. Pursuing our interest in the unsymmetric azine-based ligand, we attempted to synthesize the corresponding nickel(II) complexes bearing 2-pyridylmethylidenehydrazono-(4-imidazolyl)methane (HL^H) and 2-pyridylmethylidenehydrazono-(2-methyl-4-imidazolyl) methane (HL^{Me}).

2. Results and Discussion

2.1. Preparation of Unsymmetric (2-Pyridyl)(4-Imidazolyl)azines and Their Nickel(II) Complexes

The most intuitive and possibly simplest method for preparation of an unsymmetric (2-pyridyl)(4-imidazolyl)azine is a reaction of stoichiometric amounts of 2-pyridinecarboxaldehyde, 1H-imidazole-4-carboxaldehyde (or its 2-methyl derivative), and hydrazine (Scheme 1) [31,32]. However, as was expected, all attempts to prepare the compounds HL^R with this method failed to isolate the desired compounds, because the reaction gave a complicated mixture of the products (Figure S1 in the Supplementary Materials) which were hard to be separated by any purification method. In a previous study [30], we serendipitously found that the reaction in the presence of iron(II) salts gave selectively the crystals of a Fe^{II} complex bearing unsymmetric azine, [Fe(HL^{Me})$_2$](PF$_6$)$_2$·1.5H$_2$O. To clarify the role of transition-metal salts in the selective formation of a certain complex, we used a nickel(II) chloride for the preparation of (2-pyridyl)(4-imidazolyl)azine complexes.

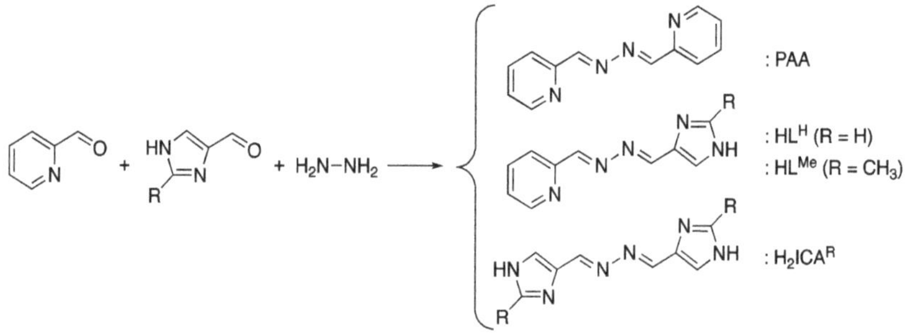

Scheme 1. Preparation of (2-pyridyl and/or 4-imidazolyl)azine compounds.

A mixture of 2-pyridine carboxaldehyde, 4-formylimidazole (or 2-methyl-4-formylimidazole), and NiCl$_2$·6H$_2$O in a 2:2:1 molar ratio in methanol was reacted with a stoichiometric amount of hydrazine monohydrate, followed by the addition of an aqueous solution of NH$_4$PF$_6$, which gave an obvious color change of the reaction solution to deep reddish orange. From the reaction mixture, air-stable deep reddish orange crude product (**1** from 4-formylimidazole or **2** from the 2-methyl derivative) was obtained by evaporation of the solvent in a relatively high yield (80% and 83% for compounds **1** and **2**, respectively). The crude products are soluble in common polar organic solvents and recrystallized from acetonitrile by vapor diffusion of methanol to deposit block-shaped deep reddish orange crystals of **1**·MeOH. For compound **2**, platelet single-crystals of **2**·H$_2$O·Et$_2$O suitable for X-ray diffraction study were deposited by vapor diffusion of diethyl ether into a methanol solution. In the FT-IR measurement of both compounds, the crude and recrystallized products gave almost identical spectra, which showed ν(C=N) stretching bands at 1619 and 1603 cm^{-1} for **1** and 1625 and 1603 cm^{-1} for **2** (Figure S2). This suggests that like the above-mentioned Fe^{II} complex [30], a certain Ni^{II} complex was selectively formed among several possible products. The elemental analyses of the vacuum-dried (partially efflorescent) samples suggested the empirical composition of [Ni(HL^H)$_2$](PF$_6$)$_2$·0.5MeOH and [Ni(HL^{Me})$_2$](PF$_6$)$_2$·MeCN·1.5MeOH for **1** and **2**, respectively.

The $\chi_M T$ values of **1** and **2** at 300 K are 1.14 and 1.21 cm^3 K mol^{-1}, respectively. These values are almost constant down to 20 K, then, decrease sharply below 20 K due to magnetic anisotropies of them (Figure S3). No significant magnetic interactions between complex cations were observed. In addition, magnetizations at 1.9 K for both complexes (Figure S3) did not reach the saturation values at 5 T, indicating the existence of magnetic anisotropies for both complexes. They are common behavior for magnetically isolated octahedral mononuclear nickel(II) complexes.

Absorption spectra of complexes **1** and **2** recorded in acetonitrile at room temperature were shown in Figure S4. Both complexes displayed two absorption bands in the region of 200–550 nm. The absorption bands in the higher energy region around 200–330 nm can be assigned to ligand-centered (LC) π–π* and n–π* transitions, respectively. The lowest energy absorption band for the complexes around 450–550 nm can be ascribed as the metal-to-ligand charge transfer (MLCT) band.

2.2. Crystal Structures of the Nickel(II) Complexes

The molecular and crystal structures of **1**·MeOH and **2**·H$_2$O·Et$_2$O was confirmed by the single-crystal X-ray analysis at 188(2) K. Compound **1**·MeOH crystallized in the orthorhombic system and a non-centrosymmetric space group $P2_12_12_1$ with Z = 4 (Table 1), indicating conglomerate crystallization (spontaneous resolution of the enantiomers). The asymmetric unit consists of one [Ni(HL^H)$_2$]$^{2+}$ cation, two PF$_6$$^-$ anions, and a methanol molecule of crystallization. An ORTEP drawing of **1** is shown in Figure 1. The Ni^{II} center was coordinated by two HL^H ligands in a pseudo-octahedral coordination geometry. Each HL^H ligand has an *E,Z* configuration (mode (i) in Scheme 2) serving as tridentate coordination to a Ni^{II} center in a meridional fashion via pyridyl-*N*, imidazolyl-*N*, and one of the azine-*N* atoms close to the pyridyl substitution group. This coordination mode forms a five-membered chelate ring on the pyridine side and a six-membered one on the imidazole side. It is noted that the other azine-*N* atom remains uncoordinated and the imidazole-NH group remains protonated.

Table 1. Crystallographic data of compounds **1**·MeOH and **2**·H$_2$O·Et$_2$O.

Compound	**1**·MeOH	**2**·H$_2$O·Et$_2$O
Chemical formula	C$_{21}$H$_{22}$F$_{12}$N$_{10}$NiOP$_2$	C$_{26}$H$_{34}$F$_{12}$N$_{10}$NiO$_2$P$_2$
Formula weight	779.10	867.25
T/K	188(2)	188(2)
Crystal color and shape	orange, block	orange, platelet
Size of specimen/mm	0.30 × 0.26 × 0.25	0.45 × 0.30 × 0.29
Crystal system	Orthorhombic	Monoclinic
Space group, Z	$P2_12_12_1$, 4	$P2_1/n$, 4
a/Å	8.6212(4)	12.5812(12)
b/Å	10.3267(6)	14.6221(12)
c/Å	33.20894(17)	19.9359(18)
β/°	90	98.664(3)
U/Å^3	2956.6(2)	3625.6(6)
D_{calc}/g cm^{-3}	1.750	1.589
μ(Mo Kα)/mm^{-1}	0.8783	0.7270
R_{int}	0.0547	0.0628
No. reflns/params.	6767/426	8299/478
R1 [F^2: $F_o^2 > 2\sigma(F_o^2)$]	0.0522	0.0903
wR2 (F^2: all data)	0.1351	0.2966
GoF	0.872	1.084
Flack param.	0.022(7)	—

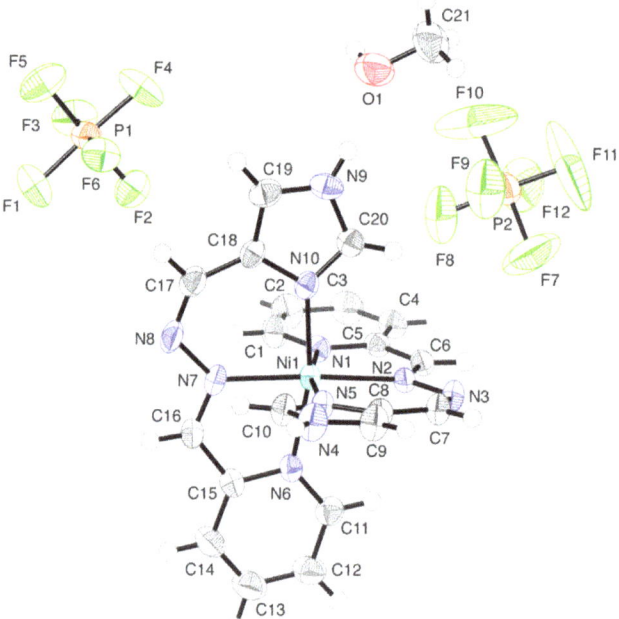

Figure 1. An ORTEP drawing of [Ni(HL^H)$_2$](PF$_6$)$_2$·MeOH (**1**·MeOH) with an atom-numbering scheme (ellipsoids are drawing at a 50% probability level).

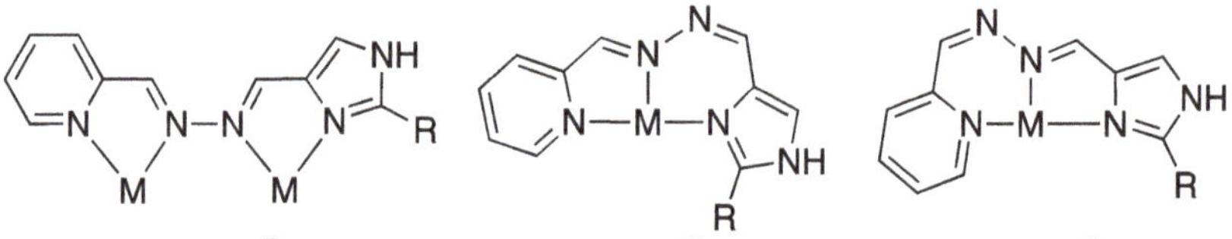

Scheme 2. Possible configuration and their bridging or tridentate coordination modes for HL^R azines.

The coordination bond lengths and angles of **1** are summarized in Table 2, which indicates the nearly ideal octahedral coordination geometry around the Ni center, with minor deviations. The Ni–N bond lengths are in the range of 2.039(5)–2.095(5) Å, which are typical for Ni^{II}–N(imine) coordination bonds [33,34]. The five-membered chelate bite angles (N1–Ni1–N2 and N6–Ni1–N7) are smaller by ca. 10° than the six-membered chelate bite angles (N2–Ni1–N5, N7–Ni1–N10), as expected. The mutually *trans* bond angle of N2–Ni1–N7 for the azine-*N* donors (175.7(2)°) is close to the ideal value.

The packing structure of **1**·MeOH was illustrated in Figure S5. In the crystal structure, an explicit hydrogen-bond was observed between one of the imidazole N–H group and the O atom of the methanol molecule of crystallization: N9(–H)···O1 2.710(8) Å (Figure 1), but no other intermolecular interactions were found. In a previous study, we reported the crystal structure of the analogous Fe^{II} complex, [Fe(HL^{Me})$_2$](PF$_6$)$_2$·1.5H$_2$O [30], in which a noble kryptoracemate resulted from a formation of a one-dimensional helical polymer by an intermolecular hydrogen-bonding interaction. In the present Ni^{II} complex **1**·MeOH, although the compound was crystallized in a non-enantiogenic (Sohncke) space group, $P2_12_12_1$, the complex cation was crystallized in a discrete form (Figure S5) and did not show the kryptoracemate phenomenon. We have tried to measure the solid-state CD spectra of a piece of single-crystal of **1**·MeOH (in a KBr disk), but no CD signal was observed.

Table 2. Selected bond lengths (l/Å) and angles (ϕ/°) around the Ni^{II} center in compounds **1**·MeOH and **2**·H_2O·Et_2O.

Compound	**1**·MeOH	**2**·H_2O·Et_2O
Ni1–N1, Ni–N6	2.094(4), 2.095(5)	2.110(4), 2.126(4)
Ni1–N2, Ni–N7	2.084(4), 2.082(4)	2.078(4), 2.093(4)
Ni1–N5, Ni–N10	2.047(5), 2.039(5)	2.068(4), 2.065(4)
N1–Ni1–N2, N6–Ni1–N7	79.22(19), 78.55(19)	78.80(16), 78.94(17)
N2–Ni1–N5, N7–Ni1–N10	87.99(18), 88.63(19)	88.95(16), 89.00(16)
N1–Ni1–N5, N6–Ni1–N10	167.20(18), 166.57(18)	166.81(16), 167.93(16)
N1–Ni1–N6, N5–Ni1–N10	88.85(18), 91.3(2)	91.91(18), 89.59(16)
N2–Ni1–N7	175.74(19)	165.77(16)
N1–Ni1–N7, N2–Ni1–N6	99.83(19), 97.26(17)	90.25(15), 92.25(16)
N1–Ni1–N10, N5–Ni1–N6	89.45(19), 93.27(19)	87.64(16), 93.46(16)
N2–Ni1–N10, N5–Ni1–N7	95.50(19), 92.97(19)	99.48(16), 102.60(16)

The compound, **2**·H_2O·Et_2O, crystallized in the monoclinic system and centrosymmetric space group $P2_1/n$ with Z = 4 (Table 1). The molecular structure of the Ni^{II} complex cation in **2** (Figure 2) is very similar to that in **1**, except for the large deviation of the bond angles, e.g., N2–Ni1–N7 and N5–Ni1–N7 (Table 2), which resulted from steric congestion from the substituted methyl group at the imidazole ring.

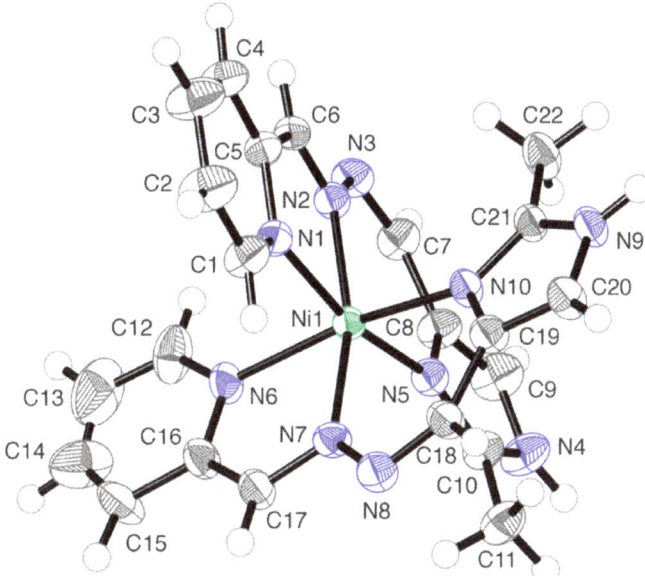

Figure 2. An ORTEP drawing of the complex cation $[Ni(HL^{Me})_2]^{2+}$ in **2**·H_2O·Et_2O with an atom-numbering scheme (ellipsoids are drawing at a 30% probability level).

In the crystal structure, the intermolecular hydrogen-bonding interaction was observed between the imidazole N–H and azine-N groups: N9(–H9)···N8 2.714(6) Å, forming one-dimensional coordination polymers (Figure 3). In contrast to the corresponding Fe^{II} complex [30], this Ni^{II} complex **2**·H_2O·Et_2O crystallized in a centrosymmetric space group $P2_1/n$, indicating the crystal consists of the racemic mixture.

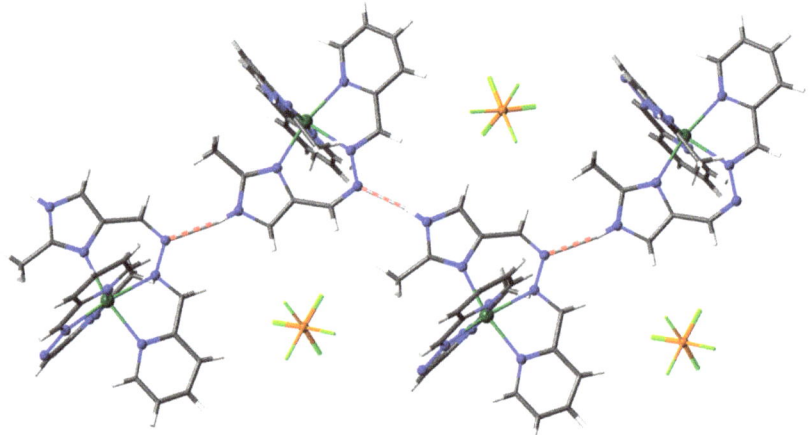

Figure 3. Intermolecular hydrogen-bonding interaction gives a one-dimensional polymer chain in 2·H_2O·Et_2O. Color code: Ni, teal; P orange; F, yellow-green; N, blue; C, gray; H, silver.

3. Materials and Methods

3.1. Chemicals and Physical Methods

All chemicals and solvents used for syntheses of azine compounds and Ni complexes were reagent grade and used without further purification. First, 2-pyridinecarboxaldehyde, 1H-imidazole-4-carboxaldehyde, 2-methyl-1H-imidazole-4-carboxaldehyde, nickel(II) chloride hexahydrate, and ammonium hexafluorophosphate were purchased from FUJIFILM (Tokyo, Japan). All reactions were carried out under aerobic conditions. Infrared spectra (KBr pellets; 4000–400 cm^{-1}) were recorded on a JASCO FT-001 Fourier transform infrared spectrometer (JASCO, Tokyo, Japan). Absorption spectra were recorded on a Shimadzu UV/Vis-1650 spectrophotometer (Kyoto, Japan) in the range of 200–600 nm at room temperature in acetonitrile. The ^{1}H NMR spectra were acquired on a Varian 400-MR spectrometer (Los Angeles, CA, USA); the chemical shifts were referenced to residual ^{1}H NMR signals of solvents and are reported versus TMS. Elemental analyses were conducted at Advanced Science Research Center, Okayama University. Magnetic susceptibilities were measured on a Quantum Design MPMS XL5 SQUID magnetometer (Tokyo, Japan) in a 1.9–300 K temperature range under an applied magnetic field of 0.1 T at the Okayama University of Science. Corrections for diamagnetism were applied using Pascal's constants [35].

3.2. Preparation of Nickel(II) Complexes

3.2.1. [Ni(HL^H)$_2$](PF$_6$)$_2$ (1)

A methanol solution (30 mL) of NiCl$_2$·6H$_2$O (0.477 g, 2.00 mmol) was added to a methanol solution (60 mL) containing 2-pyridinecarboxaldehyde (0.432 g, 4.00 mmol) and 1H-imidazole-4-carboxaldehyde (0.387 g, 4.00 mmol), followed by additions of hydrazine monohydrate (0.207 g, 4.00 mmol) in methanol (30 mL) and NH$_4$PF$_6$ (0.652 g, 4.00 mmol) in water (20 mL). The mixture was stirred at ca. 60 °C for 3 h. The resulting solution was concentrated by a rotary vacuum evaporator to give a deep reddish-orange precipitate. The crude product was dissolved in methanol and acetonitrile and recrystallized by slow evaporation to deposit deep reddish-orange microcrystals. Crystals suitable for the SC-XRD study were obtained from a mixture of acetonitrile and methanol. Yield: 1.72 g (80%). Anal. Found: C, 32.41; H, 2.36; N, 18.38%. Calcd for C$_{20.5}$H$_{20}$F$_{12}$NiN$_{10}$O$_{0.5}$P$_2$ (for **1**·0.5MeOH: C, 32.27; H, 2.64; N, 18.36%. IR (KBr pellet): ν$_{C=N}$ (imine) 1619, 1603 cm^{-1}; ν$_{P-F}$ (PF$_6^-$) 840 cm^{-1}.

3.2.2. [Ni(HL^{Me})$_2$](PF$_6$)$_2$ (2)

Complex **2** was obtained in a similar manner using 2-methyl-1*H*-imidazole-4-carboxaldehyde instead of 1*H*-imidazole-4-carboxaldehyde. Yield: 83%. Anal. Found: C, 35.45; H, 3.43; N, 17.95%. Calcd for C$_{25.5}$H$_{31}$F$_{12}$NiO$_{1.5}$P$_2$ (for **2**·CH$_3$CN·1.5CH$_3$OH: C, 35.44; H, 3.62; N, 17.83. IR (KBr pellet cm^{-1}): $\nu_{C=N}$ (imine) 1635, 1609 (fs) ν_{P-F} (PF$_6^-$) 845(s). Deep reddish-orange platelet crystals (**2**·H$_2$O·Et$_2$O) suitable for SC-XRD were obtained from a mixture of methanol and diethyl ether.

3.3. Structure Determination by X-ray Crystallography

The single-crystal X-ray diffraction data for compounds **1**·MeOH and **2**·H$_2$O·Et$_2$O were collected at 188(2) K using a Rigaku RAXIS RAPID II imaging plate area detector employing graphite monochromated Mo Kα radiation (λ = 0.71073 Å). The structures were solved by the direct method, employing the SIR2014 software packages [36], and refined on F^2 by full-matrix least-squares techniques using the SHELXL2014 program package [37]. All non-hydrogen atoms were refined anisotropically, and hydrogen atoms were included in the calculations with riding models. All calculations were performed using the Crystal Structure software package [38]. The crystal parameters, data collection procedure, and refinement results for the two compounds **1**·MeOH and **2**·H$_2$O·Et$_2$O are summarized in Table 1.

4. Conclusions

In this study, we attempted to prepare transition-metal(II) complexes of an unsymmetrical azine-type ligand, HL^R, having 2-pyridyl and (2-methyl-)1*H*-imidazol-4-yl substituent groups. The desired azine could not be isolated in pure form from a simple stoichiometric reaction of hydrazine and respective aldehydes. However, in our previous study using Fe^{II} salts, a highly selective formation of [Fe(HL^{Me})$_2$](PF$_6$)$_2$·1.5H$_2$O was observed, and the complex was found to be a kryptoracemate as a result of a one-dimensional helical chain structure by hydrogen-bonding interaction. At present, we have studied another two cases with nickel(II) salts: [Ni(HL^H)$_2$](PF$_6$)$_2$·MeOH (**1**·MeOH) and [Ni(HL^{Me})$_2$](PF$_6$)$_2$·H$_2$O·Et$_2$O (**2**·H$_2$O·Et$_2$O). In both cases, a highly selective formation of the unsymmetrical azine complex was observed among other possible symmetrical and/or unsymmetrical complexes.

In the crystal of **1**·MeOH, the compound was crystallized in a non-enantiogenic (Sohncke) space group, $P2_12_12_1$, but the complex cation, [Ni(HL^H)$_2$]$^{2+}$, was only hydrogen-bonded to the solvated methanol molecule. In the crystal structure of **2**·H$_2$O·Et$_2$O, there observed a one-dimensional hydrogen-bonded polymer chain made from [Ni(HL^{Me})$_2$]$^{2+}$, but it was crystallized in a centrosymmetric space group, $P2_1/c$. Thus, it can be concluded that the reason for the selective formation of an unsymmetric azine ligand in [Fe(HL^{Me})$_2$](PF$_6$)$_2$·1.5H$_2$O was not solely the formation of the characteristic hydrogen-bonded chain. The suitable tridentate chelate formation of E,Z-HL^R with mode (i) (in Scheme 2), which gives a five-membered chelate ring at the pyridyl coordination site and a six-membered chelate ring at the imidazolyl one, would probably be the most stable among the other coordination modes of symmetrical and unsymmetrical azine derivatives.

Supplementary Materials: The following supporting information can be downloaded at: https://www.mdpi.com/article/10.3390/molecules27206788/s1, Text for Syntheses of the symmetrical and unsymmetrical azine compounds; Figure S1: ^{1}H NMR Spectra of the azines; Figure S2: FT-IR spectra of [Ni(HL^H)$_2$](PF$_6$)$_2$ (**1**) and [Ni(HL^{Me})$_2$](PF$_6$)$_2$ (**2**); Figure S3: The $\chi_M T$ vs T plots of **1** and **2** and magnetizations of **1** and **2** at 1.9 K; Figure S4: Absorption spectra of **1** and **2** in acetonitrile at room temperature; Figure S5: Packing diagram of **1**·MeOH viewed along the crystallographic a axis; Figure S6: Packing diagrams of **2**·H$_2$O·Et$_2$O viewed along crystallographic a and b axes.

Author Contributions: Conceptualization, Y.S. and T.S.; methodology, K.M.H. and Y.S.; validation, K.M.H., Y.S. and T.S.; formal analysis, K.M.H., Y.S. and T.S.; investigation, K.M.H. and Y.S.; resources, Y.S. and T.S.; data curation, K.M.H.; writing—original draft preparation, K.M.H.; writing—review and editing, T.S.; visualization, K.M.H., Y.S. and T.S.; supervision, T.S.; project administration, T.S.; funding acquisition, T.S. All authors have read and agreed to the published version of the manuscript.

Funding: This research was supported by JSPS KAKENHI Grant No. 21K05084.

Data Availability Statement: Crystallographic data for compounds 1·MeOH and 2·H$_2$O·Et$_2$O have been deposited with the Cambridge Crystallographic Data Centre, CCDC 2209932, and 2209933, respectively. These data can be obtained free of charge from The Cambridge Crystallographic Data Centre via www.ccdc.cam.ac.uk/data_request/cif (accessed on 28 September 2022).

Acknowledgments: K.M.H. thanks to RIIS, Okayama University for generous support.

Conflicts of Interest: The authors declare no conflict of interest.

References

1. Safari, J.; Gandomi-Ravandi, S. Structure, synthesis and application of azines: A historical perspective. *RSC Adv.* **2014**, *4*, 46224–46249. [CrossRef]
2. Chourasiya, S.S.; Kathuria, D.; Wani, A.A.; Bharatam, P.V. Azines: Synthesis, structure, electronic structure and their applications. *Org. Biomol. Chem.* **2019**, *17*, 8486–8521. [CrossRef]
3. Hopkins, J.M.; Bowdridge, M.; Robertson, K.N.; Cameron, T.S.; Jenkins, H.A.; Clyburne, J.A.C. Generation of Azines by the Reaction of a Nucleophilic Carbene with Diazoalkanes: A Synthetic and Crystallographic Study. *J. Org. Chem.* **2001**, *66*, 5713–5716. [CrossRef]
4. Padwa, A. *1,3-Dipolar Cycloaddition Chemistry*; John Wiley and Sons: New York, NY, USA, 1984; Volumes 1 and 2.
5. El-Alali, A.; Al-Kamali, A.S. Reactions of 1,3-dipolar aldazines and ketazines with the dipolarophile dimethyl acetylenedicarboxylate. *Can. J. Chem.* **2002**, *80*, 1293–1301. [CrossRef]
6. Meth-Cohn, O.; Smalley, R.K. Heterocyclic chemistry. *Annu. Rep. Prog. Chem. Sect. B Org. Chem.* **1976**, *73*, 239–277. [CrossRef]
7. Godara, M.; Maheshwari, R.; Varshney, S.; Varshney, A.K. Synthesis and characterization of some new coordination compounds of boron with mixed azines. *J. Serb. Chem. Soc.* **2007**, *72*, 367–374. [CrossRef]
8. Sheng, R.; Wang, P.; Liu, W.; Wu, X.; Wu, S. A new colorimetric chemosensor for Hg^{2+} based on coumarin azine derivative. *Sens. Actuators B Chem.* **2008**, *128*, 507–511. [CrossRef]
9. Kim, S.-H.; Gwon, S.-Y.; Burkinshaw, S.M.; Son, Y.-A. The synthesis and proton-induced spectral switching of a novel azine dye and its boron complex. *Dye. Pigment.* **2010**, *87*, 268–271. [CrossRef]
10. Bodtke, A.; Pfeiffer, W.-D.; Ahrens, N.; Langer, P. Horseradish peroxidase (HRP) catalyzed oxidative coupling reactions using aqueous hydrogen peroxide: An environmentally benign procedure for the synthesis of azine pigments. *Tetrahedron* **2005**, *61*, 10926–10929. [CrossRef]
11. Dolezal, M.; Kralov, K. Synthesis and Evaluation of Pyrazine Derivatives with Herbicidal Activity. In *Herbicides, Theory, and Applications*; Larramendy, M.L., Soloneski, S., Eds.; IntechOpen Limited: London, UK, 2011.
12. Moreland, D.E. Biochemical Mechanisms of Action of Herbicides and the Impact of Biotechnology on the Development of Herbicides. *J. Pestic. Sci.* **1999**, *24*, 299–307. [CrossRef]
13. Manigandan, S.; Muthusamy, A.; Nandhakumar, R.; David, C.I.; Anand, S. Synthesis, characterization, theoretical investigations and fluorescent sensing behavior of oligomeric azine-based Fe3+ chemosensors. *High Perform. Polym.* **2022**, *34*, 321–336. [CrossRef]
14. Sawminathan, S.; Munusamy, S.; Manickam, S.; Jothi, D.; KulathuIyer, S. Azine based fluorescent rapid "off-on" chemosensor for detecting Th^{4+} and Fe^{3+} ions and its real-time application. *Dye. Pigment.* **2021**, *196*, 109755. [CrossRef]
15. Irmi, N.M.; Purwono, B.; Anwar, C. Synthesis of Symmetrical Acetophenone Azine Derivatives as Colorimetric and Fluorescent Cyanide Chemosensors. *Indones. J. Chem.* **2021**, *21*, 1337–1347. [CrossRef]
16. Acker, P.; Speer, M.E.; Wössner, J.S.; Esser, B. Azine-based polymers with a two-electron redox process as cathode materials for organic batteries. *J. Mater. Chem. A* **2020**, *8*, 11195–11201. [CrossRef]
17. Lyu, H.; Sun, X.G.; Dai, S. Organic Cathode Materials for Lithium-Ion Batteries: Past, Present, and Future. *Adv. Energy Sustain. Res.* **2021**, *2*, 2000044. [CrossRef]
18. Hager, M.D.; Esser, B.; Feng, X.; Schuhmann, W.; Theato, P.; Schubert, U.S. Polymer-Based Batteries—Flexible and Thin Energy Storage Systems. *Adv. Mater.* **2020**, *32*, 2000587. [CrossRef]
19. Tantardini, C.; Kvashnin, A.G.; Gatti, C.; Yakobson, B.I.; Gonze, X. Computational Modeling of 2D Materials under High Pressure and Their Chemical Bonding: Silicene as Possible Field-Effect Transistor. *ACS Nano* **2021**, *15*, 6861–6871. [CrossRef]
20. Singh, A.; Kociok-Köhn, G.; Chauhan, R.; Muddassir, M.; Gosavi, S.W.; Kumar, A. Ferrocene Appended Asymmetric Sensitizers with Azine Spacers with phenolic/nitro anchors for Dye-Sensitized Solar Cells. *J. Mol. Struct.* **2021**, *1249*, 131630. [CrossRef]
21. Dalapati, S.; Jin, S.; Gao, J.; Xu, Y.; Nagai, A.; Jiang, D. An Azine-Linked Covalent Organic Framework. *J. Am. Chem. Soc.* **2013**, *135*, 17310–17313. [CrossRef]

22. Konavarapu, S.K.; Biradha, K. Luminescent Triazene-Based Covalent Organic Frameworks Functionalized with Imine and Azine: N_2 and H_2 Sorption and Efficient Removal of Organic Dye Pollutants. *Cryst. Growth Des.* **2019**, *19*, 362–368. [CrossRef]
23. Sęk, D.; Siwy, M.; Małecki, J.G.; Kotowicz, S.; Golba, S.; Nowak, E.M.; Sanetra, J.; Schab-Balcerzak, E. Polycyclic aromatic hydrocarbons connected with Schiff base linkers: Experimental and theoretical photophysical characterization and electrochemical properties. *Spectrochim. Acta Part A Mol. Biomol. Spectrosc.* **2017**, *175*, 168–176. [CrossRef] [PubMed]
24. Yang, Q.-F.; Cui, X.-B.; Yu, J.-H.; Lu, J.; Yu, X.-Y.; Zhang, X.; Xu, J.-Q.; Hou, Q.; Wang, T.-G. A series of metal–organic complexes constructed from in situ generated organic amines. *CrystEngComm* **2008**, *10*, 1534–1541. [CrossRef]
25. Stratton, W.J.; Busch, D.H. The Complexes of Pyridinaldazine with Iron(II) and Nickel(II). *J. Am. Chem. Soc.* **1958**, *80*, 1286–1289. [CrossRef]
26. Stratton, W.J.; Rettig, M.F.; Drury, R.F. Metal complexes with azine ligands. I. Ligand hydrolysis and template synthesis in the iron(II)-2-Pyridinaldazine system. *Inorg. Chim. Acta* **1969**, *3*, 97–102. [CrossRef]
27. Stratton, W.J. Metal complexes with azine ligands. II. Iron(II), cobalt(II), and nickel(II) complexes with 2-pyridyl methyl ketazine. *Inorg. Chem.* **1970**, *9*, 517–520. [CrossRef]
28. Sunatsuki, Y.; Kawamoto, R.; Fujita, K.; Maruyama, H.; Suzuki, T.; Ishida, H.; Kojima, M.; Iijima, S.; Matsumoto, N. Structures and Spin States of Bis(tridentate)-Type Mononuclear and Triple Helicate Dinuclear Iron(II) Complexes of Imidazole-4-carbaldehyde azine. *Inorg. Chem.* **2009**, *48*, 8784–8795. [CrossRef]
29. Sunatsuki, Y.; Maruyama, H.; Fujita, K.; Suzuki, T.; Kojima, M.; Matsumoto, N. Mononuclear Bis(tridentate)-Type and Dinuclear Triple Helicate Iron(II) Complexes Containing 2-Ethyl-5-methylimidazole-4-carbaldehyde Azine. *Bull. Chem. Soc. Jpn.* **2009**, *82*, 1497–1505. [CrossRef]
30. Sunatsuki, Y.; Fujita, K.; Maruyama, H.; Suzuki, T.; Ishida, H.; Kojima, M.; Glaser, R. Chiral Crystal Structure of a P212121 Kryptoracemate Iron(II) Complex with an Unsymmetric Azine Ligand and the Observation of Chiral Single Crystal Circular Dichroism. *Cryst. Growth Des.* **2014**, *14*, 3692–3695. [CrossRef]
31. Safari, J.; Gandomi-Ravandi, S. Highly Efficient Practical Procedure for the Synthesis of Azine Derivatives Under Solvent-Free Conditions. *Synth. Commun.* **2011**, *41*, 645–651. [CrossRef]
32. Safari, J.; Gandomi-Ravandi, S.; Monemi, M. Novel and selective synthesis of unsymmetrical azine derivatives via a mild reaction. *Monatsh. Chem.* **2013**, *144*, 1375–1380. [CrossRef]
33. Tang, B.; Ye, J.-H.; Ju, X.-H. Computational Study of Coordinated Ni(II) Complex with High Nitrogen Content Ligands. *ISRN Org. Chem.* **2011**, *2011*, 920753. [CrossRef] [PubMed]
34. Colpas, G.J.; Kumar, M.; Day, R.O.; Maroney, M.J. Structural investigations of nickel complexes with nitrogen and sulfur donor ligands. *Inorg. Chem.* **1990**, *29*, 4779–4788. [CrossRef]
35. Bain, G.A.; Berry, J.F. Diamagnetic Corrections and Pascal's Constants. *J. Chem. Educ.* **2008**, *85*, 532–536. [CrossRef]
36. Burla, M.C.; Caliandro, R.; Carrozzini, B.; Cascarano, G.L.; Cuocci, C.; Giacovazzo, C.; Mallamo, M.; Mazzone, A.; Polodori, G. Crystal structure determination, and refinement via SIR2014. *J. Appl. Crystallogr.* **2015**, *48*, 306–309. [CrossRef]
37. Sheldrick, G.M. Crystal structure refinement with *SHELXL*. *Acta Crystallogr. Sect. C Struct. Chem.* **2015**, *C71*, 3–8. [CrossRef]
38. Rigaku Co., Ltd. *CrystalStructure*; Rigaku Co., Ltd.: Akishima, Tokyo, 2000–2016.

Article

Synthesis and Characterization of Lanthanide Metal Ion Complexes of New Polydentate Hydrazone Schiff Base Ligand

Izabela Pospieszna-Markiewicz, Marta A. Fik-Jaskółka, Zbigniew Hnatejko, Violetta Patroniak and Maciej Kubicki *

Faculty of Chemistry, Adam Mickiewicz University, Uniwersytetu Poznańskiego 8, 61-614 Poznań, Poland
* Correspondence: mkubicki@amu.edu.pl

Abstract: The new homodinuclear complexes of the general formula [$Ln_2L_3(NO_3)_3$] (where **HL** is newly synthesized 2-((2-(benzoxazol-2-yl)-2-methylhydrazono)methyl)phenol and Ln = Sm^{3+} (**1**), Eu^{3+} (**2**), Tb^{3+} (**3a**, **3b**), Dy^{3+} (**4**), Ho^{3+} (**5**), Er^{3+} (**6**), Tm^{3+} (**7**), Yb^{3+} (**8**)), have been synthesized from the lanthanide(III) nitrates with the polydentate hydrazone Schiff base ligand. The flexibility of this unsymmetrical Schiff base ligand containing N_2O binding moiety, attractive for lanthanide metal ions, allowed for a self-assembly of these complexes. The compounds were characterized by spectroscopic data (ESI-MS, IR, UV/Vis, luminescence) and by the X-ray structure determination of the single crystals, all of which appeared to be different solvents. The analytical data suggested 2:3 metal:ligand stoichiometry in these complexes, and this was further confirmed by the structural results. The metal cations are nine-coordinated, by nitrogen and oxygen donor atoms. The complexes are two-centered, with three oxygen atoms in bridging positions. There are two types of structures, differing by the sources of terminal (non-bridging) coordination centers (group A: two ligands, one nitro anion/one ligand, two nitro anions, group B: three ligands, three anions).

Keywords: Schiff base complexes; lanthanides; X-ray structures; spectroscopy; luminescence studies

Citation: Pospieszna-Markiewicz, I.; Fik-Jaskółka, M.A.; Hnatejko, Z.; Patroniak, V.; Kubicki, M. Synthesis and Characterization of Lanthanide Metal Ion Complexes of New Polydentate Hydrazone Schiff Base Ligand. *Molecules* **2022**, *27*, 8390. https://doi.org/10.3390/molecules27238390

Academic Editor: Hiroshi Sakiyama

Received: 24 October 2022
Accepted: 29 November 2022
Published: 1 December 2022

Publisher's Note: MDPI stays neutral with regard to jurisdictional claims in published maps and institutional affiliations.

Copyright: © 2022 by the authors. Licensee MDPI, Basel, Switzerland. This article is an open access article distributed under the terms and conditions of the Creative Commons Attribution (CC BY) license (https://creativecommons.org/licenses/by/4.0/).

1. Introduction

For decades, the coordination chemistry of lanthanides has been a subject of great interest. Lanthanides and their complexes present attractive physical (e.g., spin crossover, single molecule magnetism, luminescence) [1–14], chemical [15], biological (antimicrobial, antiradical, anticancer) [16–22] and catalytic [23,24] properties. The versatility and robustness of lanthanides complexes with N, O donor ligands like hydrazones and their analogues caused the formation of the diverse classes of compounds.

Compounds containing polydentate hydrazone ligands exhibit a huge variety of molecular architectures. A flexible ligand may provide more possibilities of unique, thermally and kinetically stable, coordination modes with the lanthanide ions of various ionic radii. Lanthanide contraction and the electron configuration details, very important factors affecting the structures of the complexes, may be responsible for the difficulties in controlling the coordination environment around the metal ions. Additionally, lanthanides show the low stereochemical preference and the coordination compounds with these cations usually have high coordination numbers, e.g., typically 8 or 9 for the heavy lanthanides [25–27].

It is generally assumed that the complexes are formed in self-assembly process in which a disordered system of substrates is organized to supramolecular crystal structure. In such a process the role of solvent molecules, which can be and often really are the important parts of solid state structures, is also very important.

As a part of our ongoing research on the coordination properties of hydrazone ligands [28,29], we reported the study concerning the chelating abilities of the potentially polydentate new hydrazone Schiff base ligand (**HL**) 2-((2-(benzoxazol-2-yl)-2-methylhydrazono)methyl)phenol. A series of the lanthanide complexes of general formula [$Ln_2(L)_3(NO_3)_3$]·solvent (Ln = Sm^{3+} (**1**),

Eu^{3+} (**2**), Tb^{3+} (**3a, 3b**), Dy^{3+} (**4**), Ho^{3+} (**5**), Er^{3+} (**6**), Tm^{3+} (**7**), Yb^{3+} (**8**)) has been reported, and their structural and luminescent properties were studied in detail. Photoluminescence studies show that complexes Eu(**2**) and Tb(**3a, 3b**) show characteristic emission in the visible region while Ho(**5**), Er(**6**), Tm(**7**) and Yb(**8**) in the near-infrared region. Photoluminescence spectra demonstrated that Er(**6**) and Yb(**8**) in these compounds can be effectively excited at 369 nm.

The results from this study show an interesting binding trend between N,N,O-donor ligand and selected Ln(III), as different coordination environments for the Ln atoms in the same molecule are observed. In the solid-state structures all the complexes are two-centered, lanthanide cations are always 9-coordinated in a distorted tricapped trigonal prism geometry, however, the metal coordination patterns are different depending on the crystallization conditions and/or metal cation. According to these differences, the complexes can be divided into two groups (**A** and **B**, Scheme 1). The donor atoms of functional groups from the chelating ligands play here an important role. The coordination environments of both central cations in each complex are different and range from N_6O_3, through N_4O_5, N_2O_7 to O_9 donor atom set.

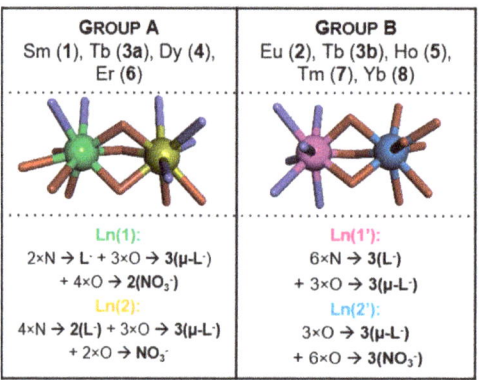

Scheme 1. Schematic representations of the two types (groups A and B) of a coordination patterns of metal ion centers.

2. Results and Discussion

2.1. Synthesis, ESI-MS and IR Spectroscopy, Thermal Analysis of the Complexes

The polydentate hydrazone Schiff base type metal ion complexes Sm^{3+} (**1**), Eu^{3+} (**2**), Tb^{3+} (**3a**), Tb^{3+} (**3b**), Dy^{3+} (**4**), Ho^{3+} (**5**), Er^{3+} (**6**), Tm^{3+} (**7**) and Yb^{3+} (**8**) containing ligand **HL** (Scheme 2 and Scheme S1) 2-((2-(benzoxazol-2-yl)-2-methylhydrazono)methyl)phenol were prepared.

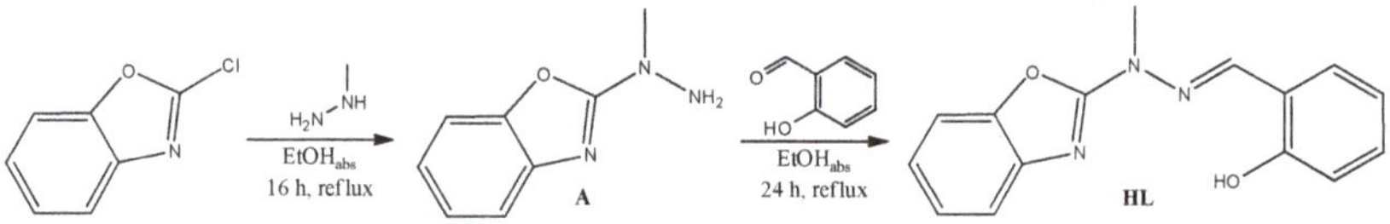

Scheme 2. Synthesis of Ligand **HL**.

Analytical data for the new compounds indicated that all complexes are in 2:3 metal to ligand stoichiometry and correspond to the general formula (in the solid state) [Ln$_2$**L**$_3$(NO$_3$)$_3$]·solvent. The metal ions are coordinated by organic ligands in a helical-like fashion, and two general coordination patterns can be identified (Scheme 1). It needs to be noted, that deprotonation of the ligand **HL** is an important step to obtain products in good quality and yields.

The ligand **HL** was prepared from the condensation of semi product A and salicylaldehyde in molar ratio 1:1. The ligand structure was identified by elemental analysis, IR, ESI-MS and ^{1}H, (Figure S1) ^{13}C NMR and was further confirmed by X-ray diffraction of a single crystal. The complexes **1–8** were characterized by ESI-MS and IR spectral analysis. The ESI-MS of complexes showed mass fragmentation pattern which fits well with the formula suggested by elemental and X-ray analyses. The peaks corresponding to the free ligand are also present in the MS as a result of demetalation. The mass spectrum of ligand **HL** showed molecular ion peaks at m/z = 266 [C$_{15}$H$_{13}$N$_3$O$_2$-H]$^-$, 268 [C$_{15}$H$_{13}$N$_3$O$_2$+H]$^+$ which match the C$_{15}$H$_{13}$N$_3$O$_2$ ligand formula.

The IR spectra of the complexes provide some information regarding the bonding mode of the ligand **HL** and nitrate counterions. All IR spectra of complexes **1–8** exhibit three bands at 3322–3014 cm^{-1} region attributable to vibrations of >N-CH$_3$ groups. The IR spectrum of ligand **HL** exhibits the bands at 3041 cm^{-1} and 2557 cm^{-1} characteristic of O-H stretching vibration and intramolecular OH$\cdots$N=C moiety, respectively. All IR spectra of compounds exhibit characteristic absorption band at around 1630 cm^{-1} attributed to C=N stretching mode confirming the Schiff base formation. The shift of this band to higher frequency, observed in the spectra of the complexes compared to the free ligand confirms the coordination of the nitrogen atom of azomethine group to the metal ions. The spectra of the complexes display vibrations indicative of coordinated nitrate groups. The ν(N–O) stretching frequencies are observed at 1475–1283 cm^{-1} region. The range of splitting confirms the bidentate coordinating behavior of the nitrate groups. The ν(C–O) phenolic frequency observed as a band at 1241 cm^{-1} for the ligand **HL**, is shifted in the complexes thus suggesting the participation of the oxygen atom of the deprotonated hydroxyl group in the formation of the M−O bonds. The metal complexes are also characterized by appearance of new bands at 519–501 cm^{-1} and 441–437 cm^{-1}, which are assigned to ν(M–O) and ν(M–N) bending frequencies, respectively.

Thermogravimetric analysis was used to study the thermal decomposition and to confirm the melting point results. The thermal stability of representative complexes was investigated under a nitrogen atmosphere by thermogravimetric analysis (TGA) in the temperature range of 30–800 °C on the powder samples. The TG curves of representative compounds (Figure S3) showed that they exhibit very similar thermal decomposition processes. The DTA curves show a high exothermic peak corresponding to total decomposition of complex in one step, at 267–332 °C temperature range.

2.2. Crystal Structures

Figure 1 shows the perspective views of **HL**. The perspective views of the chosen representative molecules of complexes are shown in Figures 2 and 3 and similar representations of all molecules are attached as Supplementary Information. Table S1 lists the relevant geometric characteristics of these molecules.

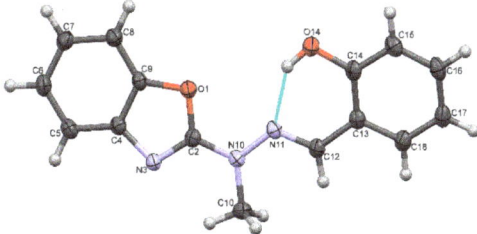

Figure 1. A perspective view of the Schiff base ligand **HL**. Ellipsoids are drawn at the 50% probability level, hydrogen atoms are represented by spheres of arbitrary radii, blue dashed line denotes the intramolecular hydrogen bond. Hydrogen atom has been localized on the basis of difference Fourier map (Figure S2) and successfully refined.

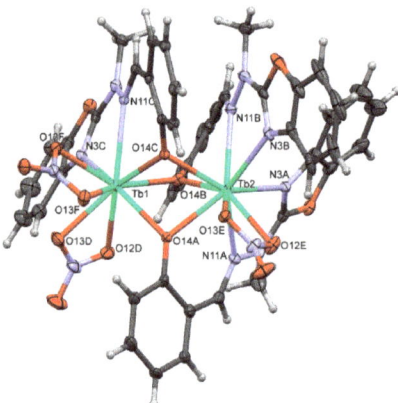

Figure 2. An example of the first of the two different types of structures **A**. A perspective view of the [Tb$_2$(C$_{15}$H$_{12}$N$_3$O$_2$)$_3$(NO$_3$)$_3$] (**3a**).

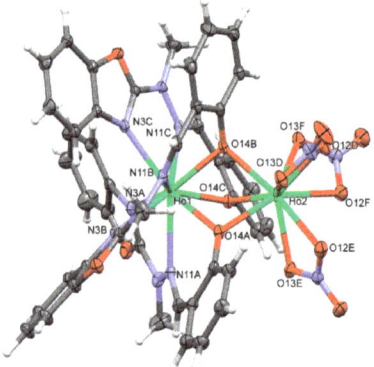

Figure 3. An example of the Second of two different types of structures **B**. A perspective view of the [Ho$_2$(C$_{15}$H$_{12}$N$_3$O$_2$)$_3$(NO$_3$)$_3$] (**5**).

Basically, there are two structurally different types of the complexes. In both cases, the complexes are non-symmetric (with an exception of (Tb)**3b**, which lies on the three-fold symmetry axis in the quite rare cubic space group P2$_1$3), of Ln$_2$**L**$_3$(NO$_3$)$_3$ composition, two-centered, triple bridged by ligand O$_{14}$ oxygen atoms, with two lanthanide cations coordinated in the most typical way with the coordination number 9. However, the details are quite different. In the group **A** ((Sm)**1**, (Tb)**3a**, (Dy)**4** and (Er)**6**) one metal ion is coordinated (besides three bridging oxygen atoms, which come from three ligand molecules) by four nitrogen atoms from two ligand molecules and two oxygen atoms from one nitrate ion (N$_4$O$_5$), while the other—by two nitrogen atoms from the third ligand molecule, and four oxygen atoms from two nitrates, so it is overall N$_2$O$_7$. In the group **B** ((Eu)**2**, (Tb)**3b**, (Ho)**5**, (Tm)**7**, (Yb)**8**) one of the Ln ions is coordinated (again, besides bridging oxygens) solely by ligands (N$_6$O$_3$), and the second one only by nitrates (O$_9$). It might be noted, that the structural differences are smaller in group **A**, with three out of four structures, **1**, **4**, and **6** closely isostructural (cf. Figure 4, the isostructurality indices are close to ideal values) than in group **B** (only (Tm)**7** and (Yb)**8**, i.e., two out of five, are isostructural).

In the Cambridge Structural Database [30] there are quite a lot examples of two-centered Ln complexes bridged by three oxygen atoms (1549 hits, of which 470 with both centers 9-coordinated), but we have found only few groups of complexes resembling relatively closely our new structures: tris(μ$_2$-(E)-2-(2-pyridylmethyleneamino)phenolato-N,N′,O,O)-tris(nitrato-O,O′)-di-lanthanide(III) [31], tris(μ$_2$-2-(2-hydroxypropyliminomethyl)phenoxo)-tris(nitrato-O,O′)-di-lanthanide(III) [32], and (μ-2,2′-[ethane-1,2-diylbis{[(pyridin-2-yl)methyl]azanediyl}methylene)]bis (6-formyl-4-methylphenolato))-(m-methoxo)-trinitrato-di- lanthanide (III) [33].

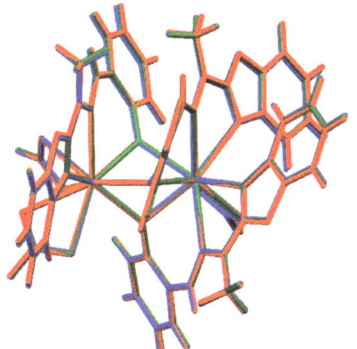

Figure 4. A perspective view of the isostructural complexes of group **A**.

Such a complicated coordination uses the flexibility of the ligand. The molecule of the free **HL** is almost planar (dihedral angle between terminal ring systems is 11.21(6)°, cf. Table S1, while all ligands in the complexes are significantly twisted. Such a conformation change allows the ligands for bridging two Ln ions, while at the same time for serving with two nitrogen atoms as coordination centers.

Each of the crystal structures of complexes contain varieties of different solvent molecules: acetonitrile, methanol, toluene, diethylether (Table S1 Crystal data). Interestingly, also in above listed examples of similar complexes, the crystal structures contained solvent molecules. This seems to suggest that the presence of appropriate—relatively small—solvent molecules is crucial for the possibility of obtaining single crystals, these molecules apparently help by filling the voids between the awkwardly shaped complex molecules. In general, weak C-H···N, C-H···O and C-H···π hydrogen bonds are predominant specific interactions which can be found in the structures.

2.3. Electronic Absorption Spectra and Luminescence Properties

The absorption spectrum of the Schiff base ligand **HL** (Figure 5) in acetonitrile showed three strong absorption bands in the range 220–450 nm (at about 239, 306 and 331 nm), that could be attributed to the π-π* or n-π* transitions [34–36]. The comparison of the absorption spectra of the ligand **HL** and the Ln^{3+} complexes in acetonitrile (Figures 5 and S4a,b) gave the numerical values of the maximum absorption wavelength and molar extinction coefficients (ε), which are listed in experimental data. Bathochromic shift of the absorption maxima and appearance of new bands were detected. These changes indicate formation of complexes between the lanthanide cations and ligand **HL**.

The intraconfigurational *f-f* transitions [37] are observed in the absorption and emission spectra of the Ln^{3+} ions. It should be noted that the ε values (molar absorption coefficient) of the forbidden *f-f* transitions are in general smaller than 10 dm^3·mol^{-1}·cm^{-1} [38].

Because of the strong ligand absorption, only a limited number of lanthanide ions show observable *f-f* transitions in the Schiff base complexes [39].

The Ho^{3+} absorption spectra (these ions are characterized by the highest values of molar absorption coefficients in the spectral range 500–800 nm) showed an increase in the absorption and bathochromic shift of the absorption bands, as presented in Figures 6 and S5. The largest changes of the bands $^5I_8$-$^5F_4$,$^5S_2$ (λ_{max}~535 nm) and $^5I_8$-$^5F_5$ (λ_{max}~640 nm) in comparison with free HL are observed for (Ho)**5**. The changes of the absorption spectra are attributed to complexation of the ligand to the Ho^{3+} ion.

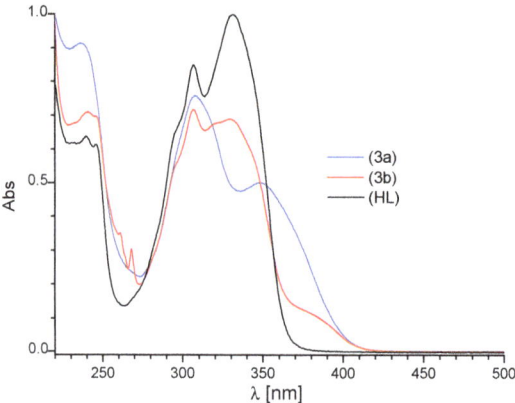

Figure 5. The absorption spectra of **HL**, (Tb group A) **3a** and (Tb group B) **3b** in acetonitrile solution; c = 2 × 10^{-5} M.

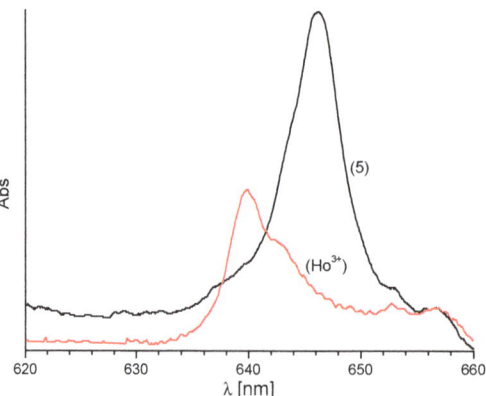

Figure 6. The absorption spectra of Ho^{3+} ion in acetonitrile solution in Ho(OTf)$_3$ and (Ho)**5** in range $^5I_8$-$^5F_5$ transitions, c = 2 × 10^{-3} M.

Photoluminescence study of the ligand **HL** and its corresponding complexes (Sm)**1**, (Eu)**2**, (Tb)**3a** and (Dy)**4** have been carried out in solution and solid state in the visible region. The exemplary excitation spectra of solid-state complexes from both analyzed groups are shown in Figure S6a,b. For the ligand and complexes of group **A**, two bands located at about 270 and 370 nm are observed, while for the complexes of group **B** only one band is observed at 370 nm. The emission spectra of all solid samples display free ligand emission bands and are blue-shifted compared to the emission spectrum of the free ligand **HL** (Figure 7). The (Sm)**1**, (Tb)**3a** and (Dy)**4** (group **A** of complexes), at wavelength excitation λ_{ex} = 368 nm, exhibit broad emission bands with maximum at 453, 442 and 450 nm respectively, while (Tb)**3b** and (Eu)**2** (group **B**) two or three emission bands located at 410, 438 nm and 412, 440, 469 nm. The main emission peak of free **HL** ligand, in solid state is located at 512 nm for λ_{ex} = 370 nm. The very high intra ligand luminescence observed in the 400–650 nm spectral range, in the solid compounds, is indicative of nonefficient intramolecular energy transfer from the ligand to the excited states of Ln(III) ions and the lack of their bands in emission spectra.

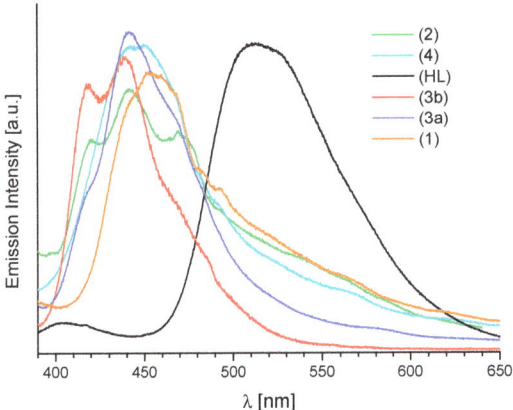

Figure 7. Normalized emission spectra of solid samples of **HL** ligand (λ_{ex} = 370 nm) and compounds (Sm)**1**, (Eu)**2**, (Tb)**3a 3b**, (Dy)**4** (λ_{ex} = 368 nm).

The similarity in the excitation spectra of **HL** and **1–4** (Sm, Eu, Tb, Dy) complexes suggests that the absorption is ligand–based. The blue- or red-shifts with respect to **HL** may be ascribed to the coordination of Ln^{3+} ions with the ligand. The reduction of the emissions of these samples, in comparison with **HL**, was also observed.

Moreover, the luminescence emissions of the compounds **5–8** (Ho, Er, Tm, Yb) in the solid state were recorded at room temperature under 369 nm LED pumping from 900 to 1700 nm (Figures 8, S7 and S8).

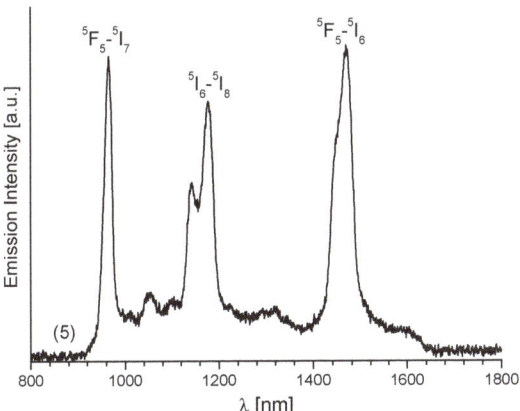

Figure 8. The emission spectrum of **5**(Ho) in NIR region.

As shown in Figure 8, the luminescence spectrum of Ho^{3+} consists of three intense NIR bands which may be ascribed to the following transitions: $^5F_5$-$^5I_7$ (966 nm), $^5I_6$-$^5I_8$ (1178 nm) and $^5F_5$-$^5I_6$ located at 1468 nm. The low intensity signals observed at about 1050 and 1300 nm are attributed to the $^5S_2$-$^5I_6$ and $^5S_2$-$^5I_5$ transitions observed in this ion. The peak position of $^2F_{5/2}$-$^2F_{7/2}$ transition in Yb^{3+} is centered at 967 nm (Figure S8). The Tm^{3+} ion in **7** exhibits near-infrared and mid-infrared emissions at 795 nm ($^3H_4$-$^3H_6$) and 1450 nm ($^3H_4$-$^3F_4$) when **6** show a main band Er^{3+} with maximum at 1509 nm, which corresponds to the $^4I_{13/2}$-$^4I_{15/2}$ transition. The intense photoluminescence in the near-IR region was observed for Er(**6**) and Yb(**8**) complexes.

The luminescence spectra of newly synthesized complexes **1–4** (Sm, Eu, Tb, Dy) were also measured in acetonitrile solution, Figure 9. The luminescence properties of (Sm)**1** and

(Dy)**4** in solutions are similar to those in the solid state. Additionally, the complexes from group A display broad emission bands located around 440 nm, previously observed in the emission spectra of solid samples, and absence of *f-f* emission bands of Sm^{3+} and Dy^{3+} ions. In contrast, the ligand emission is attenuated in the spectrum of complex **2**. Moreover, in this spectrum there are two weak bands related to the electronic *f-f* transitions in the Eu^{3+} ion (Figure 9a), i.e., transitions $^5D_0$-$^7F_1$ and $^5D_0$-$^7F_2$.

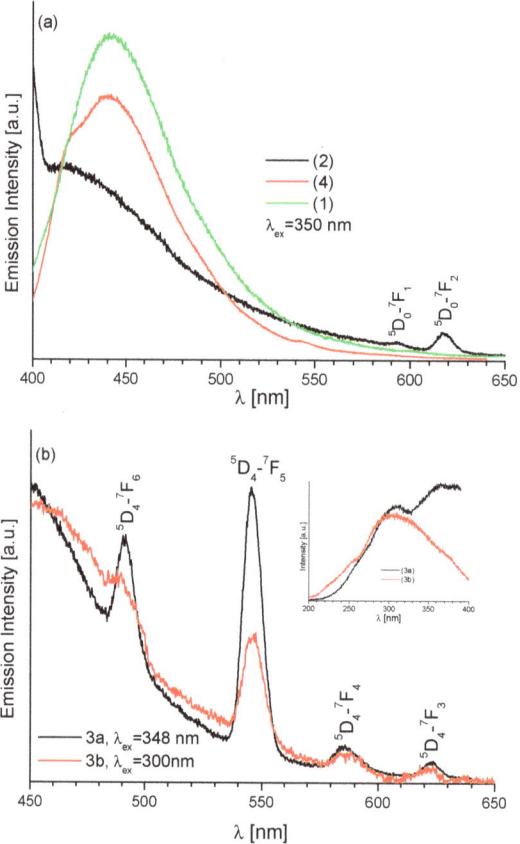

Figure 9. The emission spectra of acetonitrile solution (**a**) (Sm)**1**, (Eu)**2** and (Dy)**4** and (**b**) (Tb)**3a** and (Tb)**3b**, c = 2×10^{-5} M.

The observed differences in the emission spectra of the solid complexes Tb(III) of two groups) **3a** and **3b** were also reflected in the spectra of acetonitrile solutions (Figure 9b). In solution of compound **3a**, in addition to the ligand band located at about 440 nm, on its slope four bands reflecting the presence of Tb(III) ions, were also observed. These bands are at 491, 545, 585 and 623 nm and correspond to the $^5D_4$-^7F_j i.e., $^5D_4$-$^7F_6$, $^5D_4$-$^7F_5$, $^5D_4$-$^7F_4$ and $^5D_4$-$^7F_3$ transitions, respectively. Of them, the most intense transition is the first one, $^5D_4$-$^7F_5$. Unlike **3a**, the spectrum of **3b** displays weak emission band corresponding to $^5D_4$-^7F_j transition in Tb^{3+} ion.

The hydrazone Schiff base ligand transfers the excitation energy to the Eu^{3+} and Tb^{3+} ions. The observed ligand-centered π-π* emission band at about 440 nm which indicates that the LMET in these complexes is not complete. The low intensity of the emission bands in **2** and **3b** (group complexes) proves that this process is inefficient. This process is more efficient for compound **3a**, in which Schiff base ligands are attached to both Tb^{3+} ions.

3. Materials and Methods

3.1. Materials

2-(1-methylhydrazinyl)benzoxazole was prepared according to the procedure described in [29]. Salicylaldehyde, 2-chlorobenzoxazole, methylhydrazine and $Sm(NO_3)_3 \cdot 6H_2O$, $Eu(NO_3)_3 \cdot 5H_2O$, $Tb(NO_3)_3 \cdot 6H_2O$, $Dy(NO_3)_3 \cdot 5H_2O$, $Ho(NO_3)_3 \cdot 5H_2O$, $Er(NO_3)_3 \cdot 5H_2O$ $Tm(NO_3)_3 \cdot 6H_2O$, $Yb(NO_3)_3 \cdot 5H_2O$ were used as received from Sigma-Aldrich.

3.2. Physical Measurements

The compounds were characterized using microanalyses (CHN), IR, ESI-MS, UV/Vis, ^1H NMR, ^{13}C NMR, and single crystal X-ray structural analysis.

IR spectra were recorded in the range of 4000–400 cm^{-1}. Powders of sample were analyzed on a Nicolet iS50 FT-IR spectrometer, ATR technique was used. Mass spectra were recorded using electrospray ionization (ESI) techniques. Electrospray mass spectra were determined in methanol using a Waters Micromass ZQ spectrometer. Microanalyses (CHN) was obtained using a Perkin-Elmer 2400 CHN micro analyzer. NMR spectra were recorded in dimethylsulfoxide-d_6 on a Bruker Ultrashield 300 spectrometer model operating at 300 MHz with chemical shifts (ppm) referenced to the deuterated solvent. NMR solvent was purchased from Deutero GmbH. Melting points were determined with a EZ-MeltA automated Melting Point Apparatus, Stanford Research Systems product, apparatus and are uncorrected. Thermogravimetric analyses (TGA) were carried out on a Setsys 1200 Setaran thermogravimetry under nitrogen atmosphere with a heating rate of 5 °C/min. Ultraviolet–visible (UV–Vis) spectra of the compounds in acetonitrile (at concentration of 2×10^{-5} M) were measured using Shimadzu UV PC 2401 spectrophotometer. Luminescence spectra in Vis range were recorded on a Hitachi F7000 spectrofluorometer at room temperature with a 1 cm quartz cell and filters 320 or 395, while in the NIR range were recorded by using an Andor Shamrock 500 spectrometer (300 L/mm–blaze 1200 nm) equipped with CCD camera iDus 420 and spectrometer QuantaMasterTM 40s (Photon Technology Instrumental, Birmingham, AL, USA) equipped with Photomultiplier Tubes H10330C-75 (800–1800 nm). The samples were excited by the use a LED revolver with high-power 360–370 nm LED.

3.3. Synthesis of the 2-((2-(benzoxazol-2-yl)-2-methylhydrazono)methyl)phenol, HL=$C_{15}H_{13}N_3O_2$

Ligand HL was synthesized within two subsequent steps starting from commercially available 2-chlorobenzoxazole according to the Scheme S1. Formation of product thus formed was established by 1H NMR, 13C NMR, IR, ESI-MS spectroscopy and the X-ray diffraction methods. Ligand HL was synthesized according to the protocol described below.

To the colorless solution of **A** [29] (400.3 mg, 2.2 mmol) in EtOH$_{abs}$ (8 mL) an equimolar amount of salicylaldehyde (268.7 mg, 234.5 µL, 2.2 mmol) was added. The mixture was allowed to react for 20 h under inert atmosphere in reflux and for another 4 h at room temperature. The white precipitate was observed. The product was filter under reduced pressure. The white needle shaped monocrystals suitable for the X-ray analysis were obtained by the slow evaporation of CH$_3$CN. Yield: 513.8 mg, 1.9 mmol, 78.3%. Anal.: Calcd. for $C_{15}H_{13}N_3O_2$ (267.28 g mol^{-1}): C, 67.40; H, 4.90; N, 15.72. Found C, 67.86; H, 4.87; N, 15.67%. Melting point: 182 °C. ^{1}H NMR δ_H(300 MHz, DMSO-d_6) 10.86 (s, 1H), 8.26 (s, 1H), 7.65 (d, J = 7.8 Hz, 1H), 7.58 (d, J = 7.7 Hz, 1H), 7.48 (d, J = 7.7 Hz, 1H), 7.27 (dd, J = 15.6, 7.9 Hz, 2H), 7.16 (t, J = 7.2 Hz, 1H), 6.94 (dd, J = 7.4, 5.6 Hz, 2H), 3.69 (s, 3H). ^{13}C NMR (DMSO-d_6, 300 MHz): δ (ppm) 159.81; 156.61; 148.94; 141.97; 140.86; 139.75; 130.83; 128.95; 124.46; 121.86; 119.45; 117.03; 116.38; 109,62; 32.38. ESI-MS: m/z = 266 [$C_{15}H_{13}N_3O_2$-H]$^-$, 267 [$C_{15}H_{13}N_3O_2$+H]$^+$. IR (ATR, cm^{-1}): ν = 3041 w (OH), 2575 w (OH⋯N), 1630 s (C=N), 1582 s (C=C), 1241 s (C–O) cm^{-1}. UV-Vis (CH$_3$CN): λ_{max}/nm (ε/dm^3 mol^{-1} cm^{-1}) 239.5 (1.9 × 10^4), 306.5 (2.5 × 10^4), 331.0 (3.0 × 10^4).

3.4. Synthesis of the Complexes. General Procedures

All complexes were prepared under similar conditions. To the colorless solution of ligand HL (20 mg, 40 µmol) in CH$_3$OH (5 mL) the solution of appropriate lanthanide(III) nitrate salt [(60 µmol: 26.7 mg Sm(NO$_3$)$_3$·6H$_2$O for **1**, 25.7 mg Eu(NO$_3$)$_3$·5H$_2$O for **2**, 27.2 mg Tb(NO$_3$)$_3$·6H$_2$O for **3**, 26.3 mg Dy(NO$_3$)$_3$·5H$_2$O for **4**, 26.5 mg Ho(NO$_3$)$_3$·5H$_2$O for **5**, 26.6 mg Er(NO$_3$)$_3$·5H$_2$O for **6**, 26.7 mg Tm(NO$_3$)$_3$·6H$_2$O for **7**, 26.9 mg Yb(NO$_3$)$_3$·5H$_2$O for **8**] in CH$_3$CN (5 mL) was added. No color change was observed. After 1 h the mixtures were treated with Et$_3$N (5.6 µL, 40 µmol), which resulted in formation of clear yellow solutions. The final mixture was stirred at room temperature for 48 h under normal atmosphere. The solution volume was then reduced to 5 mL by roto-evaporation. Precipitation was carried out by addition of diethyl ether/methanol (1:1.5 mL). The solids were filtered off and dried in air. For the purpose of elemental analysis, the precipitates were dried under vacuum for 72 h in order to remove the solvent molecules [40,41]. The complexes obtained are microcrystalline variously colored powders, whose melting points (decomposition) are higher than that of the free ligand. Single crystals suitable for X-ray diffraction analysis were formed by slow diffusion of toluene (group **A**—**1**, **3a**, **4**, **6**) and diisopropyl ether (group **B**—**2**, **3b**, **5**, **6**, **7**) into sample to obtain by recrystallization of the solid from a minimum volume of CH$_3$OH/CH$_3$CN (1:1) at 4 °C over a period of 6–8 weeks. Single crystals were filtered under reduced pressure and air dried. The complexes were able to turn the solvent molecules into the crystal by recrystallization from mixed solvents. Solvents used in the processes of crystallization lead to the formation of different crystal structures and compositions, and thus affect their properties. Observed structural changes are characterized by a slight modification of the number of non-coordinated solvent molecules [42–44]. For the luminescence and UV-Vis studies, single crystals of the product were used [41].

(1) [Sm$_2$(C$_{15}$H$_{12}$N$_3$O$_2$)$_3$(NO$_3$)$_3$]·CH$_3$CN **group A**

Light yellow precipitate, Yield: 24.36 mg, 0.0184 mmol, 63%. Analytical data Calculated for C$_{45}$H$_{36}$N$_{12}$O$_{15}$Sm$_2$ (1285.57 g mol^{-1}): *Calc.* C, 42.04; H, 2.82; N, 13.07. *Found:* C, 42.18; H, 2.84; N, 13.11%. Melting point: 324 °C (decomp.). ESI-MS: m/z = 684 [Sm(C$_{15}$H$_{12}$N$_3$O$_2$)$_2$]$^+$, 604 [SmC$_{15}$H$_{12}$N$_3$O$_2$(NO$_3$)$_3$]$^-$, 268 [C$_{15}$H$_{13}$N$_3$O$_2$+H]$^+$. IR (ATR, cm^{-1}): ν = 3432 w (OH), 3058 w, 2936 w (N-CH$_3$), 1639 s (C=N), 1603 m (C=C), 1460–1287 s, 835 m (NO$_3$$^-$), 1253 m (C–O), 1028 m (N–N) 503 w (Sm–O), 476 w (Sm–N) cm^{-1}. UV-Vis (CH$_3$CN): λ$_{max}$/nm (ε/dm^3 mol^{-1} cm^{-1}) 308.4 (4.0 × 10^4), 351.2 (2.6 × 10^4).

(2) [Eu$_2$(C$_{15}$H$_{12}$N$_3$O$_2$)$_3$(NO$_3$)$_3$]·CH$_3$CN·[(CH$_3$)$_2$CH]$_2$O **group B**

Orange precipitate, Yield: 21.53 mg, 0.0167 mmol, 58%. Analytical data Calculated for C$_{45}$H$_{36}$N$_{12}$O$_{15}$Eu$_2$ (1288.78 g mol^{-1}): *Calc.* C, 41.94; H, 2.82; N, 13.04. *Found:* C, 41.82; H, 2.97; N, 13.09%. Melting point: 308 °C (decomp.). ESI-MS: m/z = 683 [Eu(C$_{15}$H$_{12}$N$_3$O$_2$)$_2$]$^+$, 605 [EuC$_{15}$H$_{12}$N$_3$O$_2$(NO$_3$)$_3$]$^-$, 268 [C$_{15}$H$_{13}$N$_3$O$_2$+H]$^+$. IR (ATR, cm^{-1}): ν = 3416 w (OH), 3108 w, 2975 w (N-CH$_3$), 1642 s (C=N), 1603 m (C=C), 1457–1294 s, 831 m (NO$_3$$^-$), 1247 m (C–O), 1032 m (N–N) 505 w (Eu–O), 439 w (Eu–N) cm^{-1}. UV-Vis (CH$_3$CN): λ$_{max}$/nm (ε/dm^3 mol^{-1} cm^{-1}) 307.6 (3.5 × 10^4), 349.6 (2.2 × 10^4).

(3a) [Tb$_2$(C$_{15}$H$_{12}$N$_3$O$_2$)$_3$(NO$_3$)$_3$]·CH$_3$OH·CH$_3$CN·C$_6$H$_5$CH$_3$ **group A**
(3b) [Tb$_2$(C$_{15}$H$_{12}$N$_3$O$_2$)$_3$(NO$_3$)$_3$]·CH$_3$CN·(C$_2$H$_5$)$_3$NHNO$_3$ **group B**

Light yellow precipitate, Yield: 21.18 mg, 0.0162 mmol, 54%. Analytical data Calculated for C$_{45}$H$_{36}$N$_{12}$O$_{15}$Tb$_2$ (1302.70 g mol^{-1}): *Calc.* C, 41.49; H, 2.79; N, 12.90. *Found:* C, 41.38; H, 2.69; N, 12.67%. Melting point: 332 °C (decomp.). ESI-MS: m/z = 691 [Tb(C$_{15}$H$_{12}$N$_3$O$_2$)$_2$]$^+$, 611 [TbC$_{15}$H$_{12}$N$_3$O$_2$(NO$_3$)$_3$]$^-$, 487 [TbC$_{15}$H$_{12}$N$_3$O$_2$NO$_3$]$^+$, 268 [C$_{15}$H$_{13}$N$_3$O$_2$+H]$^+$. IR (ATR, cm^{-1}): ν = 3666 w (OH), 3057 w, 2975 w (N-CH$_3$), 1637 s (C=N), 1608 m (C=C), 1460–1283 s, 828 m (NO$_3$$^-$), 1242 m (C–O), 1027 m (N–N) 506 w (Tb–O), 440 w (Tb–N) cm^{-1}. UV-Vis (CH$_3$CN): λ$_{max}$/nm (ε/dm^3 mol^{-1} cm^{-1}) 307.2 (4.5 × 10^4), 346.0 (3.1 × 10^4).

(4) [Dy$_2$(C$_{15}$H$_{12}$N$_3$O$_2$)$_3$(NO$_3$)$_3$]·CH$_3$CN **group A**

Light yellow precipitate, Yield: 18.67 mg, 0.0142 mmol, 53%. Analytical data Calculated C$_{45}$H$_{36}$N$_{12}$O$_{15}$Dy$_2$ (1309.85 g mol^{-1}): *Calc.* C, 41.26; H, 2.77; N, 12.83. *Found:* C, 41.06; H, 2.84; N, 12.61%. Melting point: 322 °C (decomp.). ESI-MS: m/z = 696 [Dy(C$_{15}$H$_{12}$N$_3$O$_2$)$_2$]$^+$, 616 [DyC$_{15}$H$_{12}$N$_3$O$_2$(NO$_3$)$_3$]$^-$, 268 [C$_{15}$H$_{13}$N$_3$O$_2$+H]$^+$. IR (ATR, cm^{-1}): ν = 3665 w (OH), 3066 w, 2937 w (N-CH$_3$), 1639 s (C=N), 1606 m (C=C), 1464–1289 s, 825 s (NO$_3$$^-$), 1249 m

(C–O), 1035 m (N–N) 513 w (Dy–O), 438 w (Dy–N) cm^{-1}. UV-Vis (CH$_3$CN): λ_{max}/nm (ε/dm^3 mol^{-1} cm^{-1}) 307.2 (3.7 × 10^4).

(5) [Ho$_2$(C$_{15}$H$_{12}$N$_3$O$_2$)$_3$(NO$_3$)$_3$]·3CH$_3$CN·CH$_3$OH **group B**

Light pink precipitate, Yield: 25.28 mg, 0.0192 mmol, 64%. Analytical data Calculated for C$_{45}$H$_{36}$N$_{12}$O$_{15}$Ho$_2$ (1314.71 g mol^{-1}): C, 41.11; H, 2.76; N, 12.78. Found: C, 41.07; H, 2.69; N, 12.53%. Melting point: 267 °C (decomp.). ESI-MS: m/z = 697 [Ho(C$_{15}$H$_{12}$N$_3$O$_2$)$_2$]$^+$, 617 [HoC$_{15}$H$_{12}$N$_3$O$_2$(NO$_3$)$_3$]$^-$, 493 [HoC$_{15}$H$_{12}$N$_3$O$_2$NO$_3$]$^+$, 268 [C$_{15}$H$_{13}$N$_3$O$_2$+H]$^+$. IR (ATR, cm^{-1}): ν = 3660 w (OH), 3060 w, 2962 w (N-CH$_3$), 1644 s (C=N), 160 m7 (C=C), 1475–1285 s, 829 m (NO$_3$$^-$), 1245 m (C–O), 1028 m (N–N) 501 w (Ho–O), 441 w (Ho–N) cm^{-1}. UV-Vis (CH$_3$CN): λ_{max}/nm (ε/dm^3 mol^{-1} cm^{-1}) 308.0 (4.4 × 104), 352.0 (2.8 × 10^4).

(6) [Er$_2$(C$_{15}$H$_{12}$N$_3$O$_2$)$_3$(NO$_3$)$_3$]·CH$_3$CN **group A**

Light yellow precipitate, Yield: 25,72 mg, 0.0193 mmol, 65%. Analytical data Calculated for C$_{45}$H$_{36}$N$_{12}$O$_{15}$Er$_2$ (1319.37 g mol^{-1}): Calc. C, 40.97; H, 2.75; N, 12.74. Found: C, 40.89; H, 2.81; N, 12.63%. Melting point: 310 °C (decomp.). ESI-MS: m/z = 700 [Er(C$_{15}$H$_{12}$N$_3$O$_2$)$_2$]$^+$, 557 [ErC$_{15}$H$_{13}$N$_3$O$_2$(NO$_3$)$_2$]$^+$, 268 [C$_{15}$H$_{13}$N$_3$O$_2$+H]$^+$. IR (ATR, cm^{-1}): ν = 3677 m (OH), 3067 w, 2944 w (N-CH$_3$), 1643 s (C=N), 1605 m (C=C), 1460–1283 s, 828 s (NO$_3$$^-$), 1251 m (C–O), 1037 m (N–N), 512 w (Er–O), 437 w (Er–N) cm^{-1}. UV-Vis (CH$_3$CN): λ_{max}/nm (ε/dm^3 mol^{-1} cm^{-1}) 307.5 (3.7 × 10^4).

(7) [Tm$_2$(C$_{15}$H$_{12}$N$_3$O$_2$)$_3$(NO$_3$)$_3$]·4CH$_3$CN **group B**

Light yellow precipitate, Yield: 25.17 mg, 0.0190 mmol, 66%. Analytical data Calculated for C$_{45}$H$_{36}$N$_{12}$O$_{15}$Tm$_2$ (1322.11 g mol^{-1}): Calc. C, 40.86; H, 2.74; N, 12.71. Found: C, 41.08; H, 2.82; N, 12.69%. Melting point: 322 °C (decomp.). ESI-MS: m/z = 701 [Tm(C$_{15}$H$_{12}$N$_3$O$_2$)$_2$]$^+$, 620 [TmC$_{15}$H$_{12}$N$_3$O$_2$(NO$_3$)$_3$]$^-$, 268 [C$_{15}$H$_{13}$N$_3$O$_2$+H]$^+$. IR (ATR, cm^{-1}): ν = 3680 m (OH), 3066 w, 2947 w (N-CH$_3$), 1639 s (C=N), 1605 m (C=C), 1465–1288 s, 826 m (NO$_3$$^-$), 1249 m (C–O), 1037 m (N–N), 518 w (Tm–O), 438 w (Tm–N) cm^{-1}. UV-Vis (CH$_3$CN): λ_{max}/nm (ε/dm^3 mol^{-1} cm^{-1}) 310.5 (4.2 × 10^4).

(8) [Yb$_2$(C$_{15}$H$_{12}$N$_3$O$_2$)$_3$(NO$_3$)$_3$]·4CH$_3$CN **group B**

Light yellow precipitate, Yield: 25.11 mg, 0.0188 mmol, 63%. Analytical data Calculated for C$_{45}$H$_{36}$N$_{12}$O$_{15}$Yb$_2$ (1330.96 g mol^{-1}): Calc. C, 40.61; H, 2.73; N, 12.63. Found: C, 40.81; H, 2.85; N, 12.72%. Melting point: 322 °C (decomp.). ESI-MS: m/z = 706 [Yb(C$_{15}$H$_{12}$N$_3$O$_2$)$_2$]$^+$, 626 [YbC$_{15}$H$_{12}$N$_3$O$_2$(NO$_3$)$_3$]$^-$, 268 [C$_{15}$H$_{13}$N$_3$O$_2$+H]$^+$. IR (ATR, cm^{-1}): ν = 3663 w (OH), 3038 w, 2950 w (N-CH$_3$), 1638 s (C=N), 1606 m (C=C), 1462–1286 s, 827 m (NO$_3$$^-$), 1247 m (C–O),1030 m (N–N), 519 w (Yb–O), 441 w (Yb–N) cm^{-1}. UV-Vis (CH$_3$CN): λ_{max}/nm (ε/dm^3 mol^{-1} cm^{-1}) 307.0 (3.7 × 10^4).

3.5. X-ray Crystallography

Diffraction data were collected by the ω-scan technique, for HL and 6 at 130(1) K, using mirror-monochromated CuKα radiation (λ = 1.54178 Å), on Rigaku SuperNova four-circle diffractometer with Atlas CCD detector, and in all other cases at 100(1) K with graphite-monochromated MoKα radiation (λ = 0.71073 Å), on Rigaku XCalibur four-circle diffractometer with EOS CCD detector. The data were corrected for Lorentz-polarization as well as for absorption effects [45]. Precise unit-cell parameters were determined by a least-squares fit of the reflections of the highest intensity, chosen from the whole experiment. The structures were solved with SHELXT [46] and refined with the full-matrix least-squares procedure on F2 by SHELXL [47]. All non-hydrogen atoms were refined anisotropically. All hydrogen atoms were placed in idealized positions and refined as 'riding model' with isotropic displacement parameters set at 1.2 (1.5 for CH$_3$) times Ueq of appropriate carrier atoms. In the structure b one of the solvent molecules (methanol) was found disordered over two positions for which half occupancies were assigned. Some restraints were applied for these molecules (DFIX, ISOR). In **3a**, weak restraints for the displacement ellipsoids were also applied. Some of the crystals were of relatively poor quality, and the attempts to obtain the better ones failed. Therefore, in some cases, alerts appeared in the checking procedure [48]. However, the basic structural features are reasonably proved.

4. Conclusions

Reactions of lanthanides with polydentate oxygen and nitrogen ligands have attracted great interest because of the ability of these sites to realize stable chelate complexes with high coordination numbers. The Schiff base ligand **HL** and its corresponding lanthanide complexes of the composition [Ln$_2$L$_3$(NO$_3$)$_3$]·solvent (Ln = Sm^{3+} (**1**), Eu^{3+} (**2**), Tb^{3+} (**3a, 3b**), Dy^{3+} (**4**), Ho^{3+} (**5**), Er^{3+} (**6**), Tm^{3+} (**7**), Yb^{3+} (**8**)), were synthesized and characterized. Their identity was further confirmed by the X-ray diffraction of single crystals. Interestingly, the details of the metal coordination patterns are different among the obtained complexes. These results have provided valuable information for possibility of designing new lanthanide complexes with structural chemistry depending upon crystallization conditions. The luminescent properties of ligand **HL** and compounds **1–8** have been also studied, both in solution and solid state. The efficient intramolecular energy transfer process from the triplet state energy level of ligand to excited energy level of lanthanide(III) ions is one of the factors influencing the luminescence properties of lanthanide(III) complexes. The occurrence of the luminescence of ligand **HL** and the absence or weak luminescence of characteristic emission bands of lanthanide ions in visible region is probably due to the large energy gap between the lowest triplet state level of the ligand and the excited state of lanthanides. The intense photoluminescence in the near-IR region was observed for Er(III) (**6**) and Yb(III) (**8**) complexes.

Supplementary Materials: The following supporting information can be downloaded at: https://www.mdpi.com/article/10.3390/molecules27238390/s1, Scheme S1: General structures of the resulting complexes (group A and B); Figure S1: 1H NMR spectrum of ligand in DMSO-d$_6$; Figure S2: The difference Fourier map for the structure HL without the OH or NH hydrogen atom; the position of this hydrogen next to oxygen is clearly seen; Table S1: Crystal data, data collection and structure refinement; Figure S3: Thermogravimetric analysis (TGA) curves of representative of complexes (a) **2**, (b) **3**, (c) **6**; Figure S4: The UV-Vis spectra of group **A** (a) and **B** (b) compounds in acetonitrile solution; Figure S5: The absorption spectrum of Ho^{3+} ion in acetonitrile solution in Ho(OTf)$_3$ and (Ho)**5** in range ^{5}I$_8$-^{5}F$_4$,^{5}S$_2$, c = 2 × 10^{-3}; Figure S6: **a–b**. The excitation spectra of solid samples: (a) **HL, 3a** and (b) **2, 3b**; Figure S7: Spectroscopic characteristics of a 369 LED; Figure S8: Photoluminescence spectra of solid samples **6, 7** and **8** in NIR region. Characterization including crystal and molecular structures, and structure date of the crystals, 1H NMR of the ligand; TGA of complexes; luminescence properties.

Author Contributions: Conceptualization, data curation, investigation, project administration, validation, writing—original draft, writing—review and editing, I.P.-M.; data curation, investigation, visualization, writing—original draft, writing—review and editing, M.A.F.-J.; data curation, investigation, writing—original draft, writing—review and editing, Z.H.; supervision, writing—review and editing, V.P.; investigation, validation, visualization, conceptualization, supervision, writing—original draft, writing—review and editing, M.K. All authors have read and agreed to the published version of the manuscript.

Funding: The research was funded by National Science Centre, Poland (grant no. 2016/21/B/ST5/00175).

Data Availability Statement: Crystallographic data for the structural analysis has been deposited with the Cambridge Crystallographic Data Centre, Nos. CCDC-1542869 (HL), 2062743 (1), 2062744 (2), 1542870 (3a), 2062745 (3b), 2062746 (4), 1542871 (5), 2062747 (6), 2062748 (7) and 2062749 (8). Copies of this information may be obtained free of charge from: The Director, CCDC, 12 Union Road, Cambridge, CB2 1EZ, UK. Fax: +44(1223)336-033, e-mail: deposit@ccdc.cam.ac.uk, or www.ccdc.cam.ac.uk (accessed on 29 November 2022).

Acknowledgments: Zuzanna Ruta is gratefully acknowledged for her assistance in crystallization of complexes 1, 3, 5 and 6. Przemysław Woźny is gratefully acknowledged for his registration of the NIR spectra.

Conflicts of Interest: The authors declare no conflict of interest.

References

1. Woodruff, D.N.; Winpenny, R.E.P.; Layfield, R.A. Lanthanide Single-Molecule Magnets. *Chem. Rev.* **2013**, *113*, 5110–5148. [CrossRef] [PubMed]
2. Patroniak, V.; Baxter, P.N.W.; Lehn, J.-M.; Hnatejko, Z.; Kubicki, M. Synthesis and Luminescence Properties of New Dinuclear Complexes of Lanthanide(III) Ions. *Eur. J. Inorg. Chem.* **2004**, *2004*, 2379–2384. [CrossRef]
3. Wang, X.; Chang, H.; Xie, J.; Zhao, B.; Liu, B.; Xu, S.; Pei, W.; Ren, N.; Huang, L.; Huang, W. Recent developments in lanthanide-based luminescent probes. *Coord. Chem. Rev.* **2014**, *273–274*, 201–212. [CrossRef]
4. Gao, X.-S.; Jiang, X.; Yao, C. Two new complexes of Lanthanide (III) ion with the N_3O_2-donor Schiff base ligand: Synthesis, crystal structure, and magnetic properties. *J. Mol. Struct.* **2016**, *1126*, 275–279. [CrossRef]
5. Demir, S.; Jeon, I.-R.; Long, J.R.; Harris, T.D. Radical ligand-containing single-molecule magnets. *Coord. Chem. Rev.* **2015**, *289–290*, 149–176. [CrossRef]
6. Zhang, P.; Zhang, L.; Tang, J. Lanthanide single molecule magnets: Progress and perspective. *Dalton Trans.* **2015**, *44*, 3923–3929. [CrossRef]
7. Hutchings, A.-J.; Habib, F.; Holmberg, R.J.; Korobkov, I.; Murugesu, M. Structural Rearrangement through Lanthanide Contraction in Dinuclear Complexes. *Inorg. Chem.* **2014**, *53*, 2102–2112. [CrossRef]
8. Shavaleev, N.M.; Eliseeva, S.V.; Scopelliti, R.; Bünzli, J.-C.G. Tridentate Benzimidazole-Pyridine-Tetrazolates as Sensitizers of Europium Luminescence. *Inorg. Chem.* **2014**, *53*, 5171–5178. [CrossRef]
9. Pospieszna-Markiewicz, I.; Radecka-Paryzek, W.; Kubicki, M.; Korabik, M.; Hnatejko, Z. Different supramolecular architectures in self-assembled praseodymium(III) and europium(III) complexes with rare coordination pattern of salicylaldimine ligand. *Polyhedron* **2015**, *97*, 167–174. [CrossRef]
10. Moreno-Pineda, E.; Chilton, N.F.; Marx, R.; Dorfel, M.; Sells, D.O.; Neugebauer, P.; Jiang, S.D.; Collison, D.; van Slageren, J.; McInnes, E.J.L.; et al. Direct measurement of dysprosium (III) dysprosium (III) interactions in a single-molecule magnet. *Nat. Commun.* **2014**, *5*, 5243. [CrossRef]
11. Gorczyński, A.; Marcinkowski, D.; Kubicki, M.; Löffler, M.; Korabik, M.; Karbowiak, M.; Wiśniewski, P.; Rudowicz, C.; Patroniak, V. New field-induced single ion magnets based on prolate Er(III) and Yb(III) ions: Tuning the energy barrier Ueff by the choice of counterions within an N3-tridentate Schiff-base scaffold. *Inorg. Chem. Front.* **2018**, *5*, 605–618. [CrossRef]
12. Suturina, E.A.; Mason, K.; Botta, M.; Carniato, F.; Kuprov, I.; Chilton, N.F.; McInnes, E.J.L.; Vonci, M.; Parker, D. Periodic trends and hidden dynamics of magnetic properties in three series of triazacyclononane lanthanide complexes. *Dalton Trans.* **2019**, *48*, 8400–8409. [CrossRef] [PubMed]
13. Sardaru, M.-C.; Marangoci, N.L.; Shova, S.; Bejan, D. Novel Lanthanide (III) Complexes Derived from an Imidazole–Biphenyl–Carboxylate Ligand: Synthesis, Structure and Luminescence Properties. *Molecules* **2021**, *26*, 6942. [CrossRef] [PubMed]
14. Swain, A.; Sen, A.; Rajaraman, G. Are lanthanide-transition metal direct bonds a route to achieving new generation {3d–4f} SMMs? *Dalton Trans.* **2021**, *50*, 16099–16109. [CrossRef] [PubMed]
15. Edelmann, F.T. Lanthanides and actinides: Annual survey of their organometallic chemistry covering the year 2016. *Coord. Chem. Rev.* **2017**, *338*, 27–140. [CrossRef]
16. Kozłowski, M.; Kierzek, R.; Kubicki, M.; Radecka-Paryzek, W. Metal-promoted synthesis, characterization, crystal structure and RNA cleavage ability of 2,6-diacetylpyridine bis(2-aminobenzoylhydrazone) lanthanide complexes. *J. Inorg. Biochem.* **2013**, *126*, 38–45. [CrossRef]
17. Fik-Jaskółka, M.; Pospieszna-Markiewicz, I.; Roviello, G.N.; Kubicki, M.; Radecka-Paryzek, W.; Patroniak, V. Synthesis and Spectroscopic Investigation of a Hexaaza Lanthanum(III) Macrocycle with a Hybrid-Type G4 DNA Stabilizing Effect. *Inorg. Chem.* **2021**, *60*, 2122–2126. [CrossRef]
18. Kulkarni, A.; Patil, S.A.; Badami, P.S. Synthesis, characterization, DNA cleavage and in vitro antimicrobial studies of La (III), Th (IV) and VO (IV) complexes with Schiff bases of coumarin derivatives. *Eur. J. Med. Chem.* **2009**, *44*, 2904–2912. [CrossRef]
19. Eliseeva, S.V.; Bünzli, J.-C.G. Lanthanide luminescence for functional materials and bio-sciences. *Coord. Chem. Rev.* **2010**, *39*, 189–227. [CrossRef]
20. Dokukin, V.; Silverman, S.K. Photosensitised regioselective [2+2]-cycloaddition of cinnamates and related alkenes. *Chem. Sci.* **2012**, *3*, 1707–1714. [CrossRef]
21. Mundoma, C.; Greenbaum, N.L. Sequestering of Eu (III) by a GAAA RNA tetraloop. *J. Am. Chem. Soc.* **2002**, *124*, 3525–3532. [CrossRef] [PubMed]
22. Ambiliraj, D.B.; Francis, B.; Reddy, M.L.P. Lysosome-targeting luminescent lanthanide complexes: From molecular design to bioimaging. *Dalton Trans.* **2022**, *51*, 7748–7762. [CrossRef] [PubMed]
23. Nitabaru, T.; Nojiri, A.; Kobayashi, M.; Kumagai, N.; Shibasaki, M. anti-Selective Catalytic Asymmetric Nitroaldol Reaction via a Heterobimetallic Heterogeneous Catalyst. *J. Am. Chem. Soc.* **2009**, *131*, 13860–13869. [CrossRef] [PubMed]
24. Mikami, K.; Terada, M.; Matsuzawa, H. "Asymmetric" catalysis by lanthanide complexes. *Angew. Chem. Int. Ed.* **2002**, *41*, 3554–3571. [CrossRef]
25. Vuillamy, A.; Zebret, S.; Besnard, C.; Placide, V.; Petoud, S.; Hamacek, J. Functionalized Triptycene-Derived Tripodal Ligands: Privileged Formation of Tetranuclear Cage Assemblies with Larger Ln (III). *Inorg. Chem.* **2017**, *56*, 2742–2749. [CrossRef]
26. Xu, C.; Sun, T.; Rao, L. Interactions of bis(2,4,4-trimethylpentyl) dithiophosphinate with trivalent lanthanides in a homogeneous medium: Thermodynamics and coordination modes. *Inorg. Chem.* **2017**, *56*, 2556–2565. [CrossRef]

27. Boland, K.S.; Hobart, D.E.; Kozimor, S.A.; MacInnes, M.M.; Scott, B.L. The coordination chemistry of trivalent lanthanides (Ce, Nd, Sm, Eu, Gd, Dy, Yb) with diphenyldithiophosphinate anions. *Polyhedron* **2014**, *67*, 540–548. [CrossRef]
28. Marcinkowski, D.; Wałęsa-Chorab, M.; Bocian, A.; Mikołajczyk, J.; Kubicki, M.; Hnatejko, Z.; Patroniak, V. The spectroscopic studies of new polymeric complexes of silver(I) and original mononuclear complexes of lanthanides(III) with benzimidazole-based hydrazone. *Polyhedron* **2017**, *123*, 243–251. [CrossRef]
29. Fik, M.A.; Löffler, M.; Kubicki, M.; Weselski, M.; Korabik, M.J.; Patroniak, V. New Fe (II) complexes with Schiff base ligand: Synthesis, spectral characterization, magnetic studies and thermal stability. *Polyhedron* **2015**, *102*, 609–614. [CrossRef]
30. Groom, C.R.; Bruno, I.J.; Lightfoot, M.P.; Ward, S.C. The Cambridge Structural Database. *Acta Cryst. B* **2016**, *72*, 171–179. [CrossRef]
31. Patroniak, V.; Stefankiewicz, A.R.; Lehn, J.-M.; Kubicki, M.; Hoffmann, M. Self-Assembly and Characterization of Homo- and Heterodinuclear Complexes of Zinc(II) and Lanthanide(III) Ions with a Tridentate Schiff-Base Ligand. *Eur. J. Inorg. Chem.* **2006**, *2006*, 144–149. [CrossRef]
32. Zhang, L.; Ji, Y.; Xu, X.; Liu, Z.; Tang, J. Synthesis, structure and luminescence properties of a series of dinuclear LnIII complexes (Ln = Gd, Tb, Dy, Ho, Er). *J. Lumin.* **2012**, *132*, 1906–1909. [CrossRef]
33. Zhang, S.; Shen, N.; Liu, S.; Ma, R.; Zhang, Y.-Q.; Hu, D.-W.; Liu, X.-Y.; Zhang, J.-W.; Yang, D.-S. Rare $CH_3O^-/CH_3CH_2O^-$-bridged nine-coordinated binuclear DyIII single-molecule magnets (SMMs) significantly regulate and enhance the effective energy barriers. *CrystEngComm* **2020**, *22*, 1712–1724. [CrossRef]
34. Lever, A.B.D. *Inorganic Electronic Spectroscopy*, 2nd ed.; Elsevier: London, UK, 1992.
35. Taha, Z.A.; Ajlouni, A.M.; Al Momani, W.; Al-Ghzawi, A.A. Syntheses, characterization, biological activities and photophysical properties of lanthanides complexes with a tetradentate Schiff base ligand. *Spectrochim. Acta A* **2011**, *81*, 570–577. [CrossRef] [PubMed]
36. Ajlouni, A.M.; Taha, Z.A.; Al-Hassan, K.A.; Abu Anzeh, A.M. Synthesis, characterization, luminescence properties and antioxidant activity of Ln(III) complexes with a new aryl amide bridging ligand. *J. Luminesc.* **2012**, *132*, 1357–1363. [CrossRef]
37. Lis, S.; Elbanowski, M.; Mąkowska, B.; Hnatejko, Z. Energy transfer in solution of lanthanide complexes. *J. Photochem. Photobiol. A Chem.* **2002**, *150*, 233–247. [CrossRef]
38. Carnall, W.T. *Handbook on the Physics and Chemistry of Rare Earths*; Gschneidner, K.A., Jr., Eyring, L., Eds.; North-Holland Publishing Company: Amsterdam, The Netherlands, 1979; Volume 3.
39. Binnemans, K.; Van Duen, R.; Görller-Walrand, C.; Collinson, S.R.; Martin, F.; Bruce, D.W.; Wickleder, C. Spectroscopic behaviour of lanthanide (III) coordination compounds with Schiff base ligands. *Phys. Chem. Chem. Phys.* **2000**, *2*, 3753–3757. [CrossRef]
40. Ullmann, S.; Hahn, P.; Blömer, L.; Mehnert, A.; Laube, C.; Abelc, B.; Kersting, B. Dinuclear lanthanide complexes supported by a hybrid salicylaldiminato/calix[4]arene-ligand: Synthesis, structure, and magnetic and luminescence properties of (HNEt3)[Ln2(HL)(L)] (Ln = SmIII, EuIII, GdIII, TbIII). *Dalton Trans.* **2019**, *48*, 3893–3905. [CrossRef]
41. Karachousos-Spiliotakopoulos, K.; Tangoulis, V.; Panagiotou, N.; Tasiopoulos, A.; Moreno-Pineda, E.; Wernsdorfer, W.; Schulze, M.; Botas, A.M.P.; Carlos, L.D. Luminescence thermometry and field induced slow magnetic relaxation based on a near infrared emissive heterometallic complex. *Dalton Trans.* **2022**, *51*, 8208–8216. [CrossRef]
42. Osawa, M.; Yamayoshi, H.; Hoshino, M.; Tanaka, Y.; Akita, M. Luminescence color alteration induced by trapped solvent molecules in crystals of tetrahedral gold(i) complexes: Near-unity luminescence mixed with thermally activated delayed fluorescence and phosphorescence. *Dalton Trans.* **2019**, *48*, 9094–9103. [CrossRef]
43. Bartyzel, A. Effect of solvents on synthesis and recrystallization of Ni(II) complex with N_2O_2-donor Schiff base. *Inorg. Chim. Acta* **2017**, *459*, 103–112. [CrossRef]
44. Su, H.; Li, Z.; Tan, J.; Ma, H.; Yan, L.; Li, H. Structural conversion of three copper(II) complexes with snapshot observations based on the different crystal colours and morphology. *RSC Adv.* **2020**, *10*, 42964–42970. [CrossRef] [PubMed]
45. Rigaku, O.D. CrysAlis PRO (Version 1.171.38.46), Rigaku Oxford Diffraction 2015. Available online: https://www.rigaku.com/en/;products/smc/crysalis (accessed on 12 April 2019).
46. Sheldrick, G.M. SHELXL-Integrated space-group and crystal-structure determination. *Acta Cryst. A* **2015**, *71*, 3–8. [CrossRef] [PubMed]
47. Sheldrick, G.M. Crystal structure refinement with SHELXL. *Acta Cryst. C* **2015**, *71*, 3–8. [CrossRef]
48. Spek, A.L. checkCIF validation ALERTS: What they mean and how to respond. *Acta Cryst. E* **2020**, *76*, 1–11. [CrossRef]

Article

Speciation of Tellurium(VI) in Aqueous Solutions: Identification of Trinuclear Tellurates by ^{17}O, ^{123}Te, and ^{125}Te NMR Spectroscopy

Alexander G. Medvedev [1], Oleg Yu. Savelyev [2], Dmitry P. Krut'ko [3], Alexey A. Mikhaylov [1], Ovadia Lev [4,*] and Petr V. Prikhodchenko [1,*]

1. Kurnakov Institute of General and Inorganic Chemistry, Russian Academy of Sciences, Moscow 119991, Russia
2. Faculty of Fundamental Medicine, M.V. Lomonosov Moscow State University, Moscow 119991, Russia
3. Department of Chemistry, M.V. Lomonosov Moscow State University, Moscow 119991, Russia
4. The Casali Center, The Institute of Chemistry and The Harvey M. Krueger Family Center for Nanoscience and Nanotechnology, The Hebrew University of Jerusalem, Jerusalem 9190401, Israel
* Correspondence: ovadia@mail.huji.ac.il (O.L.); prikhman@gmail.com (P.V.P.)

Abstract: Tellurates have attracted the attention of researchers over the past decade due to their properties and as less toxic forms of tellurium derivatives. However, the speciation of Te(VI) in aqueous solutions has not been comprehensively studied. We present a study of the equilibrium speciation of tellurates in aqueous solutions at a wide pH range, 2.5–15 by ^{17}O, ^{123}Te, and ^{125}Te NMR spectroscopy. The coexistence of monomeric, dimeric, and trimeric oxidotellurate species in chemical equilibrium at a wide pH range has been shown. NMR spectroscopy, DFT computations, and single-crystal X-ray diffraction studies confirmed the formation and coexistence of trimeric tellurate anions with linear and triangular structures. Two cesium tellurates, $Cs_2[Te_4O_8(OH)_{10}]$ and $Cs_2[Te_2O_4(OH)_6]$, were isolated from the solution at pH 5.5 and 9.2, respectively, and studied by single-crystal X-ray diffractometry, revealing dimeric and tetrameric tellurate anions in corresponding crystal structures.

Keywords: tellurates; NMR spectroscopy; trinuclear tellurate; spin-spin coupling; DFT computations; single-crystal X-ray diffraction

1. Introduction

Over the past decade, the chemistry of telluric acid and its derivatives, tellurates, has increasingly attracted the attention of researchers [1–5]. Hydroxidocompounds of Te(VI) are stable in ambient conditions, soluble in water, and less toxic compared to other tellurium compounds [6]. They can be easily reduced, and therefore, they are attractive as precursors for a wide range of tellurium-based compounds and materials [7–9]. Moreover, tellurates are being intensively studied as promising materials with nonlinear optical properties [1,10–14], high proton conductivity [15,16], and ferroelectric and magnetic properties [17,18]. Tellurate-based materials benefit from high-temperature stability and optical transparency. Structural diversity allows the use of tellurates as electrode materials in metal-ion batteries [19,20]. Tellurates with perovskite structures are fundamentally interesting as oxide-ion conductors in photovoltaic applications [21].

However, most tellurates are produced by conventional hydrothermal solid-state reactions based on the fusion of the corresponding oxides in the presence of carbonate at high temperatures, controlling the composition of the final compound by loading the initial reagents. Hydrothermal synthesis also allows for obtaining tellurates of various structures and compositions. For example, tetrameric potassium $K_2[Te_4O_8(OH)_{10}]$ and rubidium $Rb[Te_2O_4(OH)_5]$ tellurates with centrosymmetric tetranuclear $[Te_4O_8(OH)_{10}]^{2-}$

anion [12,22] and potassium tellurate K$_2$[Te$_3$O$_8$(OH)$_4$] [12] with infinite linear chains $_\infty$(Te$_3$O$_{12}$)$^{6-}$ were prepared by a hydrothermal route. Interestingly, the crystallization of tellurates from aqueous solutions has not been widely used to obtain the corresponding Te(VI) compound. The hexameric tellurate K$_{8.5}$[Te$_6$O$_{27}$H$_9$]·0.5H$_3$O·17H$_2$O and several other alkali metal hydrogen tellurates containing binuclear anions were isolated from aqueous solution [23].

In our previous work, based on ^{125}Te NMR studies, we demonstrated that the interaction of the telluric acid with alkali in aqueous solution results in the formation of tellurate anions of different nuclearity with dominant dinuclear and trinuclear species [9,24]. The formation of trinuclear tellurate anions corresponding to different ^{125}Te NMR signals was also proposed [9,24,25]. However, the equilibrium of Te(VI) species in aqueous solutions with a wide pH range has not been comprehensively studied. Based on CSD [26] and ICSD [27] databases, there is no crystal structure containing trinuclear tellurate anions. Therefore, it is necessary to provide a systematic study of the equilibrium processes in aqueous solutions. This, in turn, can open up new aspects of forming a particular tellurate structure and selecting the conditions for obtaining a compound of a specific composition.

Herein we present a comprehensive study of the equilibrium in tellurate aqueous solutions by ^{17}O, ^{123}Te, and ^{125}Te NMR spectroscopy. We also provide DFT calculations to support our assignments. Single-crystal X-ray diffraction studies of three samples obtained from cesium tellurate aqueous solutions are also presented.

2. Results and Discussion

The structural data of alkali metal tellurates obtained by crystallization from aqueous solutions can indirectly provide information on the structure of the tellurate anions in the initial solutions. Thus, mononuclear tellurate species in molecular or anionic form (Te(OH)$_6$ and [TeO$_6$H$_4$]$^{2-}$, respectively) and binuclear anions with different degrees of protonation [Te$_2$O$_{10}$H$_{4+x}$]$^{(4-x)-}$, where x = 0, 1, 2, were isolated from aqueous solutions. At lower pH values, anions of higher nuclearity (hexanuclear and tetranuclear anions) are formed. Thus, it can be assumed that, depending on pH and tellurium concentration, all these forms can coexist in solution. This stipulation is supported by some crystal structures containing both mononuclear and dinuclear species (K$_6$[Te$_2$O$_{10}$H$_4$][TeO$_6$H$_4$]·12H$_2$O, Cs$_2$[Te$_2$O$_{10}$H$_6$][Te(OH)$_6$]) [23].

In the present work, aqueous cesium tellurate solutions with different pH were investigated by ^{17}O, ^{123}Te, and ^{125}Te NMR spectroscopy. We have chosen the cesium countercation to provide a maximal concentration of tellurium in the resulting solution since, to the best of our knowledge, cesium tellurates have the highest solubility in water compared to the tellurates of other alkali metals. Solutions of a wide range of pH (2.5–15.2) were obtained by the interaction of telluric acid and cesium hydroxide. The tellurium concentration was 1 M, except for the solution at pH 9.2, which contained only 0.5 M tellurium.

The charges of tellurate anions in solution depend on pH, and it is obvious that at high pH, the charge of the anion should be higher. Therefore, we attribute the same signals at the spectrum of different solutions to the species with the same nuclearity (monomer **M**, dimer **D**, and trimer **T**, see Scheme 1) without discussing the accurate value of the charge and degree of protonation.

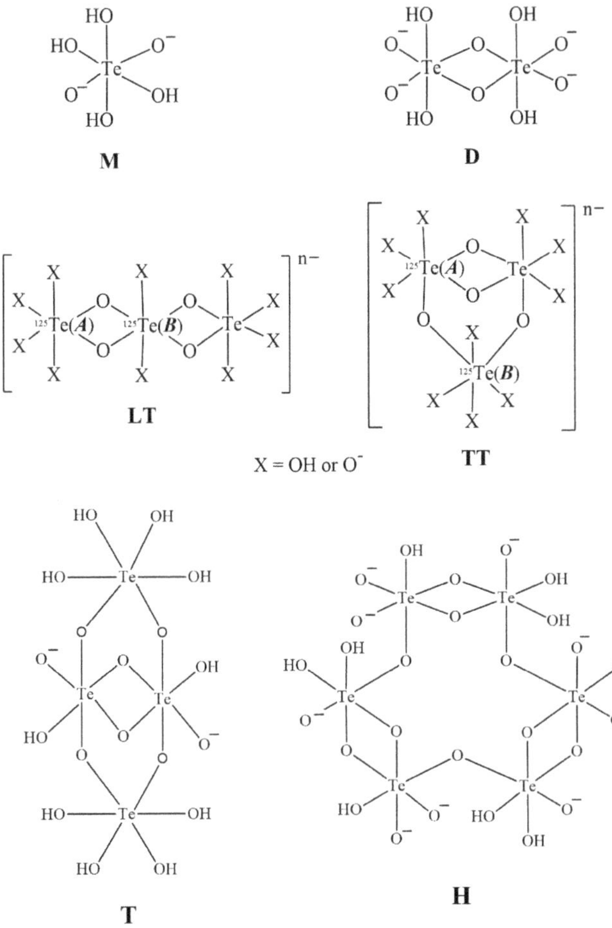

Scheme 1. The structure of tellurate anions of different nuclearity: monomer (**M**), dimer (**D**), linear (**LT**), and triangular (**TT**) trimers, tetramer (**T**), and hexamer (**H**).

2.1. ^{125}Te NMR Studies and Solution Chemistry

The signal in ^{125}Te NMR spectra with a chemical shift around 707 ppm corresponds to monomeric tellurium species (Figure 1) with different protonation degrees depending on the pH. This monomer exists as Te(OH)$_6$ molecular form at low pH (<3) (Figure 1a) and in deprotonated form as [TeO$_6$H$_4$]$^{2-}$ at higher pH (Figure 1b,c), which is confirmed by single-crystal X-ray data of the Te(VI) hydroxidocompounds isolated from corresponding aqueous solutions [28]. The spectrum of solution 2 with pH 5.2 contains an additional pair of signals around 685 and 658 ppm corresponding to a triangular trinuclear tellurate anion (Figure 1b). Previously, the pair of signals in the same region was incorrectly assigned to a linear trinuclear anion (see discussion below) [25].

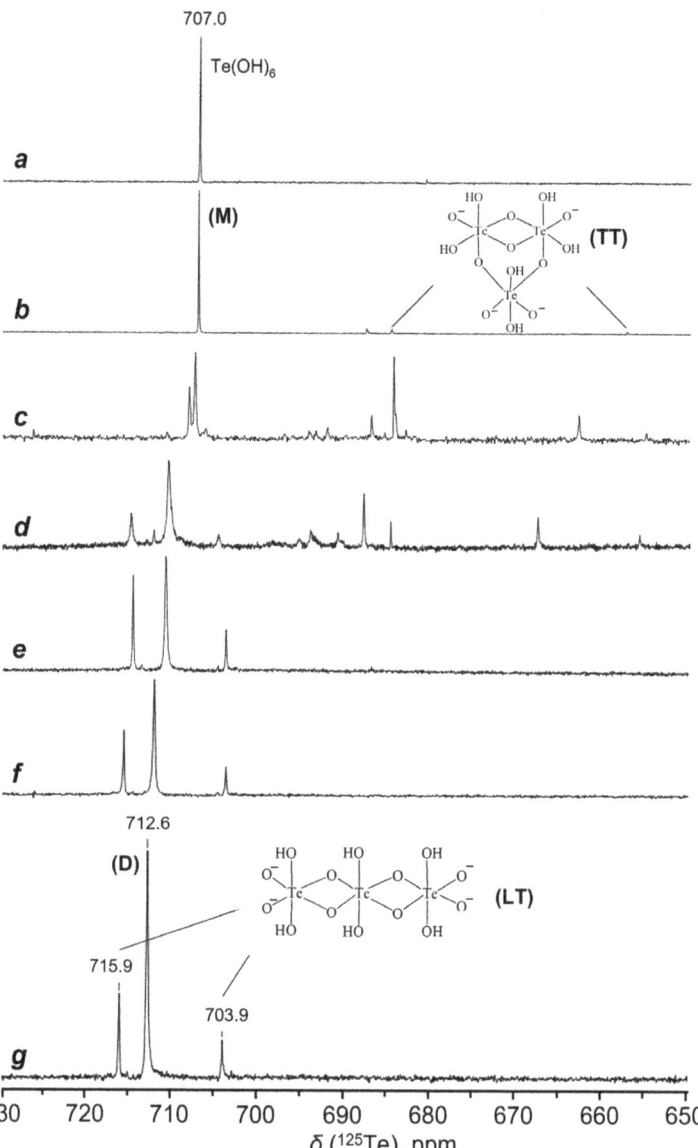

Figure 1. ^{125}Te NMR spectra of aqueous solutions with different [Te]/[Cs] ratios obtained by neutralization of telluric acid with cesium hydroxide: (**a**) solution **1** 1 M Te(OH)$_6$, [Cs] = 0, pH 2.5, (**b**) solution **2** [Cs]/[Te] = 0.1, pH 5.5, (**c**) solution **3** [Cs]/[Te] = 1, pH 9.2, (**d**) solution **4** [Cs]/[Te] = 1.5, pH 11.5, (**e**) solution **5** [Cs]/[Te] = 2.5, pH 14.3, (**f**) solution **6** [Cs]/[Te] = 3.5, pH 14.8, (**g**) solution **7** [Cs]/[Te] = 4.7, pH 15.2.

The further increase in pH by the addition of cesium hydroxide results in the formation of a precipitate. A clear solution can be obtained only at pH 9.2, even with a decrease in the tellurium concentration from 1 to 0.5 M. The ^{125}Te NMR spectra show a large number of new signals upfield from Te(OH)$_6$ signal (Figure 1c,d). However, with a further increase in pH, three intense signals with chemical shift 703.9 ppm and in the intervals 710.9–712.6 ppm and 714.6–715.9 ppm remain in the spectra (Figure 1e–g). The ratio of the relative integral intensities of the first and the third signal is 1/2 and remains unchanged. At the same time, the ratio of the integral intensities of the central signal and the two lateral ones increases with an increase in pH from ~1/1 (pH 14.3) to ~3/1 (pH 15.2). The signal at 710.9–712.6 ppm apparently corresponds to tellurium atoms in dimeric anion $[Te_2O_{10}H_4]^{4-}$ (**D**, Scheme 1) or its protonated forms, which crystallizes from these solutions [23]. This is confirmed by the fact that the ^{123}Te NMR spectrum of an aqueous solution of dimeric tellurate (see below) completely coincides with the spectra shown in Figure 1e–g. The constant ratio (1/2) of the relative integral intensities of signals at δ ~704 ppm and ~714 ppm in the ^{125}Te NMR spectra at high pH values indicates that they definitely correspond to trimeric tellurate with two equivalent tellurium atoms (Figure 1e–g).

Thus, ^{125}Te NMR studies of the solution showed the presence of two types of trimeric anions in solutions at different pH values. The structure of these trinuclear tellurate anions was unambiguously confirmed by additional NMR studies.

First, the ^{125}Te NMR spectrum of solution **5** ([Cs]/[Te] = 2.5, pH 14.3) with a high signal-to-noise ratio was collected (Figure 2).

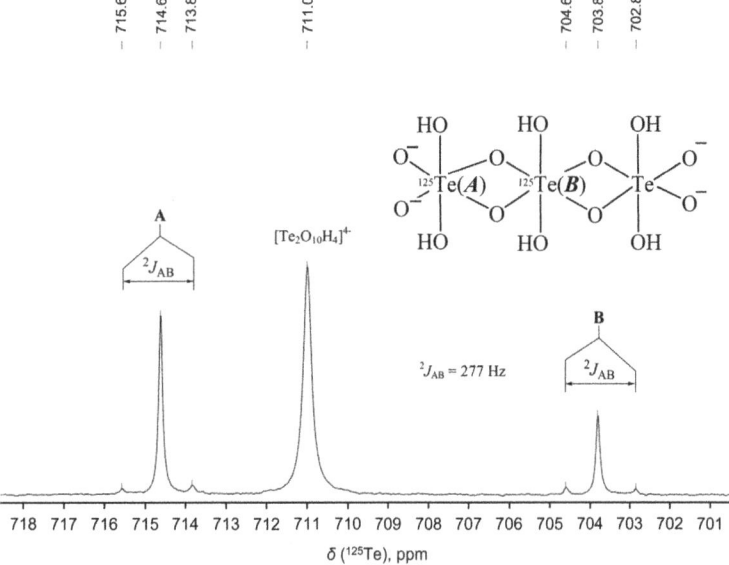

Figure 2. ^{125}Te NMR spectrum of a tellurate solution **5** ([Cs]/[Te] = 2.5, pH 14.3) (Bruker Avance-500, Bruker BioSpin GmbH, Ettlingen, Germany).

^{125}Te satellites are visible on both signals of the trinuclear anion, which form an AB-spin system ($^2J_{AB}$ = 277 Hz). The relative intensity ratio of each pair of satellites to the main signal corresponds to two neighboring tellurium atoms for one of the atoms (B, δ = 703.8 ppm) and one for the other two remaining atoms (A, δ = 714.6 ppm) (Figure 2). The chemical shift difference calculated for the AB-spin system coincides exactly with the experimental one for main signals (10.8 ppm). The linear **LT** and triangular **TT** trimers (Scheme 1) are two possible structures that satisfy the resulting ^{125}Te spectrum. Both **LT** and **TT** structures provide principally the same AB pattern in the ^{125}Te NMR spectra.

The possible existence of the triangle trimer is confirmed by single-crystal X-ray data of tetrameric tellurates $K_2[Te_4O_8(OH)_{10}]$ and $Rb[Te_2O_4(OH)_5]$ (**T**), prepared by hydrothermal route [12,22], with a similar **TT** structural fragment (Scheme 1). The same fragments were recently discovered as units of the infinite linear anionic chains in $K_2[Te_3O_8(OH)_4]$ [12].

The hexanuclear tellurate $K_{8.5}[Te_6O_{27}H_9]\cdot 0.5H_3O\cdot 17H_2O$ (**H**) is also known, in which three dimer fragments are linked by single oxygen bridges [23].

Previously, aqueous solutions of $Te(OH)_6$ at low pH (<7) were studied by ^{125}Te NMR spectroscopy using ^{125}Te enriched (up to 92.8%) telluric acid at pH 6.78 [25]. The ^{125}Te NMR spectrum, in addition to the $Te(OH)_6$ signal and several minor singlets, also contains signals related to the AB_2 spin system with J_{AB} = 682.5 Hz (δ_A = 657.5 ppm and δ_B = 682.9 ppm) incorrectly assigned to the structure of linear trimeric complex **LT**.

We reproduced this experiment using the telluric acid with ^{125}Te natural abundance (7.1%) (Figure 3), which is in complete agreement with the published data [25]. The ^{125}Te NMR spectrum with a high signal-to-noise ratio demonstrates ^{125}Te satellites from both main signals corresponding to trimeric anion. The satellites form an *AB*-spin system pattern ($^2J_{AB}$ = 682 Hz) similar to that described above.

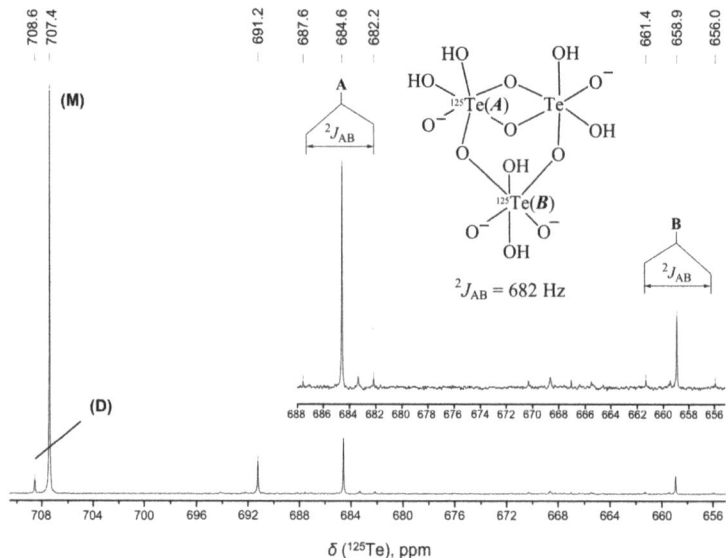

Figure 3. ^{125}Te NMR spectrum of a tellurate solution ([Te] = 0.2 M, pH 6.8).

Assignment of *AB*-spin systems depicted in Figures 2 and 3 to **LT** and **TT** trimeric structures, respectively, at first glance, can be performed based on the values of spin-spin constants $^2J_{AB}$. From general considerations, $^2J(^{125}Te-^{125}Te)$ values equal to 682 and 277 Hz should correspond to the interaction through two and one oxygen bridge, respectively. An unambiguous assignment of the two trimer structures can be performed by observing ^{123}Te (natural abundance 0.9%) satellites from ^{125}Te signals or vice versa. One set of symmetrical satellites (*AX*- and *BY*-spin systems with $^2J_{AX} = {^2J_{BY}}$) should be observed for linear trimeric tellurate **LT** corresponding to the interaction through two oxygen bridges. At the same time, two sets of satellites for a signal at 714.6 ppm, corresponding to interactions through one and through two oxygen bridges, should appear with different values of $^2J(^{125}Te-^{123}Te)$ for the triangular trimer **TT**.

2.2. ^{123}Te NMR Studies

We managed to register the ^{123}Te NMR spectrum with a sufficient level of signal-to-noise to observe ^{125}Te satellites from ^{123}Te signals at natural abundance (Figure 4). The observation of low-intensity ^{123}Te satellites in the ^{125}Te NMR spectrum is very difficult, taking into account the significant linewidths and the presence of additional signals from the *AB*-spin system. This explains the choice of the less-sensitive ^{123}Te nucleus with a lower natural content compared to ^{125}Te. A similar experiment with a natural abundance of ^{123}Te was carried out to measure indirect spin-spin coupling constants $J(^{123}$Te–^{125}Te) in organotellurium 1,8-naphthalene derivatives [29].

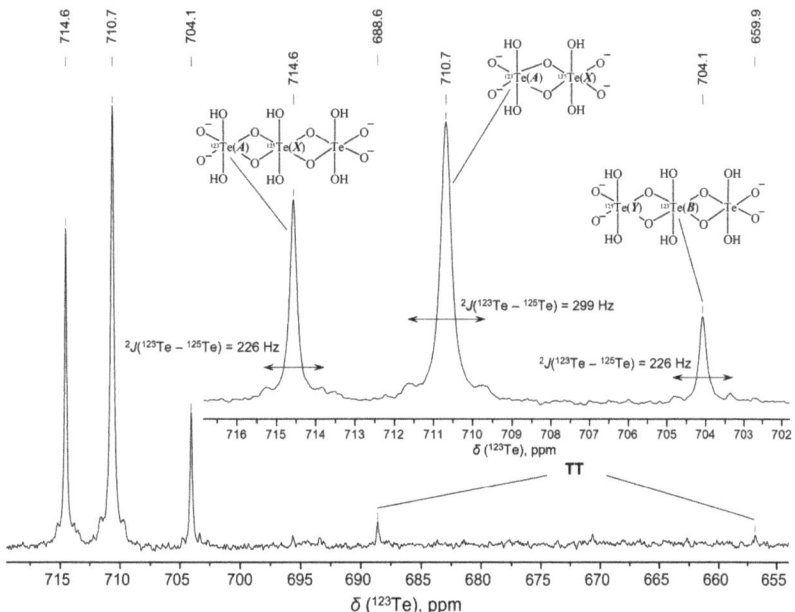

Figure 4. ^{123}Te NMR spectrum of 1 M solution of Cs$_4$[Te$_2$O$_{10}$H$_4$] in water (pH 14.4) (Bruker Avance-600, Bruker BioSpin GmbH, Ettlingen, Germany).

In the obtained spectrum, both signals corresponding to trinuclear tellurate contain only one set of satellites ^{125}Te with $^2J(^{123}$Te–^{125}Te) = 226 Hz. Multiplication by 1.206 (the ratio of resonance frequencies $\Xi(^{125}$Te)/$\Xi(^{123}$Te)) gives a value of 273 Hz, which is in excellent agreement with the value of 277 Hz obtained from the ^{125}Te spectrum (considering the linewidth of the signals, ~25–30 Hz).

The singlet of dimeric tellurate **D** also has one set of symmetrical ^{125}Te satellites with $^2J(^{123}$Te–^{125}Te) = 299 Hz. Multiplication by 1.206 gives a value of 361 Hz, which is comparable with $^2J(^{125}$Te–^{125}Te) = 277 Hz in **LT**.

The correctness of our assignment of trimeric tellurates is confirmed by the close values of chemical shifts of the **LT** and **D** signals, which is consistent with their structural similarity. At the same time, the **TT** signals are significantly shifted upfield, especially for the Te(B) atom (658.9 ppm, see Figure 3). This is also confirmed by the fact that the synthesis of tetrameric tellurate **T**, which is structurally close to trimer **TT**, was carried out at pH~3 [22].

Thus, it follows from the above data that the value of $^2J(^{125}$Te–^{125}Te) in the two-bridge Te(μ-O)$_2$Te fragment is approximately two times lower than in the Te(μ-O)Te fragment with one oxygen bridge. A possible explanation for this nontrivial fact is the significant difference in the Te-O-Te angles in these fragments. The values of these angles in Te(μ-O)$_2$Te bridging fragments for reported structures lie in the range of 100–103° [23]. At the same time, for single bridges in the **H** hexamer, their value is ~128° [23], and for the **T** tetramer, it is ~133° [22]. An increase in the Te-O-Te angle should lead to an increase in the value of $^2J(^{125}$Te–^{125}Te). To the best of our knowledge, the values of $^2J(^{125}$Te–^{125}Te) for tellurates or related tellurium compounds are currently unknown. The dependence of the geminal coupling constant on the El-O-El bond angle was studied in detail for Sn(IV) derivatives

[(R$_3$Sn)$_2$O] [30], formally an isoelectronic derivative of Te(VI). The data of this work confirm the fact that $^2J(^{119}$Sn–^{117}Sn) increases with increasing Sn ~ Sn angle. In the series of these compounds, a 1.5-fold increase in $^2J(^{119}$Sn–^{117}Sn) from 420.6 to 617.9 Hz is observed with an increase in the Sn-O-Sn angle from 137.1° (R = Ph) to 180.0° (R = Bn).

Dimeric tellurate **D** seems to be also present in the Te(OH)$_6$ solution at pH 6.8, but at a much lower concentration (Figure 3, singlet with δ (^{125}Te) = 708.6 ppm). Upfield shift with pH decrease is probably due to a change in the degree of protonation and, accordingly, the charge of the tellurate anion (obviously, protonation or deprotonation of tellurate anions cannot lead to the appearance of a new signal in the ^{125}Te NMR spectrum due to fast proton exchange in aqueous solution). Low-intensity signals of trimeric tellurate **TT** at 688.6 ppm and 656.9 ppm (Figure 4) are also present in the spectrum at high pH. At the same time, the linear trimer **LT**, which, along with the **D** dimer, dominates in solutions at pH > 14, is not observed in the spectra at low pH values (Figures 1a–c and 3). Probably, the remaining signals (see Figure 1c,d) can be attributed to the diverse polynuclear anionic forms of Te(VI), which exist in solutions with a relatively low [Cs]/[Te] ratio.

It should be noted that at high pH values (~14–15), the signal widths in the spectra are an order of magnitude larger (Δν$_{1/2}$ ~ 25–40 Hz) compared to a neutral pH 6.8 solution (Δν$_{1/2}$ ~ 4–5 Hz). This is apparently due to an increase in the rate of exchange between different tellurate anions, accompanied by the breaking of Te-O bonds in the presence of an excess of hydroxide ions. This is also confirmed by the data of ^{17}O NMR spectroscopy of aqueous solutions of tellurates (see Supplementary Materials, Section S1, Figure S1).

We can conclude that an increase in the pH of tellurate solutions leads to a shift in equilibrium toward the formation of the dimeric anion **D**, which is in equilibrium with the trimeric anion **LT** (Scheme 2). The rate of interconversion between the different forms of tellurate anions in solution increases with increasing pH.

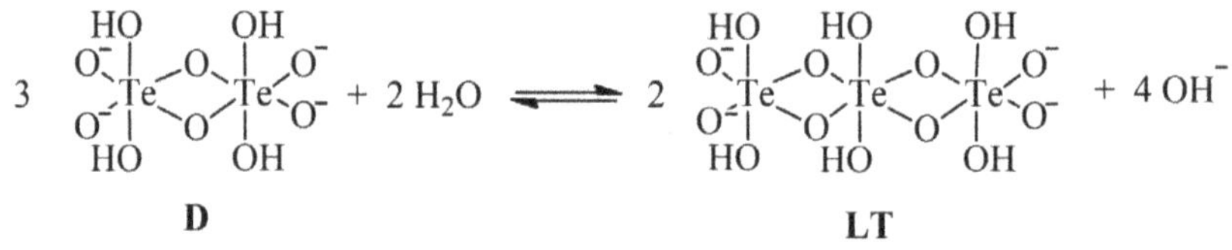

Scheme 2. Equilibrium between dimeric and linear trimeric tellurate anions in aqueous solution.

The existence of this equilibrium is confirmed by the fact that the relative content of **D** in a mixture with **LT** increases with increasing pH (see Figure 1e–g). At low pH, a number of polynuclear tellurates exist in the solution, including **TT**, whose structures apparently contain single-member oxygen bridges.

2.3. DFT Computations

In order to glean insight into the relative stability of **LT** and **TT** [Te$_3$O$_{12}$H$_6$]$^{4-}$ anions, we have conducted DFT calculations. The optimized geometries of both anions are presented in Figures 5 and 6, respectively. According to these calculations, the formation of **LT** anion in the gas phase is preferable over **TT** anion (~83 kJ/mol, Supplementary Materials Tables S1 and S2). The lower energy value for the linear anion also includes the energy of the four intramolecular hydrogen bonds with O . . . O separations lying within 2.948–3.059 Å (Figure 5). The evaluated H-bond energies are equal to 15–20 kJ/mol (Supplementary Materials Table S3). Only one hydrogen bond is found in the structure of the triangular anion (Figure 6), with a calculated energy value of 27.9 kJ/mol. Considering the observed H-bond energies, the **LT** anion is likely to be energetically preferable (by up to ~40 kJ/mol). However, this relatively small value does not exclude the possibility of **TT** anion formation.

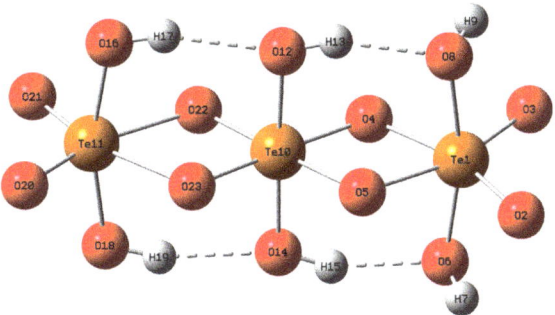

Figure 5. Optimized geometry of LT anion [Te$_3$O$_{12}$H$_6$]$^{4-}$. H-bonds are shown as dashed lines.

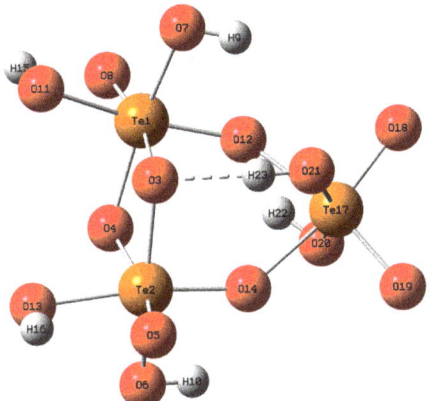

Figure 6. Optimized geometry of TT anion [Te$_3$O$_{12}$H$_6$]$^{4-}$. H-bond is shown as a dashed line.

2.4. Crystal Structures

Based on the ^{125}Te NMR data, solutions **2**, **3**, and **5** were selected as suitable candidates for the crystallization of trimeric tellurates. Three obtained crystals were collected from these solutions and characterized by single-crystal X-ray diffraction. Two new crystalline compounds, cesium octaoxidodecahydroxidotetratellurate Cs$_2$[Te$_4$O$_8$(OH)$_{10}$] **I** and cesium tetraoxidohexahydroxidoditellurate Cs$_2$[Te$_2$O$_4$(OH)$_6$] **II** were obtained from solutions **2** (pH 5.5) and **3** (pH 9.2), respectively. Previously reported cesium pentaoxidopentahydroxidoditellurate tetrahydrate Cs$_3$[Te$_2$O$_5$(OH)$_5$]·4H$_2$O (**III**, ICSD 417438) crystals [23] were collected from solution **5**. It should be noted that the charges of tellurate anions in **II** and **III** correlate with the pH of the solution. The charge of the ditellurate anion is higher in solution **5** with pH 14.3. In addition, the tetranuclear form of tellurate **T** crystallizes from a solution with a lower pH, which corresponds to the results of the NMR studies (Figure 1).

Both new crystalline hydrogen tellurates are ionic compounds containing centrosymmetric tetranuclear [Te$_4$O$_8$(OH)$_{10}$]$^{2-}$ and binuclear [Te$_2$O$_4$(OH)$_6$]$^{2-}$ anions in **I** and **II**, respectively, and Cs$^+$ cations coordinated to tellurium-containing anions. Compound **I** is isostructural to previously published potassium analog K$_2$[Te$_4$O$_8$(OH)$_{10}$] [22]. The isomorphism of the potassium and cesium salts is not typical, and it is realized only because both structures contain a bulky anion. Selected bond distances and angles of the centrosymmetric tetranuclear [Te$_4$O$_8$(OH)$_{10}$]$^{2-}$ anion in crystal **I** are given in Supplementary Materials Table S4. The tetranuclear anion can be represented by one dimeric anion bridged by four axial oxygen atoms with two Te(OH)$_6$ monomers (Figure 7). On the other hand, the [Te$_4$O$_8$(OH)$_{10}$]$^{2-}$ anion contains both linear and triangular trimeric fragments suggested by NMR studies.

The tellurium atoms have a slightly distorted octahedral coordination. The geometric parameters of $[Te_2O_4(OH)_6]^{2-}$ anion in **II** are close to those found earlier in an adduct with telluric acid $Cs_2[Te_2O_4(OH)_6][Te(OH)_6]$ (Table S4) [23]. The Te atoms have a slightly distorted octahedral environment. As in the reported adduct, the four axial and two of the four equatorial oxygen atoms are protonated (Figure 8), resulting in a total charge of −2 for the anion. There are 11 and 12 oxygen atoms in the coordination environment of the cesium cation in **I** and **II** with Cs-O distances in the range 2.995(2)–3.631(7) and 3.055(2)–3.741(2) Å, respectively. All hydrogen atoms of tellurate anions in **I** and **II** are involved in hydrogen bonding (Supplementary Materials Table S5, Figure S2).

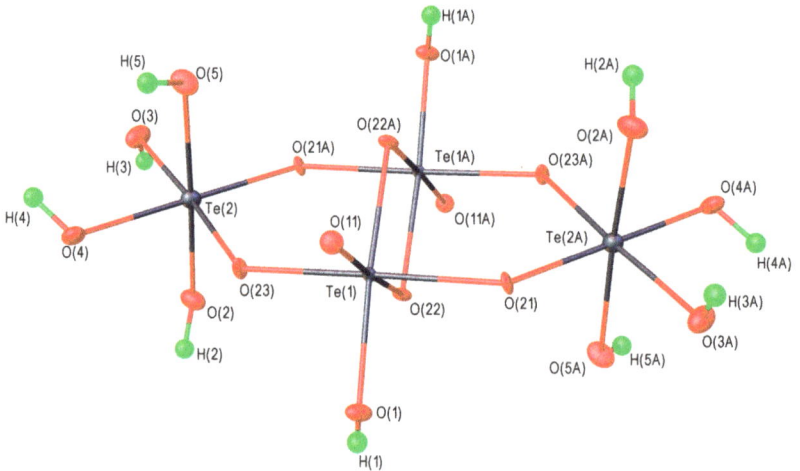

Figure 7. Structure of the $[Te_4O_8(OH)_{10}]^{2-}$ anion in compound **I**.

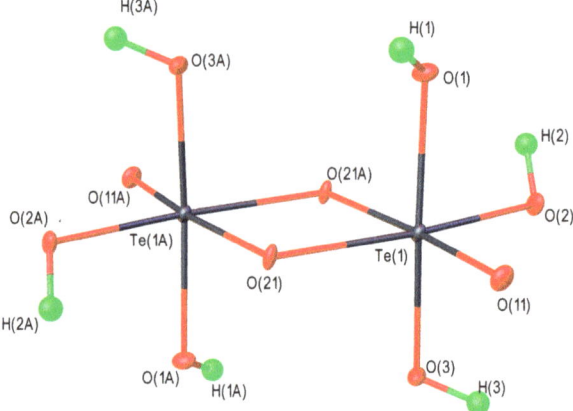

Figure 8. Structure of the $[Te_2O_4(OH)_6]^{2-}$ anion in compound **II**.

Unfortunately, we cannot isolate a sufficient amount of a single-phase product, which is confirmed by X-ray diffraction of the resulting powders. Further evaporation of solvent leads to the crystallization of different phases (See Supplementary Materials Section S2, Figure S3). Both crystalline powders contain small amounts of impurities. In this regard, we do not describe the yield of products, their vibrational spectra, or their elemental analysis.

Thus, the structural data of alkali metal tellurates obtained by natural crystallization from aqueous solutions can indirectly provide information on the structure of tellurate anions in the initial solutions. On the other hand, obtained results are consistent with NMR data.

3. Materials and Methods

3.1. Preparation of Solutions and Crystals

The solutions were obtained by neutralization of telluric acid with cesium hydroxide. The [Cs]/[Te] ratio was changed from 0 to 4.7: solution **1**—[Cs]/[Te] = 0; solution **2**—[Cs]/[Te] = 0.1; solution **3**—[Cs]/[Te] = 1; solution **4**—[Cs]/[Te] = 1.5; solution **5**—[Cs]/[Te] = 2.5; solution **6**—[Cs]/[Te] = 3.5; solution **7**—[Cs]/[Te] = 4.7. Tellurium concentration in solutions **1**, **2**, and **4**–**7**, [Te] ≈ 1 M, in solution **3**, [Te] ≈ 0.5 M (Table 1).

Table 1. Aqueous solutions with different pH obtained by neutralization of telluric acid with cesium hydroxide.

Solution	[Cs], M	[Te], M	pH
Solution 1	0	1	2.5
Solution 2	0.1	1	5.2
Solution 3	0.5	0.5	9.2
Solution 4	1.5	1	11.5
Solution 5	2.5	1	14.3
Solution 6	3.5	1	14.8
Solution 7	4.7	1	15.2

Solution **1** was prepared by dissolving telluric acid in water.

Solution **2** was prepared by adding an appropriate volume of 4.7 M cesium hydroxide solution to solution **1**.

Solutions **5**–**7** were prepared by dissolving an appropriate amount of telluric acid in a 4.7 M cesium hydroxide solution.

Solutions **3** and **4** were prepared by mixing the corresponding volumes of solutions **2** and **5**. However, it was not possible to obtain transparent solutions in the [Cs]/[Te] ratio range greater than 0.1 and less than 1. In this concentration range, a viscous precipitate is formed, which is insoluble even after prolonged stirring at moderate heating. A transparent solution with the ratio [Cs]/[Te] = 1 (solution **3**) can only be obtained by lowering the tellurium concentration to 0.5 M.

The [Cs]/[Te] ratios in solutions **1**, **2**, and **5**–**7** were calculated from the initial amounts of telluric acid and cesium hydroxide. In this case, the tellurium concentration in the resulting solution was estimated approximately (by the mass of telluric acid and the volume of water or cesium hydroxide solution, without using volumetric analytical flasks). Accordingly, in solutions **3** and **4**, the [Cs]/[Te] ratio was estimated considering the approximate tellurium concentrations in solutions **2** and **5**.

The pH of prepared solutions was measured with Edge HI 2002-02 pH meter (Hanna Instruments, Vöhringen, Germany) equipped with HI 11310 electrode with automatic temperature compensation. The accuracy of measurements is equal to ±0.01 pH.

Cesium hydrogen tellurate $Cs_4[Te_2O_{10}H_4]\cdot 8H_2O$ was synthesized according to the previously published procedure [23]. Briefly, colorless crystals were obtained by dissolution of 6.31 g $Te(OH)_6$ (0.0275 mol) in 15 mL of a 6.6 M cesium hydroxide solution. Yield 85.6% (12.89 g).

Crystalline cesium octaoxidodecahydroxidotetratellurate $Cs_2[Te_4O_8(OH)_{10}]$ **I** and cesium tetraoxidohexahydroxidoditellurate $Cs_2[Te_2O_4(OH)_6]$ **II** were obtained by slow evaporation of solvent from solutions **2** and **3**, respectively.

3.2. ^{17}O, ^{123}Te and ^{125}Te NMR Spectroscopy

^{125}Te NMR spectra were recorded on Bruker MSL-400, Bruker Avance-400, and Bruker Avance-500 (Bruker BioSpin GmbH, Ettlingen, Germany) spectrometers at resonance frequencies 126.24 and 157.81 MHz, respectively. ^{123}Te NMR spectrum was recorded on a Bruker Avance-600 spectrometer at resonance frequency 157.05 MHz. ^{17}O NMR spectra were recorded on a Bruker MSL-400 spectrometer at resonance frequency 54.24 MHz. Chemical shifts were measured using external references: aqueous solution of Te(OH)$_6$ (707.0 ppm relative to Te(CH$_3$)$_6$ [31]) for ^{125}Te and ^{123}Te and water for ^{17}O.

3.3. Computations

Gaussian09 (Gaussian Inc., Wallingford, CT, USA) was used in all computations [32]. Geometries of the linear and triangular trinuclear anion [Te$_3$O$_{12}$H$_6$]$^{4-}$ were optimized at the B3LYP/6-311G++(d,p) level using LANL2DZ basis set with effective core potential for Te atoms [33]. The normal-mode analysis did not provide imaginary frequencies for the considered species. The optimized cartesian coordinates of trinuclear anions are presented in Supplementary Materials Tables S1 and S2.

The energy of intermolecular H-bonds E_{HB} in the considered crystals (Supplementary Materials Table S3) is evaluated according to ref. [34] as:

$$E_{HB} \text{ [kJ mol}^{-1}\text{]} = 1124 \cdot G_b \text{ [atomic units]} \qquad (1)$$

where G_b is the positively defined local electronic kinetic energy density at the H⋯O bond critical point [35]. The Espinosa approach gives reasonable results for energies of intermolecular H-bonds and other non-covalent interactions [36–39].

3.4. Single-Crystal X-ray Diffraction

Experimental reflection intensity data for compounds **I** and **II** were collected on a Bruker D8 Venture diffractometer (Bruker AXS GmbH, Karlsruhe, Germany; graphite monochromatized MoK$_\alpha$ radiation, λ = 0.71073 Å) using ω-scan mode at 100 K. Absorption corrections based on measurements of equivalent reflections were applied [40]. The structures were solved by direct methods and refined by full-matrix least-squares on F^2 with anisotropic thermal parameters for all non-hydrogen atoms [41]. Hydrogen atoms were located from different Fourier syntheses and refined isotropically. Selected atom distances and bond angles are collected in Table S4. Selected crystallographic data for **I** and **II** are provided in Table 2. Atomic coordinates and anisotropic displacement parameters are given in Supplementary Materials (Tables S6–S9). CCDC 2215218 and 2215219 contain the supplementary crystallographic data for this paper. These data can be obtained free of charge via http://www.ccdc.cam.ac.uk/conts/retrieving.html (accessed on 30 November 2022) (or from the CCDC, 12 Union Road, Cambridge CB2 1EZ, U.K.; fax: +44 1223 336033; e-mail: deposit@ccdc.cam.ac.uk).

Table 2. Crystal data, data collection, and refinement parameters for **I** and **II**.

Parameter	I	II
Formula	Cs$_2$Te$_4$O$_{18}$H$_{10}$	Cs$_2$Te$_2$O$_{10}$H$_6$
CCDC	2215219	2215218
fw	1074.30	687.07
Color, habit	Colorless, prism	Colorless, prism
Cryst size, mm	0.10 × 0.05 × 0.05	0.10 × 0.10 × 0.05
Cryst syst	Monoclinic	Triclinic
Space group	P2$_1$/c	P$\bar{1}$
a, Å	5.7174(3)	6.2963(3)
b, Å	8.4698(4)	6.3962(3)
c, Å	16.6536(9)	7.3552(4)

Table 2. Cont.

Parameter	I	II
α, deg	90	67.507(2)
β, deg	97.436(2)	77.178(2)
γ, deg	90	65.661(1)
V, Å^3	799.67(2)	248.63(2)
Z	2	1
ρ_{calc}, g/cm^3	4.462	4.589
μ, mm^{-1}	11.803	13.116
$F(000)$	944	300
θ range, deg	2.47 to 30.46	3.01 to 30.48
Total no. of reflns	7894	4145
Unique reflns, R_{int}	1833, 0.0422	1504, 0.015
Reflns. with $I > 2\sigma(I)$	1632	1429
No. of variables	124	76
R_1 ($I > 2\sigma(I)$)	0.0377	0.0161
wR_2 (all data)	0.0849	0.0371
GoF on F^2	1.130	1.117
Largest diffPeak/hole, e/Å^3	1.881/−2.050	0.800/−1.034

4. Conclusions

A comprehensive study of the equilibrium in tellurate aqueous solutions was carried out by ^{125}Te, ^{123}Te, and ^{17}O NMR. The coexistence of monomeric, dimeric, and trimeric tellurate species in chemical equilibrium in the aqueous solutions of cesium tellurates at a wide pH range was proven. The formation and coexistence of trimeric tellurate anions with linear and triangular structures were shown for the first time based on NMR studies and DFT calculations. Three tellurates were crystallized from the studied solutions and studied by single-crystal X-ray analysis. Two of obtained crystals, cesium octaoxidodecahydroxidotetratellurate Cs$_2$[Te$_4$O$_8$(OH)$_{10}$] and cesium tetraoxidohexahydroxidoditellurate Cs$_2$[Te$_2$O$_4$(OH)$_6$], were isolated for the first time and shown to contain dimeric and tetrameric tellurate anions. The latter contains the triangular tellurate fragment, which was characterized by ^{125,123}Te NMR in the solution.

Supplementary Materials: The following supporting information can be downloaded at: https://www.mdpi.com/article/10.3390/molecules27248654/s1, Section S1: ^{17}O NMR studies; Figure S1: ^{17}O NMR spectra of: (a) telluric acid aqueous solution; (b) dimer D in water; (c) dimer D in water at 323 K; (d) dimer D in water, enriched by ^{17}O; Figure S2. Crystal packing in II; Table S4: Selected bond lengths and angles in the structure of I and II; Table S1: Optimized cartesian coordinates of LT anion [Te$_3$O$_{12}$H$_6$]$^{4-}$; Table S2: Optimized cartesian coordinates of TT anion [Te$_3$O$_{12}$H$_6$]$^{4-}$; Table S3: Computed values of the electron density, ρ_b, the local electronic kinetic energy density, G_b, at the O...O critical point, the H-bond energy, E_{HB}, evaluated using Equation (1); Table S5: Hydrogen-bond geometry in compounds I and II; Table S6: Fractional atomic coordinates and isotropic or equivalent isotropic displacement parameters (Å^2) for I; Table S7: Atomic displacement parameters (Å^2) for I; Table S8: Fractional atomic coordinates and isotropic or equivalent isotropic displacement parameters (Å^2) for II; Table S9: Atomic displacement parameters (Å^2) for II; Section S2: X-ray powder diffraction (XRD); Figure S3: X-ray powder diffractograms of cesium octaoxidodecahydroxidotetratellurate Cs$_2$[Te$_4$O$_8$(OH)$_{10}$] I (a) and cesium tetraoxidohexahydroxidoditellurate Cs$_2$[Te$_2$O$_4$(OH)$_6$] II (b) Calculated powder diffractograms were obtained using Mercury (CCDC) software.

Author Contributions: Conceptualization, O.L. and P.V.P.; methodology, D.P.K. and P.V.P.; software, O.L. and D.P.K.; validation, A.G.M., O.Y.S., D.P.K. and A.A.M.; formal analysis, A.G.M., D.P.K. and P.V.P.; writing—original draft preparation, D.P.K., P.V.P. and O.L.; writing—review and editing, D.P.K., P.V.P. and O.L.; visualization, A.A.M. and O.Y.S.; supervision, P.V.P. and O.L.; project administration, P.V.P. and O.L.; funding acquisition, P.V.P. All authors have read and agreed to the published version of the manuscript.

Funding: This study was supported by the Russian Science Foundation (grant no. 22-13-00426, https://rscf.ru/en/project/22-13-00426/ (accessed on 30 November 2022)).

Institutional Review Board Statement: Not applicable.

Informed Consent Statement: Not applicable.

Data Availability Statement: The data presented in this study are available on request from the corresponding authors.

Acknowledgments: X-ray diffraction study was performed at the Centre of Shared Equipment of IGIC RAS. The authors are grateful to Moscow State University (Russia) for the opportunity to use the NMR facilities of the Center for Magnetic Tomography and Spectroscopy. We thank A.V. Churakov for the helpful discussion.

Conflicts of Interest: The authors declare no conflict of interest.

Sample Availability: Samples of the compounds **I–III** are available from the authors.

References

1. Wang, D.; Zhang, Y.; Shi, Q.; Liu, Q.; Yang, D.; Zhang, B.; Wang, Y. Tellurate Polymorphs with High-Performance Nonlinear Optical Switch Property and Wide Mid-IR Transparency. *Inorg. Chem. Front.* **2022**, *9*, 1708–1713. [CrossRef]
2. Wang, D.; Zhang, Y.; Liu, Q.; Zhang, B.; Yang, D.; Wang, Y. Band Gap Modulation and Nonlinear Optical Properties of Quaternary Tellurates Li_2GeTeO_6. *Dalton Trans.* **2022**, *51*, 8955–8959. [CrossRef] [PubMed]
3. Nagarathinam, M.; Soares, C.; Chen, Y.; Seymour, V.R.; Mazanek, V.; Isaacs, M.A.; Sofer, Z.; Kolosov, O.; Griffin, J.M.; Tapia-Ruiz, N. Synthesis, Characterisation, and Feasibility Studies on the Use of Vanadium Tellurate(VI) as a Cathode Material for Aqueous Rechargeable Zn-Ion Batteries. *RSC Adv.* **2022**, *12*, 12211–12218. [CrossRef] [PubMed]
4. Song, Y.; Niu, H.; Zeng, Z.; Jiang, D.; He, X.; Liang, Y.; Huang, H.; Zhang, M.; Li, J.; He, Z.; et al. Influence of Barium Intercalated Ions on Magnetic Interaction in the Tellurate Compound $BaNi_2TeO_6$. *Inorg. Chem.* **2022**, *61*, 5731–5736. [CrossRef]
5. Chiaverini, L.; Cirri, D.; Tolbatov, I.; Corsi, F.; Piano, I.; Marrone, A.; Pratesi, A.; Marzo, T.; La Mendola, D. Medicinal Hypervalent Tellurium Prodrugs Bearing Different Ligands: A Comparative Study of the Chemical Profiles of AS101 and Its Halido Replaced Analogues. *Int. J. Mol. Sci.* **2022**, *23*, 7505. [CrossRef]
6. Gad, S.C.; Pham, T. Tellurium. In *Encyclopedia of Toxicology*; Academic Press: Amsterdam, The Netherlands, 2014; pp. 481–483.
7. Grishanov, D.A.; Mikhaylov, A.A.; Medvedev, A.G.; Gun, J.; Prikhodchenko, P.V.; Xu, Z.J.; Nagasubramanian, A.; Srinivasan, M.; Lev, O. Graphene Oxide-Supported β-Tin Telluride Composite for Sodium- and Lithium-Ion Battery Anodes. *Energy Technol.* **2018**, *6*, 127–133. [CrossRef]
8. Grishanov, D.A.; Mikhaylov, A.A.; Medvedev, A.G.; Gun, J.; Nagasubramanian, A.; Madhavi, S.; Lev, O.; Prikhodchenko, P.V. Synthesis of High Volumetric Capacity Graphene Oxide-Supported Tellurantimony Na- and Li-Ion Battery Anodes by Hydrogen Peroxide Sol Gel Processing. *J. Colloid Interface Sci.* **2018**, *512*, 165–171. [CrossRef]
9. Mikhaylov, A.A.; Medvedev, A.G.; Churakov, A.V.; Grishanov, D.A.; Prikhodchenko, P.V.; Lev, O. Peroxide Coordination of Tellurium in Aqueous Solutions. *Chem.—Eur. J.* **2016**, *22*, 2980–2986. [CrossRef]
10. Yeon, J.; Kim, S.-H.; Nguyen, S.D.; Lee, H.; Halasyamani, P.S. Two New Noncentrosymmetric (NCS) Polar Oxides: Syntheses, Characterization, and Structure–Property Relationships in $BaMTe_2O_7$ (M = Mg^{2+} or Zn^{2+}). *Inorg. Chem.* **2012**, *51*, 2662–2668. [CrossRef]
11. Lu, W.; Gao, Z.; Liu, X.; Tian, X.; Wu, Q.; Li, C.; Sun, Y.; Liu, Y.; Tao, X. Rational Design of a $LiNbO_3$-like Nonlinear Optical Crystal, Li_2ZrTeO_6, with High Laser-Damage Threshold and Wide Mid-IR Transparency Window. *J. Am. Chem. Soc.* **2018**, *140*, 13089–13096. [CrossRef]
12. Wang, D.; Gong, P.; Zhang, X.; Lin, Z.; Hu, Z.; Wu, Y. Centrosymmetric $Rb[Te_2O_4(OH)_5]$ and Noncentrosymmetric $K_2[Te_3O_8(OH)_4]$: Metal Tellurates with Corner and Edge-Sharing $(Te_4O_{18})^{12-}$ Anion Groups. *Inorg. Chem. Front.* **2022**, *9*, 2628–2636. [CrossRef]
13. Wedel, B.; Sugiyama, K.; Itagaki, K.; Müller-Buschbaum, H. Synthesis and Crystal Chemistry of New Transition Metal Tellurium Oxides in Compounds Containing Lead and Barium. *MRS Online Proc. Libr.* **2000**, *658*, 101. [CrossRef]
14. Guo, X.; Gao, Z.; Liu, F.; Du, X.; Wang, X.; Guo, F.; Li, C.; Sun, Y.; Tao, X. Optimized Growth and Anisotropic Properties of Li_2ZrTeO_6 Nonlinear Optical Crystals. *CrystEngComm* **2021**, *23*, 6682–6689. [CrossRef]
15. Dammak, M.; Khemakhem, H.; Mhiri, T. Superprotonic Conduction and Ferroelectricity in Addition Cesium Sulfate Tellurate $Cs_2SO_4·Te(OH)_6$. *J. Phys. Chem. Solids* **2001**, *62*, 2069–2074. [CrossRef]
16. Vanek, L.; Mička, Z.; Fajnor, V.Š. Thermal Dehydration and Decomposition of Oxygen-Tellurium(VI) Compounds with Alkali Metals and Ammonium. *J. Therm. Anal. Calorim.* **1999**, *55*, 861–866. [CrossRef]
17. Singh, H.; Sinha, A.K.; Ghosh, H.; Singh, M.N.; Rajput, P.; Prajapat, C.L.; Singh, M.R.; Ravikumar, G. Structural Investigations on $Co_{3-x}Mn_xTeO_6$; ($0<x\leq2$); High Temperature Ferromagnetism and Enhanced Low Temperature Anti-Ferromagnetism. *J. Appl. Phys.* **2014**, *116*, 074904. [CrossRef]

18. Augsburger, M.S.; Viola, M.C.; Pedregosa, J.C.; Carbonio, R.E.; Alonso, J.A. Crystal Structure and Magnetism of the Double Perovskites $Sr_3Fe_2TeO_9$ and $Ba_3Fe_2TeO_9$: A Neutron Diffraction Study. *J. Mater. Chem.* **2006**, *16*, 4235. [CrossRef]
19. Sathiya, M.; Ramesha, K.; Rousse, G.; Foix, D.; Gonbeau, D.; Guruprakash, K.; Prakash, A.S.; Doublet, M.L.; Tarascon, J.-M. Li_4NiTeO_6 as a Positive Electrode for Li-Ion Batteries. *Chem. Commun.* **2013**, *49*, 11376. [CrossRef]
20. Masese, T.; Yoshii, K.; Yamaguchi, Y.; Okumura, T.; Huang, Z.-D.; Kato, M.; Kubota, K.; Furutani, J.; Orikasa, Y.; Senoh, H.; et al. Rechargeable Potassium-Ion Batteries with Honeycomb-Layered Tellurates as High Voltage Cathodes and Fast Potassium-Ion Conductors. *Nat. Commun.* **2018**, *9*, 3823. [CrossRef]
21. Jin, C.; Mutailipu, M.; Jin, W.; Han, S.; Yang, Z.; Pan, S. Cation Substitution of Hexagonal Triple Perovskites: A Case in Trimetallic Tellurates $A_2A'BTe_2O_9$. *Inorg. Chem.* **2021**, *60*, 6099–6106. [CrossRef]
22. An, Y.; Mosbah, A.; Le Gal La Salle, A.; Guyomard, D.; Verbaere, A.; Piffard, Y. $K_2[Te_4O_8(OH)_{10}]$: Synthesis, Crystal Structure and Thermal Behavior. *Solid State Sci.* **2001**, *3*, 93–101. [CrossRef]
23. Churakov, A.V.; Ustinova, E.A.; Prikhodchenko, P.V.; Tripol'skaya, T.A.; Howard, J.A.K. Synthesis and Crystal Structure of New Alkali Metal Hydrogen Tellurates. *Russ. J. Inorg. Chem.* **2007**, *52*, 1503–1510. [CrossRef]
24. Ustinova, E.A.; Prikhodchenko, P.V.; Fedotov, M.A. Equilibrium in Water-Peroxide Solutions of Cesium Tellurate Studied by ^{125}Te NMR Spectroscopy. *Russ. J. Inorg. Chem.* **2006**, *51*, 608–612. [CrossRef]
25. Inamo, M. ^{125}Te NMR Evidence for the Existence of Trinuclear Tellurate Ion in Aqueous Solution. *Chem. Lett.* **1996**, *25*, 17–18. [CrossRef]
26. Groom, C.R.; Bruno, I.J.; Lightfoot, M.P.; Ward, S.C. The Cambridge Structural Database. *Acta Crystallogr. Sect. B* **2016**, *72*, 171–179. [CrossRef]
27. Belsky, A.; Hellenbrandt, M.; Karen, V.L.; Luksch, P. New Developments in the Inorganic Crystal Structure Database (ICSD): Accessibility in Support of Materials Research and Design. *Acta Crystallogr. Sect. B.* **2002**, *58*, 364–369. [CrossRef]
28. Christy, A.G.; Mills, S.J.; Kampf, A.R. A Review of the Structural Architecture of Tellurium Oxycompounds. *Mineral. Mag.* **2016**, *80*, 415–545. [CrossRef]
29. Bühl, M.; Knight, F.R.; Křístková, A.; Malkin Ondík, I.; Malkina, O.L.; Randall, R.A.M.; Slawin, A.M.Z.; Woollins, J.D. Weak Te,Te Interactions through the Looking Glass of NMR Spin-Spin Coupling. *Angew. Chem. Int. Ed.* **2013**, *52*, 2495–2498. [CrossRef]
30. Lockhart, T.P. Steric Effects in Neophyltin(IV) Chemistry. *J. Organomet. Chem.* **1985**, *287*, 179–186. [CrossRef]
31. Tötsch, W.; Peringer, P.; Sladky, F. The Solvolysis of Orthotelluric Acid in HF. *J. Chem. Soc. Chem. Commun.* **1981**, 841–842. [CrossRef]
32. Frisch, M.J.; Trucks, G.W.; Schlegel, H.B.; Scuseria, G.E.; Robb, M.A.; Cheeseman, J.R.; Scalmani, G.; Barone, V.; Petersson, G.A.; Nakatsuji, H.; et al. *Gaussian 09, Revision A.02*; Gaussian Inc.: Wallingford, CT, USA, 2009.
33. Wadt, W.R.; Hay, P.J. Ab Initio Effective Core Potentials for Molecular Calculations. Potentials for Main Group Elements Na to Bi. *J. Chem. Phys.* **1985**, *82*, 284–298. [CrossRef]
34. Mata, I.; Alkorta, I.; Espinosa, E.; Molins, E. Relationships between Interaction Energy, Intermolecular Distance and Electron Density Properties in Hydrogen Bonded Complexes under External Electric Fields. *Chem. Phys. Lett.* **2011**, *507*, 185–189. [CrossRef]
35. Bader, R.F.W. A Quantum Theory of Molecular Structure and Its Applications. *Chem. Rev.* **1991**, *91*, 893–928. [CrossRef]
36. Vener, M.V.; Churakov, A.V.; Voronin, A.P.; Parashchuk, O.D.; Artobolevskii, S.V.; Alatortsev, O.A.; Makhrov, D.E.; Medvedev, A.G.; Filarowski, A. Comparison of Proton Acceptor and Proton Donor Properties of H_2O and H_2O_2 in Organic Crystals of Drug-like Compounds: Peroxosolvates vs. Crystallohydrates. *Molecules* **2022**, *27*, 717. [CrossRef]
37. Medvedev, A.G.; Churakov, A.V.; Navasardyan, M.A.; Prikhodchenko, P.V.; Lev, O.; Vener, M.V. Fast Quantum Approach for Evaluating the Energy of Non-Covalent Interactions in Molecular Crystals: The Case Study of Intermolecular H-Bonds in Crystalline Peroxosolvates. *Molecules* **2022**, *27*, 4082. [CrossRef]
38. Buldashov, I.A.; Medvedev, A.G.; Mikhaylov, A.A.; Churakov, A.V.; Lev, O.; Prikhodchenko, P.V. Non-Covalent Interactions of the Hydroperoxo Group in Crystalline Adducts of Organic Hydroperoxides and Their Potassium Salts. *CrystEngComm* **2022**, *24*, 6101–6108. [CrossRef]
39. Churakov, A.V.; Grishanov, D.A.; Medvedev, A.G.; Mikhaylov, A.A.; Tripol'skaya, T.A.; Vener, M.V.; Navasardyan, M.A.; Lev, O.; Prikhodchenko, P.V. Cyclic Dipeptide Peroxosolvates: First Direct Evidence for Hydrogen Bonding between Hydrogen Peroxide and a Peptide Backbone. *CrystEngComm* **2019**, *21*, 4961–4968. [CrossRef]
40. Sheldrick, G.M. *SADABS, Programs for Scaling and Absorption Correction of Area Detector Data*; Version 2016/2; Bruker AXS: Karlsruhe, Germany, 2016.
41. Sheldrick, G.M. Crystal Structure Refinement with SHELXL. *Acta Crystallogr. Sect. C Struct. Chem.* **2015**, *71*, 3–8. [CrossRef]

MDPI
St. Alban-Anlage 66
4052 Basel
Switzerland
Tel. +41 61 683 77 34
Fax +41 61 302 89 18
www.mdpi.com

Molecules Editorial Office
E-mail: molecules@mdpi.com
www.mdpi.com/journal/molecules

www.ingramcontent.com/pod-product-compliance
Lightning Source LLC
LaVergne TN
LVHW070046120526
838202LV00101B/732